Advances in
Food Extrusion
Technology

Contemporary Food Engineering

Series Editor

Professor Da-Wen Sun, Director

Food Refrigeration & Computerized Food Technology
National University of Ireland, Dublin
(University College Dublin)
Dublin, Ireland
http://www.ucd.ie/sun/

Advances in Food Extrusion Technology, *edited by Medeni Maskan and Aylin Altan* (2011)

Enhancing Extraction Processes in the Food Industry, *edited by Nikolai Lebovka, Eugene Vorobiev, and Farid Chemat* (2011)

Emerging Technologies for Food Quality and Food Safety Evaluation, *edited by Yong-Jin Cho and Sukwon Kang* (2011)

Food Process Engineering Operations, *edited by George D. Saravacos and Zacharias B. Maroulis* (2011)

Biosensors in Food Processing, Safety, and Quality Control, *edited by Mehmet Mutlu* (2011)

Physicochemical Aspects of Food Engineering and Processing, *edited by Sakamon Devahastin* (2010)

Infrared Heating for Food and Agricultural Processing, *edited by Zhongli Pan and Griffiths Gregory Atungulu* (2010)

Mathematical Modeling of Food Processing, *edited by Mohammed M. Farid* (2009)

Engineering Aspects of Milk and Dairy Products, *edited by Jane Sélia dos Reis Coimbra and José A. Teixeira* (2009)

Innovation in Food Engineering: New Techniques and Products, *edited by Maria Laura Passos and Claudio P. Ribeiro* (2009)

Processing Effects on Safety and Quality of Foods, *edited by Enrique Ortega-Rivas* (2009)

Engineering Aspects of Thermal Food Processing, *edited by Ricardo Simpson* (2009)

Ultraviolet Light in Food Technology: Principles and Applications, *Tatiana N. Koutchma, Larry J. Forney, and Carmen I. Moraru* (2009)

Advances in Deep-Fat Frying of Foods, *edited by Serpil Sahin and Servet Gülüm Sumnu* (2009)

Extracting Bioactive Compounds for Food Products: Theory and Applications, *edited by M. Angela A. Meireles* (2009)

Advances in Food Dehydration, *edited by Cristina Ratti* (2009)

Optimization in Food Engineering, *edited by Ferruh Erdoğdu* (2009)

Optical Monitoring of Fresh and Processed Agricultural Crops, *edited by Manuela Zude* (2009)

Food Engineering Aspects of Baking Sweet Goods, *edited by Servet Gülüm Sumnu and Serpil Sahin* (2008)

Computational Fluid Dynamics in Food Processing, *edited by Da-Wen Sun* (2007)

Contemporary Food
Engineering Series
Da-Wen Sun, Series Editor

Advances in Food Extrusion Technology

Edited by

Medeni Maskan
Aylin Altan

CRC Press
Taylor & Francis Group
Boca Raton   London   New York

CRC Press is an imprint of the
Taylor & Francis Group, an **informa** business

CRC Press
Taylor & Francis Group
6000 Broken Sound Parkway NW, Suite 300
Boca Raton, FL 33487-2742

First issued in paperback 2016

© 2012 by Taylor & Francis Group, LLC
CRC Press is an imprint of Taylor & Francis Group, an Informa business

No claim to original U.S. Government works

ISBN 13: 978-1-138-19912-5 (pbk)
ISBN 13: 978-1-4398-1520-5 (hbk)

Visit the Taylor & Francis Web site at
http://www.taylorandfrancis.com

and the CRC Press Web site at
http://www.crcpress.com

Contents

Series Preface

Contemporary Food Engineering

Food engineering is the multidisciplinary field of applied physical sciences combined with the knowledge of product properties. Food engineers provide the technological knowledge transfer essential to the cost-effective production and commercialization of food products and services. In particular, food engineers develop and design processes and equipment to convert raw agricultural materials and ingredients into safe, convenient, and nutritious consumer food products. However, food engineering topics are continuously undergoing changes to meet diverse consumer demands, and the subject is being rapidly developed to reflect market needs.

In the development of food engineering, one of the many challenges is to employ modern tools and knowledge, such as computational materials science and nanotechnology, to develop new products and processes. Simultaneously, improving food quality, safety, and security continues to be a critical issue in food engineering study. New packaging materials and techniques are being developed to provide more protection to foods, and novel preservation technologies are emerging to enhance food security and defense. Additionally, process control and automation regularly appear among the top priorities identified in food engineering. Advanced monitoring and control systems are developed to facilitate automation and flexible food manufacturing. Furthermore, energy saving and minimization of environmental problems continue to be important food engineering issues, and significant progress is being made in waste management, efficient utilization of energy, and reduction of effluents and emissions in food production.

The *Contemporary Food Engineering Series*, consisting of edited books, attempts to address some of the recent developments in food engineering. The series covers advances in classical unit operations in engineering applied to food manufacturing as well as such topics as progress in the transport and storage of liquid and solid foods; heating, chilling, and freezing of foods; mass transfer in foods; chemical and biochemical aspects of food engineering and the use of kinetic analysis; dehydration, thermal processing, nonthermal processing, extrusion, liquid food concentration, membrane processes, and applications of membranes in food processing; shelf-life and electronic indicators in inventory management; sustainable technologies in food processing; and packaging, cleaning, and sanitation. These books are aimed at professional food scientists, academics researching food engineering problems, and graduate-level students.

The editors of these books are leading engineers and scientists from many parts of the world. All the editors were asked to present their books to address the market's need and pinpoint the cutting-edge technologies in food engineering.

All contributions are written by internationally renowned experts who have both academic and professional credentials. All the authors have attempted to provide critical, comprehensive, and readily accessible information on the art and science of a relevant topic in each chapter, with reference lists for further information. Therefore, each book can serve as an essential reference source to students and researchers in universities and research institutions.

Da-Wen Sun
Series Editor

Series Editor

Professor Da-Wen Sun, PhD, was born in Southern China, and is a world authority on food engineering research and education. He is a member of the Royal Irish Academy, which is the highest academic honor in Ireland. His main research activities include cooling, drying, and refrigeration processes and systems; quality and safety of food products; bioprocess simulation and optimization; and computer vision technology. In particular, his innovative studies on vacuum cooling of cooked meat, pizza quality inspection using computer vision, and edible films for shelf-life extension of fruits and vegetables have been widely reported in the national and international media. Results of his work have been published in over 500 papers, including about 250 peer-reviewed journal papers. He has also edited 12 authoritative books. According to Thomson Scientific's *Essential Science Indicators*^{SM} updated as of July 1, 2010, based on data derived over a period of 10 years and four months (January 1, 2000–April 30, 2010) from ISI Web of Science, a total of 2554 scientists are among the top 1% of the most cited scientists in the category of agriculture sciences, and Professor Sun is listed at the top with his ranking of 31.

Dr. Sun received first class BSc honors, and had an MSc in mechanical engineering and a PhD in chemical engineering from China before working at various universities in Europe. He became the first Chinese national to be permanently employed in an Irish university when he was appointed as a college lecturer at the National University of Ireland, Dublin (University College Dublin), in 1995. He was then continuously promoted in the shortest possible time to the position of senior lecturer, associate professor, and full professor. Dr. Sun is now professor of food and biosystems engineering and director of the Food Refrigeration and Computerized Food Technology Research Group at University College Dublin (UCD).

As a leading educator in food engineering, Dr. Sun has contributed significantly to the field of food engineering. He has guided many PhD students who have made their own contributions to the industry and academia. He has also, on a regular basis, given lectures on the advances in food engineering at international academic institutions and delivered keynote speeches at international conferences. As a recognized authority in food engineering, Dr. Sun has been conferred adjunct/visiting/consulting professorships from over ten top universities in China, including Zhejiang University, Shanghai Jiaotong University, Harbin Institute of Technology, China Agricultural University, South China University of Technology, and Jiangnan University. In recognition of his significant contribution to food engineering worldwide and for his outstanding leadership in the field, the International Commission of Agricultural and Biosystems Engineering (CIGR) awarded him the CIGR Merit Award in 2000 and again in 2006; the UK-based Institution of Mechanical Engineers named him Food

Engineer of the Year 2004; in 2008, he was awarded the CIGR Recognition Award in recognition of his distinguished achievements as the top 1% of agricultural engineering scientists around the world; in 2007, Dr. Sun was presented with the AFST(I) Fellow Award by the Association of Food Scientists and Technologists (India); and in 2010, he was presented with the CIGR Fellow Award; the title of "Fellow" is the highest honor in CIGR, and is conferred upon individuals who have made sustained, outstanding contributions worldwide.

Dr. Sun is a fellow of the Institution of Agricultural Engineers and a fellow of Engineers Ireland (the Institution of Engineers of Ireland). He has also received numerous awards for teaching and research excellence, including the President's Research Fellowship, and has received the President's Research Award from the University College Dublin on two occasions. He is editor-in-chief of *Food and Bioprocess Technology—An International Journal* (Springer) (2010 Impact Factor = 3.576, ranked at the 4th position among 126 food science and technology journals); series editor of the *Contemporary Food Engineering Series* (CRC Press/Taylor & Francis); former editor of *Journal of Food Engineering* (Elsevier); and an editorial board member for *Journal of Food Engineering* (Elsevier), *Journal of Food Process Engineering* (Blackwell), *Sensing and Instrumentation for Food Quality and Safety* (Springer), and *Czech Journal of Food Sciences*. Dr. Sun is also a chartered engineer.

On May 28, 2010, Dr. Sun was awarded membership to the Royal Irish Academy (RIA), which is the highest honor that can be attained by scholars and scientists working in Ireland. At the 51st CIGR General Assembly held during the CIGR World Congress in Quebec City, Canada, in June 2010, he was elected as incoming president of CIGR, and will become CIGR president in 2013–2014; the term of the presidency is 6 years, 2 years each for serving as incoming president, president, and past president.

Preface

Eating and drinking foods are vital habits of humans. During the centuries, different practical techniques and skills were developed by people all over the world to process foods. In recent years, extrusion cooking, which is a specialized form of processing, has become a well-established industrial technology, with a number of foods and feed applications. It is a high-temperature, short-time process that is being used increasingly in the food industries for the development of new products such as cereal-based snacks including dietary fiber, baby foods, pasta products, breakfast cereals, texturized protein food stuffs, and modified starch from cereals. In addition to the usual benefits of heat processing, extrusion offers the possibility of modifying the functional properties of food ingredients and expanding them.

Extrusion cooking as a continuous cooking, mixing, and shaping process, is a versatile and very efficient technology in food processing. In the extruder, the food mix is thermomechanically cooked to a high temperature (usually in the range 100°C–180°C), pressure and shear stress that are generated in the screw-barrel assembly. The cooked melt is then texturized and shaped in the die. Extrusion-cooked melts transit from high pressure to low (atmospheric) pressure when they exit the die. This sudden pressure drop causes part of the internal moisture and the vapor pressure to flash off, forming bubbles in the molten extrudate, thereby resulting in the expansion of the melt. The use of high temperatures reduces the processing time and allows a full transformation of raw material to its functional form in periods as short as 30–120 seconds. The extrudates generally require no further processing except for some minimal drying. Most products are extruded through round holes where the speed of the cutter determines whether it is a ball, rod, or curl. The thermomechanical action during extrusion brings about gelatinization of starch; denaturation of protein; inactivation of enzymes, micro-organisms, and many antinutritional factors; complex formation between amylose and lipids; and degradation reactions of vitamins and pigments; all these occur in a shear environment, resulting in a plasticized continuous mass. Therefore, chemical and structural transformations in foods during extrusion cooking determine the quality of extruded products.

This book was written principally as a reference for graduate and undergraduate students, researchers, and extruded product manufacturers, with special emphasis on some recent extrusion works. However, food technologists, and scientists and technicians in the extrusion processing industry may also find the contents useful. We believe this book will serve as a source of information to all involved with food and feed extrusion. For a person who is new in this area, this book will serve as a guide for understanding and properly selecting the raw material and an extruder.

Chapter 1 introduces the extrusion technology; its history, nomenclature, and working principles; and overview of various types of extruders, and parts and components of an extruder for design considerations. Chapter 2 discusses extruder selection and design, fluid flow problem with different types of raw materials, and heat transfer and viscous energy dissipation, with advantages and limitations for particular cases. Chapter 3 addresses the raw materials used in extrusion cooking with respect to chemical composition and particle size. It discusses the raw materials with specific reference to the products. Also, it discusses preparation or preconditioning of raw material for extrusion. Among these are grinding and particle size selection and moisture equilibration of raw material. Several review articles have addressed chemical and nutritional changes in extruded foods. Chapter 4 emphasizes recent research while providing an overview of trends previously reported in the literature. Chapter 5 focuses on a methodology to take a successful product (solution cast whey protein films) and create a similar extruded product. In particular, in-line viscosity measurements and extrudate rheological properties were examined to evaluate the similarities and differences between the extruded films and their solution-cast counterparts. The next four chapters present the extensive literature that currently exists on feed but mainly on food extrusion, with special emphasis on developments over the past decade. Examples are given for a number of important works such as utilization of food industry by-products for the development of extrudates, production of breakfast cereals, snack products, confectionary products, extrusion of pulses and feeds, and extrusion processing of pet foods and aquatic feed blends containing DDGS.

Chapter 10 covers the coinjection of food substances into an extruder die with the objective of creating defined colored patterns, adding internal flavors, and achieving other food injection applications into cereal-based extruded products. This chapter also deals with development of half-products or third-generation snacks (pellets). Focus is given to processes for producing corn masa– and potato-based pellets and to the identification of critical process parameters for providing precise control in hydration, as well as gelatinization levels, and overcoming variations in the characteristics of starting raw materials. Chapter 11 covers thermal and nonthermal extrusion of protein products and nonthermally extruded whey protein products.

Chapter 12 focuses on the experimental methods that are commonly used to determine extrudates' quality parameters and the most important quality properties of an extrudate. It is the objective of Chapter 13 to show the potentialities of process modeling based on continuum mechanics in the field of twin-screw extrusion for food applications through a common approach that was developed for several years. In this chapter, two main aspects are considered in twin-screw extrusion modeling. The whole process, from feeding of the raw materials in the hopper to the exit of the transformed product at the die, or just a focus on a limited portion of the extruder, is considered. Troubleshooting is the identification or diagnosis of "trouble" in an extrusion system caused by a mechanical failure or misapplication of extrusion processing parameters. Therefore, a chapter on troubleshooting the extrusion process completes the book.

This book has been prepared by authors who have extensive experience in their specialties, ranging from university members to members of the industry all over the world. We hope that the book will bring valuable information to all who are involved with the field of extrusion technology.

Dr. Medeni Maskan
Dr. Aylin Altan

Editors

Dr. Medeni Maskan is currently a professor at the University of Gaziantep Engineering Faculty Food Engineering Department. He received a BSc degree in Food Engineering from Middle East Technical University in 1988. He received his MSc and PhD degrees in Food Engineering from Gaziantep University in 1992 and 1997, respectively. He worked on "the storage stability of pistachio nuts at various atmospheric conditions" during his PhD studies. He has been a faculty member since 1998. He became a full professor in 2007. He has published more than 70 articles in national and international top class journals. His research program focuses on fats and oils, dehydration of food materials, and extrusion technology. He acts as a reviewer in several journals being published in the food science and technology area. Dr. Maskan lives in Gaziantep (Turkey) with his wife, Aysun, and their two children, Serhat and Ozan Emre.

Dr. Aylin Altan is an assistant professor of Food Engineering at the University of Mersin, Turkey. Dr. Altan received her BS and MS degrees in Food Engineering at the University of Gaziantep, Turkey, in 2000 and 2003, respectively. She received her PhD degree in Food Engineering in 2008 from the University of Gaziantep in Turkey, focusing on the extrusion process under a collaborative research with the University of California, Davis. She worked as a postdoctoral fellow at the University of California, Davis, from 2009 to 2010. She joined the faculty of Food Engineering at the University of Mersin in 2010. She is the author of 13 research papers in refereed scientific journals, 2 book chapters, and more than 20 conference presentations. She is the recipient of the third-place poster award in the Second International Congress on Food and Nutrition in 2007, best poster award in ICC International Conference in 2008, and a finalist in IFT's Food Engineering Division Graduate Paper Competition in 2008. She is a member of IFT. Dr. Altan's current research interests include processing of food materials into value-added products, modeling and optimization of food processing, utilization of agricultural processing by-products, and noninvasive imaging of food microstructure.

Contributors

Aylin Altan, PhD
Department of Food Engineering
University of Mersin
Ciftlikkoy-Mersin, Turkey

Ferouz Ayadi
Department of Agricultural and
    Biosystems Engineering
South Dakota State University
Brookings, SD

Jose De J. Berrios, PhD
USDA-ARS-WRRC
Albany, California

Françoise Berzin, PhD
INRA, UMR 614 Fractionnement des
    Agroressources et Environnement
Reims, France

Suvendu Bhattacharya, PhD
Food Engineering Department
Central Food Technological Research
    Institute
(Council of Scientific and Industrial
    Research)
Mysore, India

Mary Ellen Camire, PhD
Department of Food Science and
    Human Nutrition
University of Maine
Orono, Maine

Nehru Chevanan
Department of Agricultural and
    Biosystems Engineering
South Dakota State University
Brookings, SD

Parisa Fallahi
Department of Agricultural and
    Biosystems Engineering
South Dakota State University
Brookings, SD

Fahrettin Göğüş, PhD
Department of Food Engineering
University of Gaziantep
Gaziantep, Turkcy

Massoud Kazemzadeh, PhD
Kay's Processing LLC
Clara City, Minnesota

John M. Krochta, PhD
Department of Food Science and
    Technology
Department of Biological and
    Agricultural Engineering
University of California
Davis, California

Medeni Maskan, PhD
Department of Food Engineering
University of Gaziantep
Gaziantep, Turkey

Kathryn L. McCarthy, PhD
Department of Food Science and
    Technology
Department of Biological and
    Agricultural Engineering
University of California
Davis, California

Jorge C. Morales-Alvarez, PhD
Frito-Lay Research and Development
Plano, Texas

Sankaranandh Kannadhason
Department of Agricultural and
    Biosystems Engineering
South Dakota State University
Brookings, SD

Chinnadurai Karunanithy
Department of Agricultural and
    Biosystems Engineering
South Dakota State University
Brookings, SD

**Kasiviswanathan
Muthukumarappan, PhD**
Department of Agricultural and
    Biosystems Engineering
South Dakota State University
Brookings, South Dakota

Charles I. Onwulata, PhD
U.S. Department of Agriculture, ARS,
    Eastern Regional Research Center
Dairy Processing and Products
    Research Unit
Wyndmoor, Pennsylvania

Mohan Rao, PhD
Frito-Lay Research and Development
Plano, Texas

Murali Rai
Department of Agricultural and
    Biosystems Engineering
South Dakota State University
Brookings, SD

Daniel J. Rauch, MSc
Mattson Foods
Foster City, California

Galen J. Rokey, BSc
Wenger Manufacturing, Inc.
Sabetha, Kansas

Kurt A. Rosentrater, PhD
Department of Agricultural and
    Biosystems Engineering
Iowa State University
Ames, IA

Mahmut Seker, PhD
Department of Chemical Engineering
Gebze Institute of Technology
Gebze-Kocaeli, Turkey

Guy Della Valle, PhD
INRA, UR 1268 Biopolymères,
    Interactions & Assemblages (BIA)
Nantes, France

Bruno Vergnes, PhD
MINES ParisTech, CEMEF
Sophia Antipolis, France

Waleed A. Yacu, PhD
The Yacu Group, Inc.
Vernon Hills, Illinois

Chirag Shukla
Department of Agricultural and
    Biosystems Engineering
South Dakota State University
Brookings, SD

Sibel Yağcı, PhD
Food Engineering Department
Karamanoğlu Mehmet Bey University
Karaman, Turkey

1 Introduction to Extrusion Technology

Massoud Kazemzadeh

CONTENTS

INTRODUCTION

As technology advances and the efficacy of limited energy becomes the focus of advancing civilization, new solutions are presented for the problems that arise in the preparation and the production of food, as the world population increases to reach seven billion. Extrusion cooking and technology have been shown to be the most efficient and continuous manner by which we can break down raw food ingredients to a well cooked and predigested form so that the end product can be shelf stable and easily packaged. This process increases shelf-life from a few weeks to 9–12 months and can be consumed in a convenient, ready-to-eat form by the final consumer. Today, extruded fabricated foods are composed of cereals, starches, sugars, oils, nuts, and vegetable protein. To a limited extent, there are additional applications for some of the animal proteins mixed with cereals and grains to produce a complex matrix that would meet the needs of a specific market.

The word "extrusion" refers to a process by which a liquid to a semiliquid product is forced through a die opening of the desired cross section. The basic technology behind food extrusion systems has not changed in over 60 years. However, new applications, efficiency of use, and design have made food processing simpler, more reliable, less of an art form depending on the operator interference, and more of a science and self-correcting continuous process.

The simple design of a screw within a barrel chamber has been initially credited to Archimedes of Syracuse, a Greek mathematician and physicist who lived in 287–212 BC. His design of a wooden apparatus devised to move water from a lower level to a higher level with the turn of a screw within a round chamber, amazed the people of his time. This simple design later became the cornerstone of many different industries including material sciences such as metal fabrication, ceramics, concrete, plastic and nonplastic polymers, and, most recently, the food and feed industries.

In the last two centuries, extruders have come a long way. Joseph Bramah obtained the first extrusion patent in 1797 for making a lead pipe by having a dummy block placed in a ram type machine and forced out of a die to form a continuous profile. Today's multiplanetary screw extruders are used not only for the extrusion of profiles, but also for various other purposes from dispersive and distributive mixing, heat transfer via conductive and conversion energy, to many new designs that accommodate more varieties of raw materials and innovative applications in new concepts in the field of food processing.

NOMENCLATURE OF EXTRUDER PARTS

For a better understanding of extruder designs, we need to establish a basic nomenclature for the extruder parts. A typical extruder screw and its parts are shown in Figure 1.1. The basic nomenclature for design of the screw within a barrel is outlined below. It is the variation of this design that can generate high pressure and compression of the raw material as well as generate temperatures of over 250°C while under a pressure of over 100 bars.

Barrel opening (D_b): This dimension is usually referred to as a barrel opening in which the screw rotates. The actual screw diameter, D_s, is calculated using Equation 1.1.

$$D_s = D_b - 2\bar{d} \tag{1.1}$$

where $2\bar{d}$ is the screw clearance.

Flight height (*H*): *H* is the distance between the diameter height of the flight and the diameter of the root of the screw.

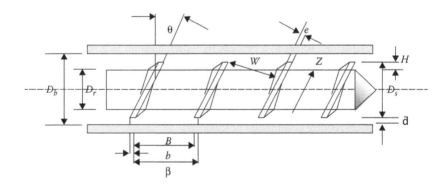

FIGURE 1.1 Schematic diagram of the extruder parts and the geometry of the single-flighted extruded screw.

Root diameter (D_r): The diameter of the root of the screw on which the flights are built is the base of the shaft that carries the rotational torque. It is calculated using Equation 1.2.

$$D_r = D_b - 2H \tag{1.2}$$

Screw clearance ($2\bar{d}$): Screw clearance is the difference between the diameter of the screw and the barrel opening. It is obtained from Equation 1.3.

$$2\bar{d} = D_b - D_r \tag{1.3}$$

Lead (β): Lead or pitch is the axial distance between the leading edge of the flight at the outside diameter and the leading edge of the same flight in front.

Helix angle (θ): The helix angle is defined as the angle of the flight with respect to the normal plane of the screw axis. It is determined using Equation 1.4.

$$\theta = \tan^{-1} \frac{\beta}{\pi D_s} \tag{1.4}$$

Channel: The helical opening that extends from the feed to the discharge end of the screw. A channel is formed by the inner side of the flight and the top of the screw root diameter and the surface of the barrel opening.

Axial channel width (W): The channel width is measured from one side of the flight to the next within the channel perpendicular to the angle of the flight. The channel width is determined using Equation 1.5.

$$W = B\cos\theta \tag{1.5}$$

where B is the axial distance between the flights.

Axial flight width (b): b is the width of a screw flight in the axial direction. The axial flight width is related to the lead and the axial distance between the flights as shown in Equation 1.6.

$$b = \beta - B \tag{1.6}$$

Flight width (e): e is the width of the screw flight measured perpendicular to the face of the flight, which is sometimes referred to as the thickness of the flight shoulder.

Channel length (Z): Z is the length of the screw channel in the Z direction, which can be one or more full turn of the screw helix.

Tip velocity of screw (V): *V* is the speed of the screw tip, which is dependent on the diameter of the screw and the rotational speed (N) of the shaft ($V = \pi D_s N \sim \pi D_b N$).

Number of flight turns (p): *p* is the total number of single flights in an axial direction. This can be single flight or multiple flights, which are commonly found when the screw enters the compression phase of the extruder.

Available volume of screw channel (V_a): V_a is the available useful volume of the screw flight, which determines the total volumetric capacity of the extruder. This value is commonly used as the comparative actual capacity of one extruder versus another based on the same horsepower and screw diameter. It is important to keep in mind that the extruder capacity is based on the volume of the product that can be extruded within a given time period, such as per hour. Many manufacturers may quote the capacity of their extruder at a much higher rate, since the base for their evaluation of the extrudate may be based on a higher bulk density of the raw material than that of the manufacturer's comparative extruder. A good example of such a misconception is to compare the extruder capacity of corn starch with a bulk density of approximately 368 kg/m³, with the extruder capacity of whole grain or a grid with 640 kg/m³. With the same screw speed and screw profile, the corn starch can extrude half of the capacity of the whole grain.

Compression ratio (CR): This is the ratio obtained by dividing the available feed volume of the screw to the next screw volume developing the pressure needed to process the raw material.

Ratio of barrel length to diameter (L/D_s): This is the ratio of the screw diameter to the length of the barrel. If the extruder barrel is in segments, one segment is usually referred as the ratio L/D_s and the total barrel set up is referred to as a number of these ratios. For example, a six barrel segment extruder with $4L/D_s$ would be $24L/D_s$.

Cone or die inlet: This is usually referred to as the end of the last section of the extruder and the entrance to the die cavity. It is usually depicted as a cone shaped addition to the end of the screw to direct the molten dough into the die area with minimum restriction and pressure development.

EXTRUDER DIE PARTS

As mentioned above, the die design of the extruder can be as elaborate and complex as the extruder screw design, or it can be simple as another stage of the extruder process. In this case, we will review the various parts of an extruder die, which in some cases may not be present as independent sections but as a combined section with the other parts of the die. Figure 1.2 shows the extruder die parts.

Screw and shaft cone (C_s): The ends of the screw shaft *S* and the screw cone C_s area are usually designed for a specific process by changing the screw end cone C_s to fit

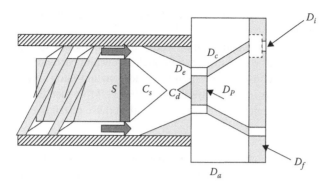

FIGURE 1.2   Schematic diagram of the extruder parts.

the pressure profile for a given mixture of raw material. For the extrusion of starchy or farinaceous raw materials, the design is mostly open and a large volume of space may be present between the die cone C_d opening and the screw cone C_s. By contrast, proteinaceous raw material would require minimum dead space between the screw cone and the die assembly and an easy flow pattern within the die assembly to the final die opening.

Die cone entrance (C_d): If the die entrance is circular with a number of holes leading into the die cavity, C_d is used to equally spread the dough between the die inlet holes with similar pressure and velocity to equilibrate the product flow throughout the final die outlet.

Die assembly (D_a): The die assembly can be located at the end of the screw of the extruder as shown in Figure 1.2. It may also be designed so that it stands by itself with its own support system while the extruder feeds the die assembly D_a via a pipe like pathway, or a chamber from a single opening, or a number of uniform openings referred to as the die entrance, D_e. Designs that require the die assembly to be remote from the end of the screw are utilized in sheeting processes or gooseneck type dies where additional space is needed to expand the die openings for coextrusion, for sheeting or laminating assembly, and for the utilization of the calendaring process. The die assembly is basically made up of the die cone C_d, the inlet pressure plate D_p, the die channel or cavity D_c for equilibration of flow into the die plate, and finally the die face plate D_f, which may have the die shaped outlets formed on the die plate or designed to utilize the die inserts D_i, to further diversify the utilization of the various shapes and designs.

COMPONENTS OF EXTRUDER DESIGN

Various designs of the extruders can be generalized into five basic design elements:

- Extruders with increasing root diameters
- Extruders with decreasing pitch, and constant root diameters

- Extruders with constant root diameters and screw pitch, but with decreasing barrel diameter
- Extruders that have a combination of the above characters such as constant root diameter, but with a decrease in pitch screw and barrel diameter
- Extruders with constant root diameter and constant pitch screw, but with restrictive pockets in a constant diameter barrel. Most such designs require a long barrel to generate high pressure within these restrictive areas of the flow

Several designs are possible for food extrusion, with single and twin-screw being the most common. Among these types of extruders, two categories can be defined:

a. Forming extruders carry much lower pressure and do not utilize thermal energy for cooking. These extruders are designed for forming dough by pressing against a die. They include pasta extruders, pretzel formers, and any of the vast variation of these types that form a dough-like consistency through a die hole to form a shape that will later be cooked to a final consistency.
b. Cooking and high-pressure extruders are designed to bring thermal energy into the raw material under high pressures of 3.4×10^3–13.8×10^3 kPa before exiting a die hole.

High pressure extruders can have various screw designs in order to fulfill a specific purpose. These designs are classified as: a) corotating and b) counterrotating screws. The corotating and counterrotating screw designs can be in a nonintermeshing or intermeshing form as shown in Figure 1.3 (Kazemzadeh 2001).

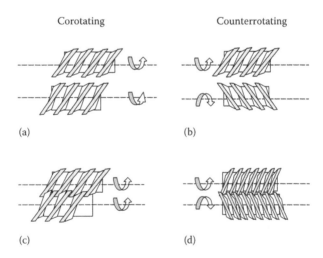

FIGURE 1.3   Screw design for the twin-screw extruder: (a) corotating and nonintermeshing, (b) counterrotating and nonintermeshing, (c) corotating and intermeshing, and (d) counterrotating and intermeshing.

Each design has its own unique benefits and usage. The corotating, non-intermeshing design is used mostly in gently pushing a low viscosity noncooked dough through a die. A more effective pumping of the dough can be achieved with little shear generation by a nonintermeshing counterrotating design. The intermeshing designs of the screw are mostly designed for their pumping action and ability to generate a great deal of pressure with high viscous products. A good application of counterrotating, intermeshing screws is found in the rubber industry or very high viscous plastic and elastomeric processing. Although they achieve a great deal of pumping action and can generate a great deal of pressure, they also have excellent mixing properties. Since they can generate a great deal of pressure between the two screws, which will cause the screws to separate and ride against the barrel of the extruder, these extruders are usually very short barrel extruders. This type of extruder design can generate a great deal of wear and tear on the barrel and screws and are recommended only for applications where the extruder L/D is very small and there are supporting bearings on both sides of the screw at 90 degrees to the line of extrusion exit die and cutter.

Among all of the extruders' screw designs, the twin-screw extruders, which are co-rotating, intermeshing, and self-wiping, have the most advanced control panels and are the most commonly utilized in the food industry. These extruders can utilize all of the important components and characteristics of the above designs of the screw configuration for the processing of the most varied raw materials common in food products, from high sugar, fat, starch, and proteins, to raw materials, from a very low viscosity dough to a very high viscous mass. The introduction of segmented screws in the early 1990s made it possible for the processing engineer to design his own screw configuration for a specific application or process. These segmented screw elements can be of various compression and conveying designs. The various pitch sections, including even a reverse pitch, can be designed and placed on a shaft that would distribute the torque and shear throughout the processing screw. The cross-sectional design of a kneading element can yield a path for two channels of screws that is constantly in a given pitch. These flight screws are referred to as lobs (Figure 1.4). The distance between the lowest part of the screw channel and the shaft, which is shown in the cross section of the kneading element as t (Figure 1.4), is the limiting factor for the maximum torque that can be realized by the screw and eventually by the extruder.

It is with this knowledge of advancement in extrusion science that we try to concentrate on discussing specific designs of the twin-screw extruder and its innovative application in the food processing industry.

There are a number of other extruder designs, which are elaborate and may have been introduced as prototypes, but have not entered the main market. Such a design is for a specific process and a specific purpose. Examples of these limited purpose designs are planetary extruders with multiple screws, split barrel extruders, and divergent extruders (where the extruder is fed from both ends with screws designed to push the material to the middle of the barrel; the exit die is located at the middle of the barrel, fed by both sides of the screws and then cut at the die). However, none of these designs have proven to be popular and none are in predominant use.

The type of extruder most commonly purchased by the food industry in the last few years is the twin-screw, which is intermeshing, corotating, and self-wiping. It has

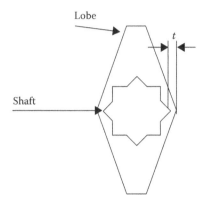

FIGURE 1.4   Schematic diagram of the kneading element.

control panels that are designed to protect the extruder and the operator from over pressurization and hazardous conditions that would result from equipment mishap. The control panels are usually designed in a manner to continuously monitor and take action to shut down the extruder, based on inconsistency in feed rate, overtorque of the motor, over-the-limit pressure at the die, feed throat backup, temperature limit of the motor and the gearbox, cutter overtorque, or the backup or blockage of the takeaway system. There are a number of other parameters that are monitored but are not controlled. They relate specifically to the operator's discretion based on product quality. These include the steam input to the conditioner or into the extruder, the barrel temperature, the die temperature, the temperature profile of the extruder barrel, and a number of other peripheral equipment items, such as pumps, takeaway systems, and exhaust. Although these are not essential to the basic design and operation of the extruder, they can be incorporated into the control panel of a computer, display, and monitor, and an emergency procedure is put in place by the control system at any point when any of the parameters exceeds the critical point. The major differences between the available intermeshing, corotating twin-screw extruders are the screw designs that can provide a true self-wiping surface with the designs that are sinusoidal cross profile, versus more of a wide shoulder hexagonal screw design profile (Figure 1.5).

Both designs can be self-wiping, where one screw wipes the next one clean and in return the dough portion is rotated and mixed as it is transferred to the adjacent screw. The wider the shoulder design of the screw, the more the shear generated and the higher the pressure introduced into the dough during processing. If the recipe is mostly starch based or contains starch and fiber, the product within the extruder with a hexagonal screw profile will have greater built-in shear factor than if it was processed using a screw with a sinusoidal screw profile. Within all of the extruder designs, a common denominator can be outlined as follows:

- The drive section where electrical or hydraulic power is converted into rotating energy, which drives the extruder shaft or shafts. This part of the extruder requires specific attention, since it determines the correct

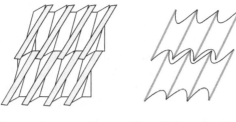

Hexagonal screw profile   Sinusoidal screw profile

FIGURE 1.5   Schematic diagram of different screw design profiles.

functionality of the extruder as well as the life span of a well-designed extruder. It contains the thrust bearings and the gear box equipped with controlled varying rotational speed of the screw. The transmission and the gear box with thrust bearings should be designed to carry three times the maximum force of the extrusion process and have an estimated life span of about 30,000–60,000 hours.

- In the feed section, the extruder is fed with raw, dry materials at a constant feed rate. It is then mixed with a liquid or a plasticizer to form a dough-like consistency before it is introduced to shear, heat, and pressure. This is an essential parameter of the extruder, which needs to be kept constant and convey the product forward while removing any air that might be present in the feed. The raw, dry materials feeder must be capable of adapting to various raw materials and bulk densities in order to keep the feed rate constant in pounds or kilograms per hour. The liquid feed, which is introduced in a chamber or port adjacent to the raw, dry materials feeder, also needs to be tested for constant delivery of the liquid. Loss-in-weight feeders are commonly used for this purpose and have self-reprograming capabilities to adjust to various bulk densities during processing in order to keep the feed rate constant in pounds or kilograms per hour.
- The compression section of the extruder is the next section after the feed section with a tighter screw pitch, and it can be varied in length and duties depending on the material and process on hand. This section accomplishes a number of tasks such as compression of the floury material to remove air while it mixes liquid with the flour to develop continuous dough-like consistency, which can have a high viscosity of over a million centipoises. If this section is well designed for the process, the dough consistency is well developed and the textural consistency of the final product will be uniform.
- The metering section is the final section before the dough enters the die area. It is the section of the extruder that consumes most of the electrical or rotational energy and is capable of generating thermal energy of over 250°C at very high pressures of 6.9×10^3–17.2×10^3 kPa. The residence time of dough in this section is not more than 10–30 s. The deformation and reconstruction of the matrix of the raw material to the finished product can easily be seen in this section of the extruder.

- The die section is the last section. Depending on the process, it can be as simple as a single outlet hole or as elaborate as the extruder itself with various chambers and pathways to accommodate the fluid dynamics of the molten dough within the extruder. In most cases, the die plate and the die pressure equilibration plate are all that is needed. In other processes, the sheeting and profile development may require more elaborate designs. It is also at this stage that the fast thermal and pressure sensors can be most accurate in providing a window to the process outline.

According to Harper and Holoy (1979), certain parameters such as the temperature and viscosity of the material became the focus of study and later became the backbone of today's extrusion technology studies on which the new designs and ideas of extrusion are developed. In order to understand the material behavior within the extrusion system, we must consider that the extruders are a rheometer by which the raw material behavior changes as it is sheared, heated, pressurized, and finally extruded through a die opening.

In food extrusion, the raw materials are considered to be of three distinct categories: (a) thermoplastic, (b) thermosetting, and (c) nonfunctional.

a. Thermoplastic raw materials are commonly defined as those materials that can be heated to a melting point and cooled to solidify. This process may repeat itself over and over. Such raw materials are metals, glasses, most plastics, starches, and sugars.
b. Thermosetting raw materials are those that, once they have been heated and go through the glass transition phase, become cured. Once this curing process is complete, the molecules tend to be very strongly set. As a result, if they are heated again after cooling, they will not follow the original pathway and flow. Thermosetting raw materials include elastomers, certain elastomic polymers, plastics, rubbers, and proteins.
c. Nonfunctional raw materials are those that do not go through a glass transition phase at temperatures introduced within the extrusion system and do not play a role in the building of the cross sectional matrix of the final product. If heated to a high enough thermal energy, they will bypass the amorphous phase and will decompose. These are materials that are made up of products such as minerals, fibers, and cellulose.

EXTRUSION COOKING

Webster's dictionary describes cooking as, among other things, the heating process of the food for consumption. According to Caldwell and Fast (1990), this indicates that: (1) cooking or the introduction of heat to the food is considered as the processing of food giving rise to the term "processed food," and (2) it takes into account the thermal energy required to bring the food to a given temperature in order for a chemical reaction to take place.

According to Yousria, William, and Kazemzadeh (1995), the most controversial and important aspect of cereal processing, for example, is the conversion of the batch cooking process into a continuous process. More specifically, it is the conversion of batch cooking to extrusion cooking without a change in texture or quality as defined. In order to closely examine the cooking process, we must review the elements of an equation developed to specify and define the term "cooked." In reviewing the total cooking process, four components play a major role: moisture content, granule size of the particulate making the dough, residence time, total energy input, and the type of energy input. Hydration of the product plays a major role in the gelatinization of starch granules, in protein texturization and denaturation, and in the elasticity of such components in the presence of shear. Moisture also serves as a plasticizing agent, which allows the starch granules to change phases from the crystalline stage, which can be easily fractured to the amorphous stage during extrusion. Residence time is a key factor in providing conduction of heat energy, and also hydration of particulates, because very narrow residence time distribution provides a homogeneous balance so that all particulates have access to the hydration, conduction, conversion, and conversion energies simultaneously. The smaller the particulates or the finer the flour is ground, the quicker is the hydration, thereby requiring lower thermal energy to gelatinize the starch granule.

Twin-screw extruders are well known to have much narrower residence time than single-screw extruders, due to the complete displacement and forward conveying of the product toward the die. The narrow residence time also provides a more uniform cooking in the cross section of the dough and a reduction in the nodules of over cooked dough and of burned dough at the cross section of the extrudate. In most cases, where the extruder is designed for 20/1 *L/D,* the residence time at 50% rotational speed is not more than 1.2 min, which makes the cooking process extremely efficient and fast, compared to any other means of cooking. Time/temperature/moisture history becomes an important issue in low- or no-shear processes, including low-shear cooking extrusion. For high-shear processes, such as high-shear cooking extruders, the energy of the cooking process is supplied dominantly by mechanical energy or conversion energy, which is shear-dependent. The conversion energy is proportional to the heat energy generated instantaneously at the site of the screw/dough/barrel, and is directly dependent on the viscosity of the dough, the rotational speed of the screws, and the pressure at the chamber. This phenomenon is well known in other industries, such as the grinding or milling industry, as hot spots. There are other variables that we do not bring into this equation since they are independent variables, such as wear and tear of the screw that results in the increase in distance between the screws and the barrel.

High pressure cooking extruders are considered as one of the most efficient ways to cook a given raw food material. The energy is supplied to the food matter from three sources of energy: (1) conductive, (2) convective, and (3) conversion.

1. Conductive energy: Conductive energy can be defined as the energy that is inputted into the system by heating the barrel and screw. Conductive energy is direct, externally applied heat energy via the jacketed surface of the barrels. Also, minor conduction can be placed from the screw surface to the product. Due to the short residence time of only 1–1.5 min, the product within the extrusion system receives a very minimum amount of

the conductive energy based on its coefficient of conductivity. There are a number of patents for twin-screw extruders with straight barrels and twin-screws that push the product forward in a long chamber with dividers that are designed to maximize the conductive energy of the barrel. These extruders, while they have achieved their purpose, have not yet been commonly seen to perform well under wider, raw material parameters. The extruder barrels can be heated by various means such as circulating steam, hot water, hot oil, and hot air, or by electrical energy via heating elements. Although all these are valid ways of heating the barrels, each has definite shortcomings and benefits. The steam energy cannot reach very high temperatures of 204°C unless it is under high pressure. Hot air also has similar problems in reaching lower temperatures. On the other hand, hot oil and electrical energy are capable of reaching temperatures of well over 176°C and can burn the extrudate to a crisp with limited exposure time.

2. Convective energy: Convective energy is defined as energy input in the form of gas or steam that is at a higher thermal energy than the food material within the extruder, and thus this energy gets incorporated into the extrudate within the barrel and is dissipated into the extrudate in a continuous format, thereby elevating its thermal energy. This energy can also be in negative form whereby, for example, through the use of liquid nitrogen, the total output of the system lowers the thermal energy of the extrudate to form a low temperature or even a frozen product. This is the energy supplied by steam injection into the extruder barrel or steam applied to the raw material at the point of conditioning before entering the extruder. This energy can also be obtained by turning hot water into steam during the process and thus passing that energy from the steam bubble to the raw material on a continuous basis. This has been demonstrated by Kazemzadeh (1985) by examination within the cross section of the extrudate matrix at different sections of the barrel. Such an approach increases the efficiency of the input of thermal energy into the product. Therefore, it not only increases the extruder capacity, but also provides direct thermal energy to hydrated grain particles to gelatinize. This, in turn, makes the starch or protein molecules more elastic and less susceptible to shear damage. Even though such an approach is a good solution to most extrusion problems, it is limited by how much steam can be injected into the system due to a limited extruder volume. In an extreme case, only 20% of the total energy can be inputted via this method since the residence time in the conditioner is usually not more than 3 min and in the extruder not more than 1.2 min.

3. Conversion energy: Conversion energy is the third and most important source of energy. It takes place when electrical energy is converted into heat energy by the rotation of the screw and shear is generated within the dough by screw rotation and pressure formed. This action is similar to generating heat energy by rubbing our hands together fast with the application of some pressure. Conversion energy usually refers to the amount of energy harvested from mechanical energy as a result of the electrical or hydraulic energy input by the drive via the gearbox. The harvesting of conversion energy is directly proportional to the conveying, pressurizing, back mixing,

and kneading provided within a screw profile design. This energy input can be easily disruptive and damaging to the starch granule. It provides thermal energy through shear stress and shear rate based on the apparent viscosity expressed in the following power law model for viscosity:

$$\eta = \frac{T}{Y} \tag{1.7}$$

$$T = KY^n \tag{1.8}$$

where η is the apparent viscosity (Ns/m^2), τ is the shear stress (N/m^2), γ is the shear rate (s^{-1}), K is the consistency index, and n is the flow index. In the case of dough viscosity and flow behavior within the extruder, the n value is estimated to be $n - 0.36$ and the consistency K is estimated by Equation 1.9.

$$K = 0.73 \exp(-14M) \exp\left(\frac{7884}{T}\right) \tag{1.9}$$

where M represents moisture on a dry basis and T is the absolute temperature of the product. It should be noted that the flow index and the flow consistency values are merely estimates, and that during the extrusion process the product goes from a flour consistency to a high, viscous, dough-like consistency as the raw material is mixed with the plasticizer, cooked, and sheared.

Therefore, it is apparent that not only the viscosity behavior changes, but also the flow index is shifted. If the starch granules and the protein globules are not well hydrated and elastic, the convective energy can easily damage both the starch granules and protein globules, resulting in a popcorn-type texture for starch with a high concentration of dextrin and a metallic off-taste in proteinaceous snacks. In extreme cases, damaged starch-based products exhibit rapid water solubility, gumminess in the mouth, and a very short bowl life for cereals. It can be concluded that within the extrusion system, thermal energy measured by a thermocouple inserted into the product within the extruder is not the true measure of total energy input. This is because damage occurring from the grinding action of the starch granules and protein globules by the screws and barrel surface does not generate enough energy to be reflected in this measurement.

$$ET_s \neq Et_{convertive} + Et_{raw} + Et_{conductive} + Et_{convective} \tag{1.10}$$

where ET_s is the sensible energy measured using the temperature of the extrudate within the extruder using thermocouple, $Et_{convertive}$ is the energy converted from the electrical drive shaft, Et_{raw} is the temperature of the raw material entering the system, $Et_{conductive}$ is the energy conducted from the wall and the screw of the extruder, and $Et_{convective}$ is the energy inputted from steam or hot gas injected into the extruder.

To better understand the effects of conversion energy, according to Kazemzadeh (1985), it is important to evaluate how the energy is absorbed by the ideal product. Energy is absorbed by the ideal product in three distinct pathways:

a. Pressurization of the product results from the pumping and efficient conveying action of the screw toward the restricted die. Highly proteinaceous products (greater than 58%) usually require higher pressure of 1×10^4 kPa buildup at the die area in order for the moisture and thermal energy to totally infiltrate into the protein globules and allow the protein matrix to be formed and elongated into fibrous texture at the die. With respect to the starch granules, very low levels of pressure are required in order to achieve developed matrix and expansion. Achievement of the specific point of pressurization for the plasticized dough is essential in order to utilize the high-pressure steam bubbles within the continuous medium to totally infiltrate and affect all the starch molecules within the matrix. A total infusion of moisture under high pressure and temperature will result in a cross section of the extrudate under the polarized light of a microscope, showing no presence of a Maltese cross. Otherwise, variations in the extent of cooking may occur where the outside of the pellet may be well sheared to form a continuous texture due to the addition of shear at the die wall, while the inner matrix of the extrudate may be undersheared and undercooked, thus forming a harder inner core texture.
b. Sensible heat is the energy used to increase the temperature of the product to the point where it is measured by a thermocouple. This reaction can easily be determined by placing a thermocouple, which has a fast reactive time, while the thermocouple is insulated in order to isolate the influence of the barrel and the housing temperature from the product temperature.
c. The heat of reaction is the energy required to drive a chemical reaction from one point to the next. Since cooking of most starchy and proteinaceous foods is an endothermic reaction, it refers to the amount of energy absorbed by the product components to proceed with the reaction. This part of the heat absorption has been described by Harper and Holoy (1979) in the following equation:

$$\frac{E_t}{m} E_t/m = \int_{T_1}^{T_2} C_P dt + \int_{P_1}^{P_2} \frac{dp}{Q} dp/Q + \Delta H^{\circ} H^{\circ} + \Delta H_{st} \tag{1.11}$$

where E_t/m is the total energy (E_t) absorbed per unit of mass (m), T is the product temperature, C_P is the specific heat of the product, Q is the density of the product at pressure P, ΔH° is the heat absorbed by cooking reactions, ΔH_{st} is the heat of fusion of lipids, and subscripts 1 and 2 refer to the start and end of the process. More simply put by McCabe and Smith (1956):

$$\frac{E_t}{m} E_t/m = \int_{T_1}^{T_2} C_p dt + \Delta H^{\circ} H^2 \tag{1.12}$$

In cereal cooking, the main reaction is starch gelatinization and protein denaturation. The endothermic heat of reaction has been reported to be 90–100 kJ/kg for protein (Harper, Rhodes, and Wanninger 1971). When no real data is available for the heat of reaction, the total heat of reaction can be estimated from the average heat of reaction.

$$H = 14X_s + 95X_p \tag{1.13}$$

where X_s is the starch fraction and X_p is the protein fraction of the formula. The specific heat of a product is a function of temperature as well as composition. This becomes significant in a continuous extrusion cooker where high temperatures are often reached without real need. Shepherd and Bhardwaj (1986) present a more accurate formula based on experimental data for cereal grains as follows:

$$C_p = 1.424X_c + 1.549X_p + 1.675X_f + 0.837X_a + 4.187X_m \tag{1.14}$$

where C_p is the specific heat of the product (kJ/(kg · K)) and X is the weight of fractions of the formula designated by carbohydrate c, protein p, fat f, ash a, and moisture m. This can be further simplified to yield the following:

$$C_p = 1.71\left[X_m\left(4.187C_d\right)\right] + 0.00304T - 0.292 \tag{1.15}$$

where T is the temperature in degrees Centigrade and C_d is the specific heat of the dry matter (kJ/(kg · K)).

It is obvious from the flow and dynamic values of product viscosity that all three sources of energy input are not interchangeable and that they contribute to the total energy balance in a different manner. Both mechanical energy and conducted heat energy are transferred to the product by the surfaces of the cooking extruder barrels and are associated with the heat transfer coefficient, which, in turn, is influenced by wiping and resurface generation of the product at the heat exchange point. From the above outline, it is apparent that starch gelatinization can occur at temperatures of 82°C–93°C. The additional heat energy provided to the product is to fully distribute the thermal energy within the product medium. When this energy is over supplied, it can cause earlier plasticization of the starchy product, thus providing an increase in the shear level of the system.

It is with these limitations in mind that the extruder cooker should be designed to increase mixing; increase heat exchange by conduction and convection energy; and provide conversion energy at viscosity levels where the product is pressurized, plasticized, and worked to utilize this energy in a low-shear environment. This is due to the viscosity, as well as the index of flow behavior of the extrudate, changes during the extrusion process due to the evolution of the extrudate from the raw floury particulates phase to a continuous pressurized viscous dough-like consistency.

DOUGH RHEOLOGY

Parameters such as input of thermal energy and its effects on viscosity and flow properties of material in various polymers and nonpolymers of food material were put to use in the extrusion in order to design new and innovative systems to accommodate a greater variety of processes and raw materials and their various flow properties within the extrusion system. The use of intermeshing, corotating, self-wiping twin-screw extruders can influence the extent and role of conduction energy; however, it is still limited to approximately 284–315 W/m^2, depending on the extruder size and type.

Rheology can be defined as the study of deformation and flow properties of materials. In our focus, the material that is most commonly used is dough. Dough is a combination of a number of elements such as starches, proteins, sugars, and fibers. The defining point of the study of dough within the extrusion system is the isolation of the following parameters in order to study the effects of each parameter within the flow properties of the given dough. Of these parameters, some are dependent variable and some are independent variable, including: velocity profile of the dough material after it has been mixed with the plasticizer within the screw channel with respect to the surface of the screw and the barrel chamber; total energy requirement for the extrusion process as was outlined in the previous section; pressure and shear profiles along the extrusion screw; residence time distribution of dough material passing through the extruder; heat transfer and thermal energy generated via mechanical energy within the screw profile; and finally, flow and pressure drops through the die assembly and die opening.

It is obvious that studying the rheological properties of a complex dough material made up of various ingredients with each following specific flow properties can be difficult, and, in some cases, impossible. According to Harper, Rhodes, and Wanninger (1971), the mathematical model of flow behavior used to better understand the flow properties and directional flow of the dough within the extruder and more specifically within the screw channel needs some assumptions to be considered. The basic assumption is that the dough within the extruder behaves as a Newtonian fluid. To further simplify the equation, we have to base our calculations on the apparent viscosity of the Newtonian fluids. This flow behavior is further elaborated based on the shear stress and shear rate in an idealized situation.

The definition of shear in this context can be explained as the movement of two parallel plates with an area of A, with a velocity of V, at a distance of Y, and a force between the plates of F (Figure 1.6), filled with an ideal fluid with no leakage. Two parameters that are proportional to the viscosity of the ideal fluid can be evaluated. Shear stress (τ_{yx}) is determined by the force exerted on the two plates with an area of A. Since the force and movement are in two directions, the values must be identified in the x and y axes. A linear velocity gradient in the fluid between the two plates and no slippage at the walls are assumed. The slope of the gradient is $-(V_x/Y)$ and outlined as shear rate. The relationship can be shown as:

$$\frac{F}{A} - \left(\frac{V_x}{Y} \right) \tag{1.16}$$

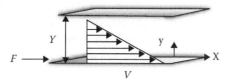

FIGURE 1.6   Velocity profile between parallel plates.

or

$$\tau_\downarrow yz = -\left(\left(dv_\downarrow x\right)/dy\right) \tag{1.17}$$

which states that the shear force per unit area is directly proportional to the negative velocity gradient or shear rate. The above equation is derived directly from the Newtonian law of viscosity and fluids, which are classified as Newtonian fluids. The negative sign is needed in order to obtain a positive τ when the shear rate is negative. When the direction of shear stress and the shear rate are known and obvious, the negative sign is usually ignored and the equation is written as:

$$\tau = \gamma \tag{1.18}$$

where τ is the shear stress (N/m²), μ is the viscosity (Ns/m²), and γ is the shear rate (s⁻¹).

A Newtonian fluid is the viscosity of fluids where the rates of shear and stress are proportional to each other. Other fluids that are not Newtonian in viscosity are Bingham plastic, pseudoplastic, and dilatant, having a wide variety of viscoelastic properties, based on their relationship between the shear stress and shear rate as shown in Figure 1.7.

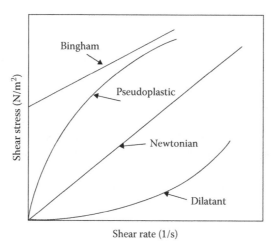

FIGURE 1.7   Curves for typical time-independent fluids.

The two classic non-Newtonian fluids are shear thinning (pseudoplastic) fluids in which the viscosity of the fluid decreases as the shear rate, γ, increases, and shear thickening (dilatant) fluids in which the viscosity increases as the shear rate increases. Extensive use of the Ostwalde and de Waele power law model (Equation 1.8) (Rossen and Miller 1973) has been employed due to its simplicity and ability to correlate observed rheological behavior over the shear rate range of 10–200 s⁻¹ found in food extruders. Clark (1978) has discussed the reasons for the acceptance of the power law model (for describing the rheological behavior of intermediated and low moisture food dough). This discussion fits the experimental data reasonably well, it is convenient to use, and it calculates the flow property of both pseudoplastic and dilatant behaviors. The power law model has its own limitations and does not account for yield stress, nor does it extrapolate well to shear rate outside of the range of the data, and can be applied only to a small range of γ. Once the correlation between τ and γ has been established, we can clearly define the concept of apparent viscosity η. Apparent viscosity is the viscosity that exists at a particular value of γ and changes with varying degree of γ values for non-Newtonian fluids. Therefore, we can define the apparent viscosity by:

$$\eta = \frac{\tau}{\gamma} \tag{1.19}$$

The symbol η is used to depict apparent viscosity and clearly distinguishes it from μ, which is a fluid property for Newtonian fluids. Substituting for τ in terms of γ using the power law equation:

$$\tau = \frac{K\gamma^n}{\gamma} = K\gamma^{n-1} \tag{1.20}$$

From the simplified equations above and with a few assumptions that the viscosity of the dough remains consistent at any moment of time within the extruder, we can realize the following statements for any extrusion system. Increased rotational speed of the screws will increase the shear rate proportionally, and its absolute value is dependent on the size of the extruder and the distance between the screw tips and the barrel. The larger the extruder size and the faster the speed of the screw, the more shear generated. The higher the pressure within the screw, the more shear stress the dough will experience. Also, the greater the gap between the screw and the barrel wall, the lesser the shear produced and a lesser maximum pressure can be achieved. The same concept can be applied to the dough with higher moisture content or a lower viscosity under the same extrusion parameters, generating less shear rate and shear stress. As the wear and tear in the extruder increase, the gap between the barrel and the screw increases. This in turn, lowers the shear rate, thereby lowering the torque and cook temperature.

CONCLUSION

It has been established that the extrusion system is presently the most efficient way of processing the foods that are low in water activity and are shelf stable for over

12 months, depending on the storage conditions and packaging outline. The extrusion process lends itself to a number of markets in the food industry. One underexplored market where extrusion offers a much needed solution is low cost mass production of highly nutritious foods for the developing countries that are in great need of such foods for survival. The low water content of most final products produced by the extrusion process, the low cost of shipping, and ready-to-eat convenience foods, allow this method to lead all other processes. We suggest that the extrusion process with its diverse applications and efficiency of use will become a key component of food processing in solving the ever-growing food demand in the developing countries.

According to Kazemzadeh (1995), among the various designs of the extruders, the one that is the most outstanding for its versatile application to the food industry and capable of processing a wider array of raw materials is the twin-screw corotating, intermeshing, and self-wiping extruder (TSE). These extruders are sold as a system including: a complete control panel, the dryer, and coaters, as well as a number of peripheral equipment pieces such as a takeaway system, pumps, and pneumatic lines.

Today's extrusion process needs to be developed and installed for a certain type of product. By this we mean that the TSE is a very versatile type of processing equipment. This versatility becomes applicable for a wide variety of raw materials and allows for inconsistencies of the raw materials. With some changes in the equipment configuration, such as the increase or decrease in the length of the barrel, or change in the screw configuration, the extrusion system can become the process that will meet all expectations. For example, a low bulk density starchy material requires a different configuration than a high density, proteinaceous recipe, a high fiber, or high fat recipe. To shift from one major ingredient composition to another, there needs to be some modifications to the screw and even to the barrel length in order to get the best and most efficient result. The extrusion process is usually referred to as a system since it needs a peripheral equipment to match the versatility of this process. A system and its controls may include the feeder for raw material; the conditioning and steam mixing chamber; the extruder with variable speed; the barrel temperature control, the cutter; the takeaway system; and even the dryer, coating system, and secondary drier to process a range of given products. TSEs are mostly utilized in pet food, breakfast cereal, and specialty markets. Their application ranges from the Dutching of cocoa, to flavor bits and cookie bits.

In order to be able to map the screw configuration of a process for a given mix of raw materials, some understanding of food rheology is needed and a more detailed understanding of the behavior of such mixtures at elevated temperatures is required. Most of this information is an art and is based on empirical data. However, one can have a great understanding of the process and how to set up a mapping of the screw configuration and set up a process outline for scale-up, if there is some understanding of how the shear rate and shear stress are affected by variation of the extruder parameters. A simple increase in moisture input will reduce the viscosity of the dough, which in turn will reduce the shear rate and shear stress. The reduction of screw revolutions per minute (RPM) can reduce the degradation of the starch or protein molecules by the reduction of shear rate. The relationship between the apparent viscosity and the ratio of shear stress and shear rate are essential for the visualization of what type of shear history and shear rate a product has experienced within few seconds of extrusion. High shear rate, which is usually proportional to the decrease in plasticizer input and the increase in speed of the screw on a starch based product, will generate a popcorn characteristic that is very water soluble and contains a high degree of dextrins. But, a high concentration of plasticizer, combined with lower RPM, will generate a more gelatinized, low water soluble, starch based product that will give a more baked consistency and mouth feel to the final product.

Finally, the use of TSE is growing worldwide, due to its versatility and ease of operation. Its application varies from nonfood industries in manufacturing adhesives, low molecular plastics, to elastomers, plastic colorants, paints, coatings, soaps, and detergents. In the food industry, its application varies from high starch whole grain chips and cereals to high starch cheese puffs, and from candy and chocolate production to coextrusion of cookies and sugary bits for inclusions.

TSE's applications can vary from fruit processing and making of fruit based leathery products to high vegetable starchy snacks, and from high protein meat analogs for vegetarians to meat and grain based high protein and fiber cereals, snack chips and bits for dieters. TSEs are also well utilized in the specialty feed and pet food industries where the process requires specific throughput of a very specialized salmon, trout, or drug delivery system for animals. High protein products are also designed for animal recovery, specifically to carry ingredients for animal recovery foods, and for specialty diets for diabetic or obese animals. Many other applications are in the works that cannot be discussed at this point due to confidentiality or patent rights that are in place. Generally, any process that works with highly viscous material and that needs mixing, heating, cooling, and shaping can use this system as a major component.

REFERENCES

Caldwell, E.F. and Fast, R.B. 1990. *Breakfast Cereals and How They Are Made.* St. Paul, MN: American Association of Cereal Chemists Inc.

Clark, J.P. 1978. Dough rheology. *Food Technol (Chicago)* 32(7): 73.

Harper, J.M. and Holoy, S.H. 1979. Optimal energy usage in food extrusion. *Transactions, ASAE paper* 79: 6508.

Harper, J.M., Rhodes, T.P. and Wanninger, L.A., Jr. 1971. Viscosity model for cooked cereal doughs. *Chem Eng Prog Symp Ser* 67: 40.

Kazemzadeh, M. 1985. Morphological and ultrastructural study of defatted soy flour at various stages in extrusion process. M.S. Thesis, University of Texas A&M College Station, Texas, USA, pp. 86–108.

Kazemzadeh, M. 1995. Extruded cereals and snacks. In *Food Process Design and Evaluation*, ed. R.K. Singh, 191–251. Lancaster, PA: Technomic Publishing Company.

Kazemzadeh, M. 2001. Baby foods. In *Extrusion Cooking: Technologies and Application*, ed. R. Guy, 182–199. Cambridge: Woodhead Publishing Limited.

McCabe, W.L. and Smith, J.C. 1956. *Unit Operation of Chemical Engineering.* New York: McGraw-Hill.

Rossen, J.L. and Miller, R.C. 1973. Food extrusion. *Technol (Chicago)* 27(8): 46.

Shepherd, H. and Bhardwaj, R.K. 1986. Thermal properties of pigeon pea. *Cereal Foods World* 31: 466.

Yousria, A.B., William, M.B. and Kazemzadeh, M. 1995. Stability of hard red spring (HRS) wheat flour and vital wheat gluten as nutritional binder in extruded aquaculture diet, Book of Abstracts, Abstract No. 891. IFT Annual Meeting, Institute of Food Technologists, Chicago, IL.

2 Extruder Selection, Design, and Operation for Different Food Applications

Waleed A. Yacu

CONTENTS

23

INTRODUCTION

Extruders are widely used in the food industry. They are used for simple forming applications, such as pasta, to more complex operations involving significant modifications of the extruded material. Their feed material often includes many solid and liquid ingredients resulting in ill-defined transformation reactions and rheology/viscosity fields.

There are two major types of extruders: single-screw and twin-screw (corotating and counterrotating). These come with a wide range of screw diameter and length/diameter (*L/D*) ratio with or without steam preconditioning. Given the current process know-how limitations, it may not always be easy to optimally select and design the extruder for a specific application. This chapter intends to break down the extrusion process to discrete sections, describe their function and operating parameters, address the process needs, and provide a logical approach to the selection and design process. The approach should improve with experience and better understanding of the material characteristics, rheology, transformation reactions, and interactions between the system design and operating parameters.

As a starting point, this chapter discusses the use of a single-screw extruder for plasticating/melting synthetic polymers. Most of the extrusion theoretical development is based on this application. Comparing plastic to food extrusion is necessary to learn from and appropriately analyze the single-screw food extrusion process. Twin-screw extruders, corotating and counterrotating, are described later and compared with single-screw extruders and with each other.

Food extruders utilize thermal and mechanical energy. Understanding the energy consumption and input requirements is very important to improved performance and economical system design. Water is a common ingredient in almost all the food extrusion formulations. A part of the water can sometimes be applied in the form of steam, thus supplementing the total extruder energy. This option can have a significant impact on the extruder selection, design, and performance as well as on the product characteristics.

Combined extrusion cooking and forming of unexpanded products is a common process that is used to make snacks, cereals, and other products. Accelerated cooling before the final forming step is critical in such operations. This is typically achieved by evaporating part of the liquid water in a venting stage either within the extruder or in between a cooker and a forming extruder. The selection and design of such a step is briefly described in a section of this chapter.

When the extruded product is developed on a small production basis, scale-up becomes part of the total extruders selection and design process. The scale-up process will be briefly described by identifying its important factors, limitations, and complimenting process options.

Finally, the extruder's important design parameters and operating variables are listed and discussed. It will become clear that these are generally interactive, thus impacting the extruder performance and the product characteristics. As such, one may be able to suggest more than one system design and operation to make the same product. The optimum selection is likely to be governed by availability, flexibility, and economics.

COMPARISON BETWEEN PLASTIC AND FOOD EXTRUSION OPERATIONS

Nonfood polymers can be divided into three main groups (Rauwendaal 1986): thermoplastics, thermosets, and elastomers. Thermoplastic materials soften when they are heated and solidify when they are cooled. Their chemical composition does not generally change during the extrusion process. Thermosets undergo a cross-linking reaction when the temperature increases above a certain limit. This causes the formation of a three-dimensional network, which remains intact when the temperature is reduced. The cross-linking reaction is irreversible and therefore, thermosets cannot be recycled as thermoplastic materials. Elastomers or rubbers are materials capable of very large deformations when subjected to shear force, but recover their original shape when the applied force is removed.

The analysis and development of the polymer extrusion technology has mainly concentrated on the use of extruders for thermoplastics. In most cases, only a single polymer ingredient is extruded. The process involves conveying, plasticating (melting), and melt metering to the die.

Two types of polymerization reactions, "addition" and "condensation," produce plastic polymers. In the first type, individual monomers such as ethylene add to each other directly without any change in composition and form a long chain molecule. In the second type, two or more monomers react with each other with the elimination of a small part of them (usually water) to form a linear chain molecule. The reacting monomers in this reaction type usually carry two functional groups, which can react with each other continuously. Examples of this type of polymers are polyamides (nylons) and polyester. All of the linear or slightly branched polymers produced by addition or condensation reactions are thermoplastic. This means that they can be repeatedly softened at an elevated temperature and solidified again by cooling. No chemical changes occur during these thermal treatment processes.

The extruder is normally used to convey the solid polymer, melt, mix, and form a finished product shape such as sheet, pipe, and rod. It is also used in compounding and in the production of plastic raw materials such as pellets for further processing operations. In addition to the continuous operations, the extruder is also used in a semicontinuous manner in the case of injection molding machines by reciprocating the screw forward when the mold filling cycle starts.

Extrusion food ingredients are mostly composed of biopolymers of carbohydrates (starch, fiber sugars, and others) and proteins. Other ingredients such as water, oils/fat, minerals, and vitamins are also included in the feed mix. The structure of the biopolymers is not well defined and varies with the sources. In food extrusion, the process usually involves feeding a mix of solid and liquid ingredients. These ingredients react in complex and ill-defined processes involving, in many cases, irreversible changes in the physical and chemical structures. Inside the extruder, the ingredients form a fluid-like material resulting from either:

1. A low temperature dough formation process in the presence of adequate fluids (water and/or oil). Water is always present and it acts as a plasticizer, reducing the glass transition and melting temperature. The components of the formed dough may undergo structural modification if adequate energy is provided. For example, the starch granule will undergo a gelatinization/melting process if a certain temperature range is exceeded. The latter can result in a significant viscosity rise due to the increase in the water absorption capacity of the gelatinized starch.
2. At low water/oil concentration, softening and melting processes of the biopolymers (carbohydrates and proteins) can be achieved by applying adequate energy. This can be in the form of solid friction and viscous heat dissipation of mechanical energy, steam energy, and/or heat transfer through the barrel. Typically, the food material in these cases also undergoes irreversible physical and chemical changes.

The heated melt in most food extrusion cases behave as non-Newtonian pseudoplastic material. Its viscosity decreases with increasing shear rate and temperature. It also decreases by increasing the concentration of water and other viscosity reducing agents.

Food extrusion has been divided into two general categories: forming and cooking. For forming applications, low shear extruders are basically used to mix and form the desired product shape with minimum energy input. The product applications include pasta, cold formed snack, and other unexpended precooked pellets. The screw typically has a deep channel (0.2–0.3 D) with no compression ratio and operates at a relatively low speed (below 50 rpm).

Cooking extrusion applications normally utilize medium and high shear single and twin-screw corotating extruders. Significant energy can be provided to the product through viscous heat dissipation, heat transfer through the barrel, and sometimes with steam injection. The screws run at relatively high speeds (>100 rpm) with shallower channel depth and a compression ratio of up to 4:1 for single-screw extruders. These extruders are used for cooking as well as forming expanded products.

Both single and twin-screw extruders are used for plastic and food extrusions. Twin-screw extruders have significant advantages over single-screw extruders in feeding, mixing, heat transfer, residence time distribution, and pumping performance. However, they are more expensive. The main areas of twin-screw application in polymer extrusion are: (i) profile extrusion of thermally sensitive material, e.g., polyvinyl chloride (PVC); and (ii) compounding and difficult polymer processing operations that require good mixing of various compounds, devolatilization, and chemical reaction.

SINGLE-SCREW EXTRUDERS

In plasticating and food extrusion applications, the extruder performs the following functions: (i) conveying and compacting of the solid feed materials, (ii) plasticating/melting or dough forming, and (iii) metering the melted product through a die assembly.

A single-screw extruder's performance is well understood within polymer processing. Quantitative analysis of the three main sections is at a relatively advanced stage. Rauwendaal (1986) presented a very good description of single-screw extrusion of polymers, and the reader is referred to that title for a detailed analysis. Figure 2.1 presents a descriptive example of a single-screw extruder. Figure 2.2 provides definitions of the screw terminology using modular screw elements.

SOLID CONVEYING SECTION

Solid Conveying Section in Plastic Extruders
Darnell and Mol (1956) were the first to carry out a comprehensive analysis of solids transport in single-screw extruders. This work has been further extended, although

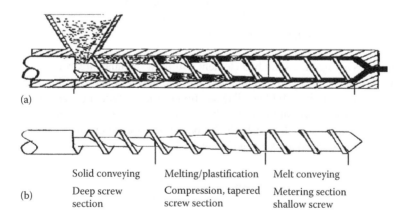

FIGURE 2.1   A descriptive example of: (a) a thermoplastic single-screw extruder, (b) a thermoplastic single-screw.

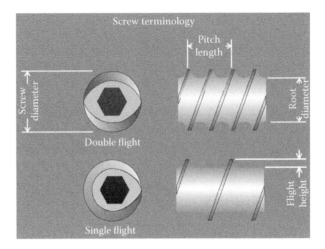

FIGURE 2.2 Description of screw terminology using a single-screw extruder screw elements. (Courtesy of Wenger Manufacturing, Inc., Sabetha, Kansas.)

the basic analysis remains unchanged. The drag-induced solids conveying process analysis assumes choke feeding and pressure development in the solid conveying zone. The analysis further assumes:

1. The compressed particulate (resin beads) behaves like a continuum solid bed.
2. The solid bed is in contact with the entire channel wall.
3. The channel depth is constant.
4. The flight clearance is negligible.
5. The solid bed moves in plug flow.
6. The pressure is a function of down channel distance only.
7. The coefficient of friction is independent of pressure.
8. Gravitational and centrifugal forces are neglected.
9. Density changes in the plug are neglected.

The forces acting on a solid bed element within the channel are evaluated. These are the frictional forces at the boundaries and the normal forces resulting from the pressure gradient in the solid bed. The solution of the analytical equations provides the angle, velocity, and output at which the solid is moving, provided the friction coefficient of the material with both the barrel and the screw surface, geometry details, and screw speed are known (Equation 2.1).

$$M_s = \rho H W n V_b \frac{\sin\theta}{\sin(\theta + \varphi)} \tag{2.1}$$

where M_s is the solid flow rate, ρ is the solid bed density, H is the channel height, W is the channel width, n is the number of parallel flights, V_b is the barrel velocity, θ is the solid conveying angle, and φ is the screw helix angle. The solids conveying angle is calculated from a momentum balance by taking into consideration the pressure gradient and the material friction coefficients with the screw and barrel.

Grooving the barrel in the solid conveying section improves the conveying performance as it increases the coefficient of friction with the barrel. Other claimed benefits of barrel grooving are improved stability and lower pressure sensitivity. Extruders fitted with a grooved solid conveying section can develop very high pressures and therefore, would require improved thrust bearing capabilities.

Solid Conveying Section in Food Extruders

The solid ingredients used in food extruders have a relatively small particle size (flour/powder). Most of the food extruders are starved fed primarily to avoid bridging problems in the feed port. The feeding device therefore determines the extruder output. In most of the extruders, barrel grooving is employed to improve conveying. The solid ingredient is conveyed in nonfilled screws similar to a "plug" flow by the drag action of the rotating screw. In completely filled screw sections, solid flow mechanism is likely to be different than that described above for thermoplastic extruders. Except in low shear forming extruders, the screw speed is generally higher in food extruders than that of thermoplastic extruders. In completely filled screw sections, solid conveying is not believed to be that of a continuum plug flow. The solid bed is sheared and dispersed as the internal locking forces between adjacent particles are weaker than the applied shear stress on the field.

The analysis of a solid conveying section for thermoplastic extrusion application may not be directly applicable to high speed food extruders. One can, however, utilize principles similar to that used in the aforementioned analysis to minimize the length of this section by optimizing the screw and barrel designs. Jasberg, Mustakas, and Bagley (1979) applied the same method of plastic materials for evaluating the friction coefficient between the food and extrusion surfaces.

PLASTICATING/MELTING SECTION

Plasticating/Melting Section in Plastic Extruders

The melting section in a choke fed single-screw plastic extruder starts as soon as the melt appears, usually after 3–5 diameters of extruder length (Rawindaal 1986). The first traces of the melt usually appear at the barrel surface because of the higher friction force between the screw flight tip and barrel surface. As melting proceeds, the initial film at the barrel surface grows in thickness. The melt flows into the screw channel, displacing the solid bed. In most cases, the solid bed is pushed against the passive flight flank and the melt starts to accumulate in the melt pool between the solid bed and the pushing flight flank. Due to screw rotation, the melt is sheared and exerts considerable friction and pressure on the solid bed. This reduces the solid bed size as more material melts flow away from the solid bed. Maddock (1959) was the first to accurately describe melting behavior in single-screw extruders. Tadmor, Dukdevani, and Klein (1967) were the first to perform a theoretical analysis of the melting process.

The melting rate is predicted from an energy balance across the channel (Rauwendaal 1986). The following assumptions are made in order to evaluate the temperature profile in the melt film:

1. The process is a steady state.
2. The polymer density and thermal conductivity are constant.
3. Convection heat transfer is neglected.
4. Conduction heat transfer is only in a direction normal to the interface.
5. The polymer melt flow is laminar.
6. Inertia and body forces are negligible.
7. There is no slip at the wall.
8. There is no pressure gradient in the melt film.
9. The temperature dependence of the viscosity is neglected.

In most of the thermoplastic extruders, the maximum (approximately 80%–90% or more) energy is supplied by the screw through viscous dissipation of mechanical energy (Rauwendaal 1986). Viscous heat dissipation is preferred to heat transfer through the barrel as it is relatively uniform and because polymers have poor thermal conductivity.

In reality, polymer melts are non-Newtonian and the melting process is nonisothermal. This complicates the analytical evaluation of the process, necessitating the use of numerical methods to solve the equations.

Klenk (1968) and Dekker (1976) suggested other melting models. Klenk observed the melting of PVC and noticed a melt pool located against the passive flight flank. This unusual melting behavior was attributed to the slippage of the PVC melt along the wall. Dekker's observation on polypropylene melting was that there was no clear melt pool at any side of the channel, but the solid bed was more or less suspended in a melt film. Lindt (1981) developed a mathematical model describing this melting behavior.

Dough Forming/Melting Section in Food Extruders

Food solid powders can be transformed into a fluid (dough) through hydration if adequate water is present without the application of heat. This process is influenced by: formula and particle size, moisture content, temperature, pressure, and mixing pattern. Water acts as a plasticizer, reducing the glass transition and melting temperature of starch (Figure 2.3). Heating the formed dough above a certain temperature will result in starch gelatinization/melting. The latter phenomena involve starch granular swelling (due to an increase in water uptake), resulting in a significant viscosity rise.

With decreasing moisture content, the food solid material glass transition and melting temperatures increase. Below approximately 30% moisture content, there is inadequate water to form dough-like material at an ambient temperature. In this case, forming a fluid-like material can be achieved by applying heat. This process is similar to plasticating thermoplastic materials. That is except for the fact that the solid particles are mostly sheared rather than being in a continuum solid bed. Interparticulate friction plays a significant role in the heating and melting processes.

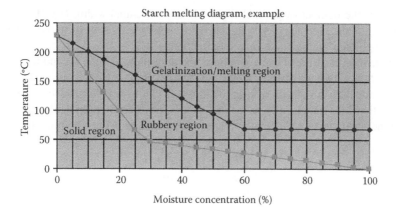

FIGURE 2.3   A general diagram describing the influence of moisture concentration on the gelatinization/melting and glass transition temperatures of starch.

The formed melt in most food extrusion processes is typically pseudoplastic or shear thinning. Theories of plasticating melt flow and mixing pattern developed for thermoplastic materials could be applied to food extrusion.

MELT CONVEYING SECTION

Melt Conveying Section in Plastic Extruders

The melt conveying zone function is to pump the polymer melt across the die, in most cases against considerable pressure. It starts at the point where the melting process is completed. This zone has been the subject of considerable engineering analysis since 1922 (Rowell and Finlayson 1928). The simplest modeling development is based on the following assumptions:

1. The process is a steady state.
2. The polymer density and thermal conductivity are constant.
3. The polymer melt flow is laminar.
4. Inertia and body forces are negligible.
5. There is no slip at the wall.
6. The melt viscosity is Newtonian and doesn't change with temperature.
7. The channel width is infinite and its curvature effect is neglected.

The solution of the equation of motion in the down channel provides the velocity profile $V_{z(y)}$. The integral across the channel of which provides the volumetric melt flow rate Q:

$$Q = \frac{1}{2} nWHV_{bz} - \frac{nWH^3}{12\mu} \frac{dP}{dz}$$

(2.2)

where n is the number of parallel flights, W is the channel width, H is the channel height, V_{bz} is the melt velocity at the barrel, μ is the melt viscosity, and dP/dz is the pressure gradient in the down channel direction.

The first term of the above equation is the drag flow, and the second term is the pressure flow. Optimum channel depth and the helix angle can be calculated by setting the flow rate derivatives with respect to these variables to zero (Rauwendaal 1986).

Equation 2.2 assumes infinite channel width. The effect of flight flanks is obtained when the shear stress acting on the plane x–z is included in the equation of motion:

$$\frac{dP}{dz} = \frac{d\tau_{yz}}{dy} + \frac{d\tau_{yz}}{dx} \tag{2.3}$$

The solution of this equation is considerably more complicated than that of the infinite channel width model. The volumetric output has been conveniently expressed in the following form:

$$Q = \frac{1}{2} nWHV_{bz}F_d - \frac{nWH^3}{12\mu} \frac{dP}{dz} F_p \tag{2.4}$$

where F_d and F_p are the drag and the pressure flow shape factors, respectively.

The development of theoretical analysis of the melt conveying zone has progressed further (Rauwendaal 1986) to account for: (i) the effect of clearance between the screw tip and the barrel; (ii) non-Newtonian melt viscosity for one-dimensional flow analysis. For two- or three-dimensional analysis with the temperature-dependent viscosity and a finite channel width, the complexity of the mathematical analysis increases significantly. In most of the cases, an analytical solution is not available and high level of sophistication may not be warranted when the boundary conditions are not well defined. In addition, the melt viscosity and other physical properties may not follow the original assumptions.

The standard single-screw design has been generally described with the following characteristics (Rauwendaal 1986):

Total length = 20–32 D, where D denotes the screw/barrel diameter. Of which, the:

Length of the solid conveying section = 4–8 D
Length of the metering section = 6–10 D
Number of parallel flights = 1
Flight pitch=1 D (helix angle = 17.66°)
Flight width=0.1 D
Channel depth in the feed section = 0.1 D–0.15 D
Channel compression measured as H_f/H_m depth ratio = 2–4

The dimensions of the majority of single-screw thermoplastic extruders fall within those listed above. However, some variations to these exist. For example, the addition of another parallel flight in the feed section has been claimed to smooth out pressure fluctuations and to balance the forces acting on the screw. Another suggested variation has been a variable pitch screw to allow for optimum helix angle in the solid conveying and melt metering sections. The optimization process also suggests that a reduction in the flight width in the melting and metering sections would result in a reduction of specific energy consumption.

The barrier flight extruder screw is a different type of single-screw that was introduced by Mailefer in 1959. The operating principles of this screw are based on separating the melt phase from the solid phase. The cross-sectional area of the solid phase channel continues to decrease while that of the melt phase channel continues to increase. At the end of the barrier section, the solids channel reduces to zero and the melt channel occupies the full channel. The barrier flight ensures that all the solid particles must be reduced in size to that below the clearance of the barrier flight.

Melt Conveying Section in Food Extruders

The behavior of food melts in the melt conveying section is not very different from that of thermoplastic melts under similar extrusion conditions. Their rheology is invariably pseudoplastic and the viscosity decreases with increasing temperature. The viscosity is, however, also affected by the moisture/oil content and by any structural/chemical changes. Therefore, the influence of mixing and time–temperature history is very significant on food melts. The engineering analysis of the plastic melt conveying section can be generally applied to food melts. Harman and Harper (1974) were the first to adopt thermoplastic melt conveying principles in evaluating forming extruders by assuming Newtonian and isothermal conditions. In cooking extruders, the melt flow needs to be described as non-Newtonian and nonisothermal. With continuing changes in the melt physical properties, cooking extrusion is a more complex process than the standard thermoplastic extrusion.

SIMPLIFIED DESCRIPTION OF MELT CONVEYING IN A SINGLE-SCREW EXTRUDER

The single-screw extruder is a "drag flow" pump. It relies on the principle that materials assume the speed of the boundary they are in contact with. This is true when there is no slip at the boundary. Inside the extruder, the material is also exposed to another form of flow and that is called the "pressure flow." This is caused by a difference in the pressures at various points within the extruder. To understand these two flow mechanisms better, let us refer to Figure 2.1a and b.

Imagine the screw displayed in Figure 2.1b as being rotated alone with no barrel surface surrounding it. Let us assume that the screw is also filled with a material that is adhering to the screw surface and will not fall off because of gravity or rotating inertia forces. Under such circumstance, the material will also rotate with the screw and will have a relative speed of zero with respect to the screw. In other words, the material will stay in its same position relative to the screw and will have no forward or backward movement. This is the same case as when someone is sitting in a moving vehicle where the speed of the person is the same as that of the vehicle and there is no relative movement between the person and the vehicle.

Let us now place the same screw inside a cylindrical barrel. The part of the material that is on the outside will now come in contact with the barrel surface. The barrel is stationary while the screw is rotating. The material in contact with the barrel will not rotate with the screw if it does not slip on the barrel surface. The relative movement of the barrel toward the screw is perpendicular to the screw axis. If the rotating part was a cylindrical shaft with no continuous helix (screw flight), the material in between would have been subjected to a radial shear field. There would have been no forward or backward movement along the shaft or barrel axis.

The presence of a helical flight changes that situation in the following manner. When the material at the barrel location is prevented from rotating with the screw, then it will be displaced by the flight in the axial direction by a distance equal to one screw pitch for every rotation. This is similar to a bolt and nut situation if the bolt is rotated and the nut is restricted. However, one needs to remember that the bolt is moving as a plug since it is slipping on the screw surface. The material will be subjected to a shear profile if it does not slip on the screw surface. Thus, the screw helix provides the material with an axial as well as a radial movement. Collectively the resultant net flow will be in the helical direction within the channel and is referred to as the "drag flow." It has a zero movement at the screw root and a maximum forward movement (one screw pitch in length per revolution) at the barrel interface as shown in Figure 2.4.

The mathematical representation of the drag flow is given by Equation 2.5.

$$Q_d = \frac{1}{2} WHV_{bz}$$

$$V_{bz} = \pi DN \cos \phi$$

(2.5)

where V_{bz} is the relative speed of the "barrel" in the helix direction, D is the screw diameter, H is the screw channel, N is the screw speed, and ϕ is the flight helix angle with the vertical coordinate.

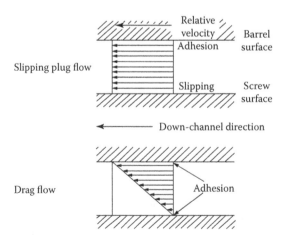

FIGURE 2.4 Drag flow representation with and without slippage on the screw surface.

From Equation 2.5, it can be seen that the drag flow is entirely dependent on the screw geometry and the screw speed. It is approximately equal to half of the plug flow. That will be the case if the material slips completely on the screw surface.

Let us now consider another type of flow within the extruder, namely the pressure flow. For a moment, let us think about the laminar flow inside a pipe and not in the extruder channel. We will refer to the diagram in Figure 2.5. In this system, a viscous fluid (dough for example) is being pumped from point "A" to point "B." To do so, the pump must generate adequate normal stress (pressure) to overcome the shear stress (friction resistance) inside the pipe. Therefore, the pressure in point "A" must be higher than that of point "B." Based on the nonslip mechanism, one might wonder how does the material flow inside the pipe when the pipe itself is stationary? The answer is that the material layer in contact with the inside surface of the tube is stationary, but due to the applied force, the material will be subjected to shear and that will result in a velocity profile as shown in Figure 2.5b.

The fluid portion in the center of the pipe will have the highest velocity. The flow profile has a bell shape, and the velocity of the adjacent layers gradually decreases until it is zero at the boundary (pipe surface). For a Newtonian fluid, the maximum velocity in the pipe center will be equal to twice the average velocity of the fluid. In such a case, the volumetric flow rate "pressure flow" can be determined from:

$$Q_p = \frac{\pi \cdot r^4}{8\mu} \frac{\Delta P}{L} \tag{2.6}$$

where ΔP is the pressure drop over the length L and radius r. The same type of flow can take place in a rectangular cross section duct pipe work. The fluid velocity will be highest in the center and zero at the duct surface boundary. Let us go back to the extruder example. The helical screw channel is a continuous rectangular duct. In a normal extruder operation, the pressure is higher at the end of the extruder (nearest the die) than at the feed end. Under such conditions, it is expected to have a pressure flow (backward direction) within the extruder channel. This is in the opposite

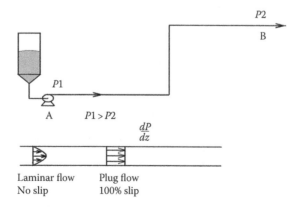

FIGURE 2.5   A general description of pressure flow in a cylindrical pipe.

direction of the drag flow explained earlier. Equation 2.7 mathematically represents the pressure flow in a rectangular or a screw channel:

$$Q_p = \frac{WH^3}{12\mu} \frac{\Delta P}{L} \tag{2.7}$$

Notice here that apart from the screw dimensions, the pressure flow is also influenced by the material viscosity and the pressure gradient.

The net flow (Figure 2.6) within a choked fed extruder is therefore the difference between the drag flow (forward) and the pressure flow (backward). As the restriction level increases, the drag flow component does not change but the pressure flow component increases, thus reducing the net flow (output) of the extruder. The screw channel depth has a much more significant effect on the pressure flow than on the drag flow. Thus with deeper screw channels, the extruder net flow decreases more rapidly with increasing flow restrictions (pressure) than in shallower screw channels (Figure 2.7).

In starved fed extruders, increasing the pressure flow does not practically decrease the extruder output since it is operating below its maximum limit. However, the percentage screw fill will increase, producing a similar effect on the response variable to that of a choked extruder. That is, a longer material retention time, more intense mixing, and greater energy consumption. At some point, the increasing pressure flow can result in a complete screw fill, creating a choked fed extruder situation.

So far our discussion has been limited to a nonslip situation. In reality, the material may slip to a varying degree on either or both the barrel and screw surfaces. If the material slips on the barrel surface, it will rotate with the screw either partially or completely, depending on the degree of slippage. In the latter case, the drag flow will be zero and no forward movement (by the flight action) is obtained. Single-screw extruder barrels are grooved to increase friction, minimize slippage, and improve

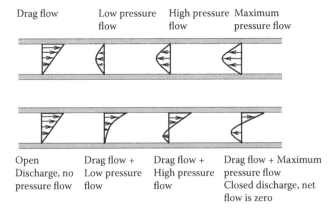

FIGURE 2.6 Description of the drag flow, pressure flow, and combined flow under increasing pressure scenarios.

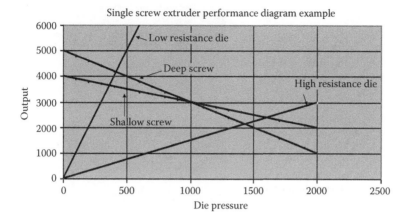

FIGURE 2.7   Influence of screw channel depth on the extruder output with increasing die pressure.

conveying. Cooling or heating the barrel surface may change the material friction factor with the barrel surface and its slippage characteristics. Slippage generally decreases with cooling and increases with heating. Adjusting the friction factor is one of the important reasons for controlling the barrel temperature in large single-screw extruders.

If the material slips on the screw surface, the drag flow will move the material like a continuous plug, potentially increasing the output by 100%. As mentioned before, this case resembles the movement of a restricted nut on a rotating bolt. Similarly, if the material slips completely on a pipe surface, the material velocity profile diminishes and a plug flow will result. The shear rate within a plug flow (in a screw channel or a pipe) is zero and that will result in lower energy dissipation by the friction action.

TWIN-SCREW EXTRUSION

The following discussion will be limited to intermeshing twin-screw extruders. There are two types depending on the manner one screw rotates with respect to the other screw: corotating and counterrotating extruders. One major difference between single-screw and twin-screw extruders is the type of transport that takes place within the extruder. Single-screw extruders depend entirely on frictional drag flow in the solid conveying zone and viscous drag flow in the melt-conveying zones. The transport in a twin-screw extruder is less dependent on the frictional properties of the material due to the action of the second screw in the intermeshing region. This provides a degree of positive displacement transport. Intermeshing counterrotating twin-screw extruders provide the most positive displacement action.

The channel in a corotating twin-screw extruder cannot be physically completely closed in the intermeshing region (Figures 2.8 through 2.10). That is dictated by the nature of such screw rotation. Due to the surface wiping of the screw surface in the intermeshing region, the material is wiped off one screw surface, changes

Double flight corotating
intermeshing screws

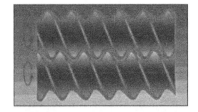

Twin double counterrotating
intermeshing screws

FIGURE 2.8   Corotating and counterrotating intermeshing screws. (Courtesy of Wenger Manufacturing, Inc., Sabetha, Kansas.)

direction, and moves to the other screw (Figure 2.11). The pressure difference within the extruder acts on the material and can cause backward movement since the channel is open in the axial direction.

In fully intermeshing counterrotating extruders, the screw channel is closed in both radial and axial directions. The material is physically prevented from moving to the other screw by the flight of the second screw in the intermeshing region. It is thus forced to move as "C"-shaped closed chambers (along each screw) by one screw

FIGURE 2.9   Intermeshing corotating double flight screws and a 45° kneading block. (Courtesy of Coperion Inc., Ramsey, New Jersey.)

Flow path in a corotating twin-screw extruder

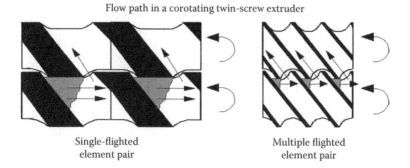

Single-flighted
element pair

Multiple flighted
element pair

FIGURE 2.10 Corotating screws showing open channel in the intermeshing region. (Courtesy of Coperion Inc., Ramsey, New Jersey.)

pitch for every screw rotation (Figures 2.8 and 2.11). These extruders are thus not vulnerable to pressure flow and are capable of generating higher die pressure than other extruders under similar circumstances. Material mixing is limited to the flow profile within the "C"-shaped closed chambers and leakage flow between the screws and barrel parts.

The material flow in a corotating twin-screw extruder follows the figure "8" profile with a corresponding change of direction and surface renewal. This results in a relatively uniform shear stress distribution around the screws (Figure 2.12). The screw configuration generally includes mixing and flow restricting elements. This creates a complex flow pattern that results in good mixing and heat transfer, large melting capacity, and good melt temperature control. The flow pattern complexity makes it more difficult to mathematically analyze and scale up. In practice, successful scale-up has been partly achieved by modular extruder design where

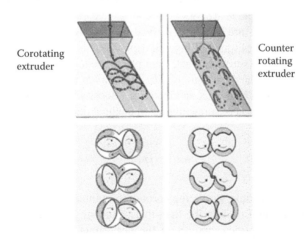

Corotating
extruder

Counter
rotating
extruder

FIGURE 2.11 General flow patterns in corotating and counterrotating twin-screw extruders (two flight screws in both cases). (Courtesy of Coperion Inc., Ramsey, New Jersey.)

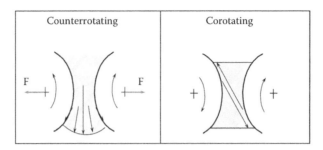

FIGURE 2.12   Shear stress profile in counterrotating and corotating twin-screw extruders. (Courtesy of Coperion Inc., Ramsey, New Jersey.)

the screw configuration can be easily adjusted to obtain the desired performance characteristics.

The shear stress distribution around the screw in counterrotating extruders is not uniform (Figure 2.12). The trapped material in the intermeshing region pushes the two screws outward toward the barrel, potentially causing excessive wear. This limits their operation to melt extrusion with relatively low viscosity and at low screw speeds. Their strength coincides in their high degree of positive displacement conveying.

Typically, intermeshing corotating screws are designed with one, two, or three starting flights. The use of three flight screws is only physically possible in relatively shallow channel extruders ($H < 0.15\ D$). Fully intermeshing single flight screws have a wider flight tip per unit screw pitch than the total of that of double or triple flight screws. This creates a greater degree of restriction in the intermeshing region per unit screw pitch, creating greater reduction in the pressure flow (Figure 2.10). Thus, they are less sensitive to downstream pressure and can therefore provide a greater degree of positive displacement and narrower residence time distribution than the same pitch double or triple lead screws.

Twin-screw extruders are generally starved fed and the feeding system controls their output. The degree of screw filling will depend on the flow rate, material viscosity, screw design, screw speed, and pressure profile generated by either the restrictive screw elements or the die assembly. The mechanical energy input thus depends on the extruder design, shear rate, and material viscosity. The shear rate within the screw channel is a function of screw diameter, channel depth, and screw speed.

$$\gamma \approx \frac{DN}{H} \tag{2.8}$$

The material flow profile within the corotating twin-screw extruders is complex and much more difficult to describe than that of the single-screw extruders. Booy (1980) presented a simplified analysis of the performance of the filled corotating twin-screw extruders, distinguishing two flow regimes, one between the screws and the barrel and another in the intermeshing region. Due to the self-wiping action, the material is assumed to flow forward at a rate of one pitch per revolution in the intermeshing region. Thus, the total net flow rate Q for a Newtonian melt would be:

$$Q = Q_d + Q_a - Q_p \tag{2.9}$$

where Q_d is the screw drag flow rate, Q_a is the forward flow rate in the intermeshing region, and Q_p is the pressure flow rate.

$$Q_d = \frac{1}{2}(2n-1)F_dHWV_{bz} \tag{2.10}$$

$$Q_a = A_a V_b \tan \varphi \tag{2.11}$$

$$Q_p = \frac{(2n-1)F_pH^3W}{12\mu}\frac{dP}{dz} \tag{2.12}$$

where n is the number of screw flights, F_d and F_p are the drag and pressure flow correction factors, respectively, V_{bz} is the melt velocity at the barrel, A_a and V_b are the channel area and melt velocity, respectively, in the intermeshing region, and φ is the flight helix angle.

Fully intermeshing counterrotating screws are designed with small tolerance (gaps) between the screws and the barrel. The trapped material in closed C-shaped chambers is conveyed in a positive displacement manner with some leakage pressure flow (Q_l) through the gaps. Janssen (1978) analyzed the flow in these extruders, distinguishing and mathematically describing four kinds of leakage flows in the extruder. Three of these take place in the intermeshing region (calendar, tetrahedron, and screw to screw), and one between the screw tip and the barrel. The volumetric rate output is simply represented as:

$$Q = 2nNV - Q_l \tag{2.13}$$

where n is the number of screw flights, N is the screw speed, and V is the C chamber volume. The total leakage flow for a new set of screws and barrel is expected to be below 10% of the total forward flow.

Twin-screw extruders were first used for food cooking applications in the mid-1970s. Their use has since increased at the expense of single-screw extruders and due to the expansion in food extrusion applications in general. As mentioned earlier, corotating twin-screw extruders have a greater positive displacement and mixing capability than single-screw extruders. This has enabled the development of improved process capability and flexibility with more consistent food product quality. They have been adopted in low and medium as well as high viscosity applications. Modular extruder design and good process control have widened their use to applications requiring multiprocess and functional zones. These include multiple liquid injections, downstream solid feeding, and venting ports. They can be designed with relatively long barrels, providing the long residence time needed for certain transformations and reactions. The flexibility in extruder design has also accelerated the development and optimization of new and more difficult products despite the fact that their exact flow behavior is not well understood. Figure 2.13 demonstrates an example of possible capabilities of a twin-screw extruder for a multiprocess extrusion system. It shows two dry mix feeders, a liquid and a direct steam injection ports, a side feeder for other solid components, a vent port with a vent stuffer, an injection port for thermally sensitive liquid ingredients, a start up/throttle valve, the die plate, and die face cutter assembly. In practice, most of the extrusion systems are much simpler and do not include some of the units shown in Figure 2.13.

Counterrotating extruders have been used for low viscosity applications that require high pressure forming capabilities such as candy and liquorish. They typically rely to a significant extent on the barrel heat transfer to heat and cook the material. The trapped material in the C-shaped chambers can be heated to high temperatures without the steam blowing back to the feed port. The pressure within these chambers can be raised and maintained to a higher level than the corresponding local steam saturation pressure.

Considerable research has been carried out using corotating twin-screw extruders. The emphasis in most of these has been the establishment of empirical relationships correlating the effect of feed composition and operating conditions on the product characteristics. The twin-screw design typically incorporates mixing discs, paddles, or reverse screw elements. The added complexity of food composition makes the task of detailed engineering analysis more difficult. Yacu (1985) presented a simulation

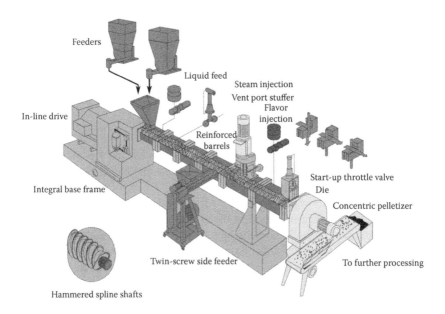

FIGURE 2.13   Example of possible arrangement of a corotating twin-screw extruder for a multiprocess food extrusion system. (Courtesy of Coperion Inc., Ramsey, New Jersey.)

analysis of a twin-screw corotating extruder. The analysis was unidirectional and accounted for non-Newtonian and nonisothermal melt rheology.

DIE FORMING

The forming process encompasses passing the melt through a specifically designed die opening. A certain pressure drop takes place across the die, depending on the die geometry, the melt viscosity, and the flow rate:

$$\Delta P \approx \frac{Q\mu}{K} \tag{2.14}$$

where K is the die conductance factor, which equals the reciprocal of die resistance. Equation 2.14 assumes a laminar flow through the die with no material slippage on the metal surface. Slippage results in a plug flow-like pattern and can be promoted by the use of smooth and low friction die material. Die heating may further decrease the friction factor, thus increasing the slippage further. Material slippage is expected to result in a lower pressure drop across the die. This may impact the total extruder fill, mixing intensity, and energy input. From a product forming point of view, the main advantage of die slippage is the formation of a smooth and intact product surface. That is why Teflon inserts are used in pasta product dies.

Food melts rheology analysis at different shear rate (volumetric output) is more complicated than that of thermoplastic materials. One needs to take account of the changes in the melt characteristics with changing feed rate as it impacts the time–temperature history in the extruder.

In most food extrusion operations, the extrudate swells after exiting the die (Figure 2.14).

The polymer's large and complex molecular structure gives rise to an elastic nature. This is believed to be largely responsible for this phenomenon (Tanner 1970; Jackopin and Herman 1978). Swelling is a form of elastic recovery from deformation the polymer gets exposed to in the die. Swelling increases with abrupt (not profiled) die flow entry and by shortening of the die land length. It also increases with increasing the shear rate within the die. In a circular die, the shear rate can be written as:

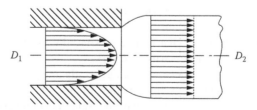

Velocity profile in the die exit region

Extrude swelling, $D_2/D_1 > 1$

FIGURE 2.14   Extrudate swelling at the die exit.

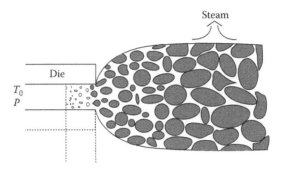

FIGURE 2.15   Product expansion after exiting the die due to vapor pressure difference.

$$\gamma = \frac{Q}{\pi \cdot r^3} \tag{2.15}$$

where Q is the volumetric feed rate within the die and r is the die radius. In cooking extrusion, the extrudate may also expand when the melt temperature behind the die significantly exceeds the boiling point of water (Figure 2.15). The expanding medium is water and the process is driven by its vapor pressure difference with the exit atmosphere.

With complex extruded food product shapes, the design and development of the dies have been based more on logical and practical methods rather than on analytical methods. These shapes are further advanced in cold extrusion such as pasta and half products. Stream lining of material flow in cooking extrusion applications is important to avoid material stagnation. Such material can cause quality defects, process instability, and potential blockage of small die areas.

MIXING

The basic mechanism of mixing in polymer extruders is a convective motion in the laminar flow region. The mixing action generally takes place by shear and elongational flow. It is called distributive mixing when the mixed components do not exhibit a yield point. Distributive mixing can be described by the extent of deformation or strain the fluid elements are exposed to. Dispersive mixing, on the other hand, is like a milling action and involves the breakdown of solid particles and agglomerates. The actual stress is important in this form of mixing, to overcome the yield stress of the solid component. Distributive mixing always exists when dispersive mixing takes place, but not necessarily vice versa.

Distributive mixing in extruders has been generally analyzed by determining the velocity profiles occurring in the screw channel. In most of the analyses, the fluid is considered Newtonian, the components have the same flow properties, and the flow through the flight clearance is neglected. The mixing performance improves with increasing pressure-to-drag flow ratio (throttle ratio). In choke fed extruders, this will reduce the extruder output. In all the cases, it will increase the average residence time and widen the residence time distribution (more back mixing).

Special mixing elements have been designed for a single-screw extruder to promote distributive mixing (Rauwendaal 1986, 1998). Distributive mixing element types include pin, dulmage, saxton, pineapple, slotted flight, and a cavity transfer mixing section. The quantitative analysis of these mixing elements is quite difficult, and development of these devices has been mostly empirical using experimental evaluation methods. When used in a thermoplastic single-screw extruder, they are typically placed at the discharge end of the metering section. Modular single-screw (such as Wenger Manufacturing) and twin-screw extruder manufacturers use kneading paddles, cut flight screws, and discs to promote distributive mixing. These elements are placed in any logical location within the extruder.

Dispersive mixing is critical in compounding and some food extrusion applications. The breakdown stress of agglomerates depends on their size, shape, and nature. The stresses acting on the agglomerate will depend on the flow and rheological properties of the polymer field. The higher the viscosity, the greater will be the dispersive mixing. High-speed solid–solid mixing can also cause a higher level of agglomerate breakdown than liquid–solid mixing. Several dispersive screw elements have been used in extruders (Rauwendaal 1986). These include the Union Carbide, Egan, Dray, and Blister rings. In all of these cases, the agglomerates must pass over a specifically designed clearance between the mixing element and the barrel. They are subjected to a high shear stress, which breaks them down and enhances mixing.

Twin-screw extruder manufacturers generally provide wide kneading paddles to promote dispersive mixing. It is not possible to create tight clearance dam areas in intermeshing twin-screw extruders with side-by-side equal diameter discs. In this case, the maximum disc diameter is equal to the center distance between the two screw shafts. The created tolerance/gap with the barrel will be equal to half of the screw channel depth. A tight clearance dam can be best created by placing two sets of intermeshing discs (following each other) on the screw shafts (Figure 2.16). Baker

Thin discs   Wide discs

FIGURE 2.16   Intermeshing orifice plugs (discs) for severe screw restriction and dispersive mixing. (Courtesy Baker Perkins, Grand Rapids, Michigan.)

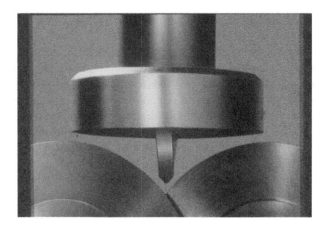

FIGURE 2.17 Intermeshing discs with a barrel valve (vane) in the open position. (Courtesy of Baker Perkins, Grand Rapids, Michigan.)

Perkins designed and developed this restrictive disc arrangement with an adjustable barrel valve (vane shape) placed at the discs location (Figure 2.17). The valve can adjust the flow resistance of the intermeshing discs, providing an added tool for controlling the degree of extruder fill in that section.

PRESSURE DEVELOPMENT WITHIN THE EXTRUDER

As stated earlier, most of the food single and twin-screw extruders are normally starved fed. Their throughput is determined by the feeding system. Accordingly, the screws in the solid conveying section remain partly full with negligible mechanical work done on the feed material. The screw filling takes place in the successive sections due to: (i) reduced conveying capacity of the forwarding screws or kneading elements, or (ii) placement of obstructions in the form of kneading elements, discs, reverse screws, or die plate. As the resistance to the flow increases, the feed material (solid and liquid) is compacted, mixed, and transformed into a fluid/melt through a hydration process accompanied by solid–solid/solid–fluid friction and dissipation of mechanical energy. This transformation/melting process can involve a significant temperature rise, depending on the generated friction field and material viscosity.

The length of the conveying screw section to generate the necessary pressure for overcoming restrictions is affected by the screw design, effective material viscosity, and the screw speed. The pressure gradient for a Newtonian fluid can be generally represented by:

$$\frac{dP}{dx} \approx \frac{K_1 DN\mu}{H^2 \tan\varphi} \tag{2.16}$$

where K_1 is a constant and is influenced by the number of screw flights, the degree of intermeshing and gaps between the screws and between the screw tips and the barrel; D is the screw diameter; N is the screw speed; μ is the material viscosity; H is the flight height; and φ is the screw flight helix angle. It can be seen from Equation 2.16 that for a given extruder, the length of the melt pumping section length decreases with increasing screw speed, melt viscosity, and decreasing screw channel depth and helix angle. This equation applies to situations where the material is in a fluid (melt) form either completely or partially (acting as the continuous phase).

The pressure gradient across the resistance sections, being made of kneading elements, discs, reverse screw, or die plate, for a Newtonian fluid can be generally represented by:

$$\Delta P \approx \frac{Q\mu}{K} \qquad (2.17)$$

where Q is the melt volumetric flow rate and K is the conductance factor of the restrictive section. For a reverse kneading/screw element section, K will be affected by the formed helix angle, the channel depth, the screw gaps, and the screw diameter. At steady state, this section will always be full and material forward flow takes place by a pressure drop within the section. The pressure flow should be greater than the drag flow component of the restrictive elements of this section.

For a discharge die obstruction such as a circular orifice, K can be represented by:

$$K = \frac{\pi \cdot r^4}{8L} \qquad (2.18)$$

where r is the die radius and L is the die land length. The pressure drop within a resistance section increases with increasing flow rate, viscosity, and the restrictive value of the section.

Figure 2.18 displays an example of a twin-screw extruder design showing the likely pressure and temperatures within the extruder.

ENERGY CONSUMPTION AND SUPPLY WITHIN THE EXTRUDER

ENERGY CONSUMPTION

The energy supplied in the extruder is mostly consumed by the product. Some inevitable losses also take place from the barrel, the gear box, and the motor. The energy consumed by the product results in the following changes:

1. Enthalpy rise, q_h, (kW):

$$q_h = m \cdot c_p \cdot \Delta T \qquad (2.19)$$

where m is the total extrudate flow rate, kg/s, c_p is the material specific heat, kJ/(kg·°C), and ΔT is the material temperature rise, °C. In cooking extrusion applications, the enthalpy increase accounts for the maximum consumed energy.

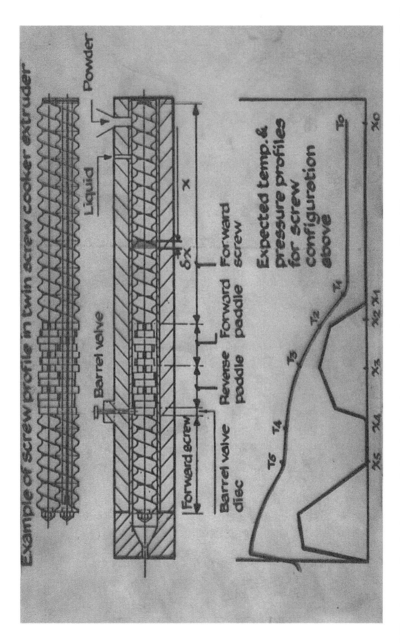

FIGURE 2.18 An example of twin-screw design showing the pressure and temperature profiles within the extruder. (Courtesy of Baker Perkins, Grand Rapids, Michigan.)

2. Heat of reaction/transformation such as heat of starch gelatinization and molecular breakdown. Gelatinization energy has been determined for different starches, and it is estimated to be in the range of 10–20 kJ/kg (Sablani 2007). These values are typically measured at excess water concentrations, accounting only for the gelatinization endotherm. In high viscosity extrusion applications (at low moisture concentration), starch and other biopolymers undergo a fractionation (depolymerization) process that is also expected to consume some energy. This energy is hard to measure and difficult to quantify.
3. Potential Energy, q_p (kW):

$$q_p = \frac{Q \cdot \Delta P}{1000} \tag{2.20}$$

where Q is the material volumetric flow rate in m³/s and ΔP is the pressure drop at the die, N/m².

The material kinetic energy is generally too small and can be ignored. Energy losses:

1. From the barrel by a circulating liquid in the barrel jacket, q_{cl}, (kW):

$$q_{cl} = m_c \cdot c_p \cdot \Delta T \tag{2.21}$$

where m_c is the coolant flow rate, kg/s, c_p is the coolant specific heat, kJ/(kg °C), and ΔT is the coolant temperature increase, °C.
2. From the barrel in the form of convection cooling by the surrounding air.

$$q_{c2} = U \cdot A \cdot \Delta T_{lm} \tag{2.22}$$

where U is the convection heat transfer coefficient, kW/(m² °C), A is the outside barrel area, m², and ΔT_{lm} is the logarithmic temperature difference between the barrel surface and air, °C.
3. From the gear box and the motor. This comes in the form of forced cooling to remove generated heat and maintain the gear box and motor at acceptable temperatures. The losses can be estimated by calculating the applied cooling load. They are generally in the range of 5%–15% of the motor energy, depending on the extruder size and operation.

ENERGY SUPPLY

Energy is supplied to the material from the extruder motor, barrel heat transfer, and in some cases through direct steam injection.

MECHANICAL ENERGY INPUT

The mechanical energy is provided by the extruder motor. It is mostly consumed by the material in the filled sections of the extruder through friction and viscous heat dissipation. For a Newtonian fluid, the mechanical energy input (E) can be calculated from:

$$\int dE \approx \int C\mu \cdot N^2 dx \qquad (2.23)$$

where C is a constant representing the screw design. The food material generally exhibits a pseudoplastic non-Newtonian flow behavior. Together with the effect of moisture (M) and fat (F) content and temperature (T), the viscosity can be empirically represented by:

$$\mu = \mu_o \cdot e^{-aM} \cdot e^{-bF} \cdot \gamma^{n-1} \qquad (2.24)$$

where μ_o is a reference (initial) viscosity and γ is the shear rate within the screws. The latter is linearly proportional to the screw speed.

It can be seen that viscous heat dissipation can be increased by increasing the length of the screw filled section, the material viscosity, and the screw speed. It is important to remember that increasing the screw speed results in a decrease of the melt pumping sections length. On the other hand, the resistance sections (discs, reverse kneading elements, and screws) will always be full and their fill length is constant, regardless of the screw speed or the feed rate.

The gross specific mechanical energy input (kWh/kg) for a direct current (DC) motor with constant maximum torque can be estimated from Equation 2.25:

$$S = \frac{\%N * \%T * P}{F} \qquad (2.25)$$

where N is the maximum extruder screw speed (rpm), T is the maximum motor torque (Nm), P is the total installed motor power (kW), and F is the feed rate (kg/h). The screw speed and torque are proportional to the motor's armature voltage and current, respectively. For an alternative current (AC) motor, the actual power should be obtained from a watt meter.

HEAT TRANSFER THROUGH THE BARREL

The rate of heat transfer (kW) to or through the extruder barrel can be estimated from:

$$q_b = U \cdot A \cdot \Delta T_{lm} \qquad (2.26)$$

where U is the convection heat transfer coefficient (kW/(m² °C), A is the internal barrel surface area (m²), and ΔT_{lm} is the logarithmic temperature difference (°C) between the internal barrel surface and the material in contact.

The heat transfer coefficient in the nonfilled sections is relatively small due to the lack of close contact with the surface. The heat transfer coefficient is affected by the material's physical properties and the mixing pattern. It generally increases with decreasing viscosity and increasing screw speed. Levine and Miller (2007) presented a review of the published research works on barrel heat transfer. The heat transfer coefficient has been estimated to be in the range of 50–300 W/m² °C.

The role of barrel heating becomes more important and significant in low viscosity extrusion cooking applications such as those containing high moisture, sugar, and/or fat concentration. The rate of mechanical energy input decreases in such applications. In high melt viscosity applications, motor mechanical energy is typically the main source.

Heat can also be transferred from the material by applying cooling to the barrel jacket. Product cooling results in melt viscosity rise, causing an increase in the rate of mechanical energy input. The net product cooling is therefore lower than the overall cooling load taken out from the barrel.

Barrel heat transfer per unit output can be greater in small extruders than in larger ones. In both cases, barrel temperature control is important even when the heat transfer rate is insignificant. As mentioned earlier, the material friction coefficient with the metal surface may be influenced by the surface temperature.

Steam Injection into the Barrel

Water is typically a natural constituent of the total food formula. A portion of the total water can be directly injected into the extruder in the form of steam. This can bring a significant amount of energy, thus supplementing and/or substituting part of the motor and barrel heating energy. The steam energy q_s (kW) can be estimated from Equation 2.27.

$$q_s = m_s \cdot \lambda_s \qquad (2.27)$$

where m_s (kg/s) and λ_s (kJ/kg) are the steam mass flow rate and enthalpy, respectively.

Steam is usually injected in the nonfilled sections of the extruder. Control of the steam flow rate is important to achieve steady state operation. It is also necessary to have a highly filled section between the injection port and the dry mix feed port to avoid steam blow back. This can be achieved by the use of reduced conveying capability screw elements (small pitch or cut flight screws) or restricting discs "shear locks."

EFFECT OF SCREW/BARREL WEAR ON THE EXTRUDER PERFORMANCE

The extruder functions as a pump and a thermomechanical reactor. These functions are influenced by its screw design including the various gaps within the extruder. With use, the screw elements and barrel material undergo a process of wear. The wear rate will depend on the abrasive, adhesive, and corrosive nature; the parts material of construction; and the operating conditions. The increase in the gap dimension of forward conveying sections decreases the extruder's pumping efficiency and increases the extent of back mixing and retention time. The total mechanical energy input to the product is likely to increase in this case. This may also reduce the extruder output if it is operating at or close to a choked feed manner.

On the other hand, the restriction sections within the extruder and die assembly may also experience wear. The resulting larger gaps in this case will reduce the flow resistance of these sections, requiring a lower pressure drop across them. That may decrease the extruder fill length of the proceeding pumping section and the total energy input to the product.

For thermally sensitive materials such as food products, larger gaps and probable material stagnation could potentially result in product deterioration and quality defects, possible die blockages, and process instabilities.

Adjustments in the product formula (particularly the added water rate) and operating conditions can in some cases compensate for the gradual wear and prolong the useful life of extruder parts. The screw profile may also be altered so as to adjust the melt filled section length. In a scenario where the extruder length is longer than necessary, one can move the filled (working) sections up or down the screw to make use of the nonworn sections of the barrel liner. This, however, needs adequate processing knowledge on the effect of screw configuration on the process and product performance. Eventually, a situation will be reached where the product and/or output specifications cannot be met and parts replacement will have to be made. With new parts, the extruder conditions may have to be adjusted again if they have been altered during the parts wear lifetime.

VENTING WITHIN THE EXTRUDER

Venting within a cooker extruder can be very useful for a number of objectives, such as accelerated cooling, removal of certain volatile components, and affecting certain changes within the product. A specifically designed screw is needed to allow for venting in a nonfilled section. Typically, a high volumetric flow capacity screw section is positioned at the venting port succeeding a restrictive/low volumetric flow capacity screw section acting as a gas-tight seal (Figure 2.19).

The melt rheology and temperature affect the behavior of the melt and the extent of volatile removal from a venting section. Relatively low viscosity melts (for example, those of a high initial moisture content) can be successfully vented. With reduced initial moisture/fat content and/or excessive vapor release, the vented melt may produce highly expanded mass that is hard to convey forward. The use of a stuffer (Figure 2.20) will help in certain circumstances.

In applications such as cooked half products, considerable product cooling is required to avoid product expansion and achieve good die face cutting. The optimum extrusion process design will depend on the nature of the formula and the desired degree of cooking. If this can be achieved at a relatively low temperature, cooking, cooling, and forming can be carried out in a single extruder with or without the use of a venting port.

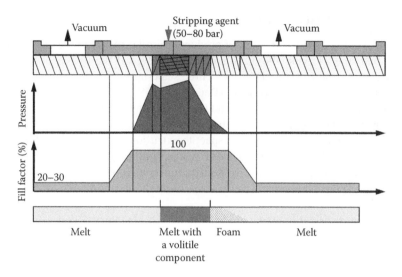

FIGURE 2.19   An example of a screw design with two venting ports for degassing of volatiles. (Courtesy of Coperion Inc., Ramsey, New Jersey.)

In applications where a relatively high temperature (say above 135°C in the cooking section) and at high production capacities, the use of "two extruders" system is typically preferred. The first extruder cooks the material. Venting and cooling are achieved between the two extruders. The second extruder compresses the expanded material and forms the cooled product to the desired shape. The two extruders can be thus designed and operated independently. In this manner, the cooking and the forming extruders are utilized to their maximum throughput and efficiency without having to compromise, as it may be the case in a single cooking/forming extrusion system. Forming extruders are typically of a single-screw type similar to those used in the pasta product applications.

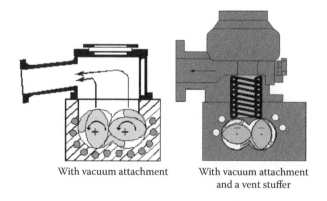

FIGURE 2.20   Examples of vent port adaptor designs. (Courtesy of Coperion Inc., Ramsey, New Jersey.)

SCALE-UP CONSIDERATION

Scale-up can be defined as the task of producing similar results in the larger system for the first time (Levine 1989). This can be achieved if the treatment regime in both scales of production is the same. Three scale-up methods are considered for thermoplastic single-screw extruders (Rauwendaal 1986). Scale-up for constant (i) "common" factors, (ii) heat transfer, and (iii) for mixing. Scale-up for constant mixing is generally the most common method used in food extrusion (Levine 1989; Yacu 1992). It is the most attractive with respect to extrusion output increase. It requires the following guidelines: (i) geometrical design similarity is maintained between the small and the large extruders, (ii) both extruders are operated at the same screw speed, and (iii) the die resistance is the same. With these conditions:

1. The volumetric throughput of the extruder is scaled up as follows:

$$V_2 = V_1 \left(\frac{D_2}{D_1} \right)^3 \tag{2.28}$$

   This relationship will also apply to the volumetric/mass holdup within the extruder. Therefore, the average residence time and distribution should be similar in both extruders.

2. The maximum installed motor power is limited by the screw speed and extruder shaft torque. Luckily, the torque (T) for a similar shaft material of construction and design can be scaled up in a similar way to that of the volumetric output.

$$T_2 = T_1 \left(\frac{D_2}{D_1} \right)^3 \tag{2.29}$$

   The specific mechanical energy input can therefore be maintained constant since the production throughput and motor power are increased in the same proportion.

3. The barrel heat transfer to or from the product scales up according to Equation 2.30.

$$q_2 = q_1 \left(\frac{D_2}{D_1} \right)^2 \tag{2.30}$$

   Equation 2.30 indicates that the available rate of heat transfer does not increase in the same proportion as that of throughput and the motor power. This makes the scale up process more complicated when heat transfer through the barrel is significant and important.

   In practice, there has been a greater degree of freedom in scaling up food extruders than plastic extruders. Substituting part of the water with steam can potentially reduce the dependence on barrel heating in large extruders. Similarly, the removal of part of the water as steam in a venting step can minimize the dependence on barrel cooling in large extruders. Barrel steam injection and removal processes can follow scale-up for mixing guidelines. Steam is also often used to preheat the formula ingredients in a mixing cylinder generally called a "preconditioner." Heat transfer through the preconditioner barrel is typically not significant. Therefore, it can be scaled up and operated using the same "constant mixing" scale-up method.

FOOD EXTRUDER SELECTION AND DESIGN

The extrusion process has been successfully utilized for a large range of food products. It covers both low energy (forming) and high energy input (cooking) applications. The extruder feed formulas are very diverse, producing a wide range of melt rheology. The extruder selection and design need to be tailored to the specific application and operation requirements. This must deliver optimum system performance and consistent product quality at the desired (preferably maximum) economic output. This can be achieved by:

- Properly understanding and delivering the necessary extruder functions (conveying, mixing, melting, metering, and others)
- Supplying the optimum balance of mechanical and thermal energy
- Providing the material the necessary retention time with a controlled and acceptable degree of residence time distribution

The above approach can be logically and practically addressed by the following factors analysis:

EXTRUDER TYPE

As discussed earlier, single-screw extruders generally cost less and have been successful in a large number of applications. The adoption of a variable screw speed drive and improved instrumentation and control has improved their performance over the early generation single-screw extruders. They are typically the extruders of choice for low shear forming applications. They are also widely used in pet food, snacks, and other extrusion cooking applications. Figure 2.21 describes an example of a single-screw food extrusion system with a vent port fitted with a vent stuffer.

Single-screw extruders remain physically inadequate for applications with difficult-to-handle formulations and functional demands. Under such challenging circumstances, they may not be able to deliver acceptable process performance, stability, and consistent product quality.

Twin-screw corotating extruders are more expensive. They have, however, demonstrated their ability, sometimes with better economic performance to make products like those typically made on single-screw extruders. Their use has been extended to the more challenging applications that single-screw extruders are not able to perform well. The use of counterrotating twin-screw extruders (operating at a relatively low screw speed) has been limited to low melt viscosity applications requiring high pressure generation capability.

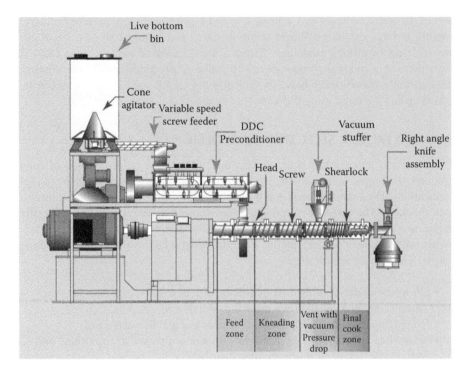

FIGURE 2.21 An example of a single-screw extrusion cooking system with a steam preconditioning cylinder. (Courtesy Wenger Manufacturing, Inc., Sabetha, Kansas.)

EXTRUDER DIAMETER

The extruder diameter is generally determined from the required production output and operating screw speed. For a specific output, the lower the screw speed, the larger would be the extruder diameter. In some cases, the die pressure and screw shaft torque may also play a role in determining the minimum extruder diameter. The choice of extruder diameter is determined by the equipment manufacturer. Typically, these may range from very small (15 mm) for bench top and pilot plant operations to very large (300 mm) for production operations.

EXTRUDER LENGTH TO DIAMETER (*L/D*) RATIO

The main factors affecting extruder length are:

- Total energy required to produce the desired product. This can be estimated from energy consumption analysis and extruder losses.
- Formula and the nature of the fluid/melt viscosity it will subsequently generate.
- Availability of steam if it can be used as a substitute of part of the required water to supplement the total energy input.

- Available extruder motor power and screw speed.
- Operating extruder pressure.
- Processing functional requirements of the extruder. Whether it is a straight forward conveying, melting, and forming operation; or the one that has other functions such as venting/cooling, additional downstream solid, liquid, and/ or gas injection steps and/or other desired operations.
- Optimum screw and barrel design. Some extruders are used to make more than one product requiring different screw designs. In these cases, the extruder length will be determined by the product/application that requires the longest functional screw design. Pilot plant and other test extruders are necessarily long to accommodate different product needs. The active extruder length can obviously be much shorter than the actual extruder length.

Generally speaking, for similar total energy input applications and production output:

- High viscosity feed formulas require relatively short extruders with the likelihood of relying mostly on motor energy. Viscous heat dissipation of mechanical energy is rapid and can heat the material in a relatively short time. The available motor energy is proportional to the screw speed. Therefore, extruders with larger motors and higher screw speeds can provide the energy at a faster rate and their required length can be even shorter. One should be careful with this since higher screw speed decreases product residence time. The input parameters (feeders, pumps, and temperature control) will require greater accuracy and control with decreasing processing time to avoid extruder instabilities and product quality fluctuation.
- Steam energy utilization reduces the demand for motor energy and barrel heat transfer and can result in shorter extruders.
- The extruder length increases with increasing numbers of process functional steps. These may require specific retention time that will increase the extruder length requirement.

EXTRUDER MOTOR SIZE, ROTATION DIRECTION, AND SCREW SPEED

The motor size is determined from the extruder output and the rate of mechanical energy input per unit feed rate. This can range between 0.015–0.180 kWh/kg, depending on the generated viscosity field and extruder operating conditions. The extruder power delivered to the screw shaft is equal to:

$$\text{Power } (P) = \text{Screw speed } (N) \times \text{Shaft torque (T)} \qquad (2.31)$$

The screw shaft strength thus plays a major and more important role in high viscosity applications.

The screw rotation can be clockwise or counterclockwise and is determined by the extruder manufacturer. This has no impact on the extruder performance and analysis approach. It will obviously determine the screws helix angle (direction) with respect to the vertical axis. A right handed screw will move the material forward if it is rotating clockwise. A left handed screw moves the material forward when it is rotating counterclockwise. The locking screw cap (placed at the end of the screw shaft) will need to be tightened in the opposite direction of the screw rotation. This is essential to avoid undoing of the cap during extruder operation. Thus, the screw cap of a clockwise rotating screw shaft has a left handed locking screw cap.

Screw Design

As discussed earlier, the extruder provides many functions. The screw and barrel designs need to appropriately address these functions. The following is a general guideline on screw design selection principles. It should apply to both single and twin-screw extruders unless otherwise stated. Figure 2.22 shows screws with different designs.

Screw Pitch

The larger the screw pitch, the greater its conveying capacity will be, resulting in a shorter retention time in the nonfilled sections. Screws with a relatively long pitch (1–2 D long) are generally employed in the nonfilled sections such as the solid conveying and venting regions of the extruder. Medium length pitch screws (0.5–1 D) are used in the filled sections where efficient melting and melt conveying are desirable. As demonstrated earlier, the screw melt pumping efficiency generally improves with reducing screw pitch. Short pitch screws (0.25–0.75 D) are used in the extruder metering section particularly close to the die to generate necessary pressure in a relatively short distance.

Twin lead screw, 1.0 D pitch   Single lead screw, 1.0 D pitch   Twin lead reverse, 1.0 D pitch screw 0.25 D long

Screw twin lead , 0.4 D pitch   Single lead Screw, 0.25 D pitch   Twin lead reverse, 1.0 D pitch screw 0.5 D long

FIGURE 2.22   Examples of corotating screw elements. (Courtesy Baker Perkins, Grand Rapids, Michigan.)

Screw Channel Depth

It was shown earlier that the drag flow is proportional to the channel depth (H). For a Newtonian viscosity melt, the pressure flow is proportional to H^3. In single-screw extruders, screw channel depth can be independently chosen to deliver the optimum performance in a specific section. The channel is typically deep in the solid conveying section (up to 0.25 D) and shallow in the metering section (as low as 0.05 D). The influence of the screw channel depth on the extruder output was described in the section entitled "Single-Screw Extruders." At high die pressure, a shallow screw provides a more stable output and performance since it is less vulnerable to pressure flow fluctuations.

Screw channel depth is not an independent variable in parallel intermeshing twin-screw extruders. It is fixed by the extruder manufacturer. The manufacturers typically supply their extruders with the same channel/diameter ratio across the range of screw diameters they offer. A screw channel depth of around 0.2 D is common in twin-screw extruders. It provides a good balance of volumetric output, shear rate, and shaft torque.

Conical single and twin-screw extruders have variable channel depths along the screw length. These extruders are not widely used in food applications.

Number of Screw Flights

The screw flight width and the number of flights are independent variables in single-screw extruders. Increasing the flight width and/or the number of screw flights reduces the channel volume and screw conveying capacity. Generally, single flight screws are used in the solid conveying zone. Double and sometimes triple flighted screws are used in the melting/kneading sections of a food extruder. They provide a greater retention time and thus result in higher energy input. Single lead screws are commonly used in the metering section. A multiple flight screw (2–4 starts) is used at the discharge end to minimize the extruder output oscillation caused by the screw rotation. This is more important with the low speed extruders such as pasta and half-product snacks.

The flight width is not an independent variable in intermeshing twin-screw extruders. In corotating extruders, the single flighted screws have a wider flight tip per unit screw pitch than the total of that of double and triple flighted screws. This reduces the open channel volume in the intermeshing region, thereby restricting the flow path for pressure flow. Thus, single flighted screws are less sensitive to downstream pressure and can therefore provide a greater degree of positive displacement and narrower residence time distribution than a same pitch twin lead screw. A special screw element that converts the single flight to a double flight should be placed between a single and a twin flighted screw when these follow each other in the screw design. Otherwise, one of the flights of the twin flighted screw will not be able to deliver to or receive material from the single flighted screw.

Mixing and Restrictive Elements

Mixing and restrictive screw elements come in different forms and shapes. They are designed to induce a change in the flow pattern, increase extruder fill degree, and improve mixing. Along with such action, the material residence time distribution and energy input are also impacted. A general discussion of the two types of mixing, distributive and dispersive, was provided in the section entitled "Mixing." The following discussion will concentrate on some of the common mixing elements used in twin-screw extruders:

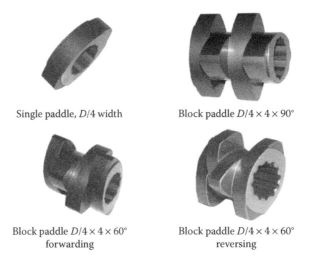

Single paddle, $D/4$ width   Block paddle $D/4 \times 4 \times 90°$

Block paddle $D/4 \times 4 \times 60°$ Block paddle $D/4 \times 4 \times 60°$
forwarding   reversing

FIGURE 2.23 Single and different paddle block arrangements. (Courtesy Baker Perkins, Grand Rapids, Michigan.) Please note that the Baker Perkins Extruder screws rotate counterclockwise.

Mixing Paddles/Kneading Block

An individual paddle represents a slice of an intermeshing screw. The paddles are made of different thicknesses (0.10–0.25 D). They are fitted to the screw shaft in a staggered fashion to produce certain angles from each other. As intermeshing elements, they are self-wiping and can provide a varying degree of conveying and mixing. They are provided by the manufacturers either as individual paddles or combined together as "blocks" (Figures 2.23 and 2.24).

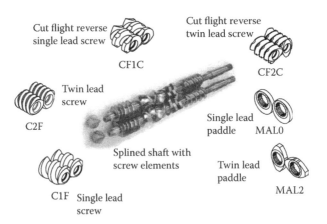

Cut flight reverse single lead screw

CF1C

Cut flight reverse twin lead screw

CF2C

Twin lead screw

C2F

Single lead paddle   MAL0

Splined shaft with screw elements

Twin lead paddle

C1F   Single lead screw

MAL2

FIGURE 2.24 Examples of corotating twin-screw elements and shaft assembly. (Courtesy of Clextral Inc., Firminy, France.) Please note that the Clextral and most other extruder manufacturers' screws rotate clockwise.

Single and twin start paddles and kneading blocks are used. The paddles can be typically staggered in steps of 15° from each other. The most common formed angles are 30°, 45°, 60°, and 90° (neutral). The formed paddle helix angle can be in the same direction (forward) or opposite direction (reverse) to that of forward conveying screws. Table 2.1 shows how twin lead paddles with different thickness and angles form.

The helix angle is calculated from Equation 2.32.

$$\tan \varphi = \frac{P}{\pi D} \tag{2.32}$$

It can be seen that the formed paddles pitch (P) and helix angle decrease with decreasing paddle thickness and increasing staggering angle. For the same paddle thickness, the leakage flow in "forward" paddles increases with decreasing pitch due to the greater open space between adjacent paddles on the same shaft. This results in a greater degree of back mixing and a reduction of the conveying capacity. The neutral paddles (angle=90°) have no conveying capacity. The following guidelines can be adapted in the use of paddles/kneading blocks in filled sections:

1. Narrow thickness forwarding paddles can provide good distributive mixing per unit distance with relatively low energy input.
2. Wide forwarding paddles can provide good dispersive mixing per unit distance with relatively high energy input.
3. Within the staggering angles listed in the above table, the smaller the staggered angle, the greater is their conveying output.
4. The above also apply to neutral and reverse paddles, taking into consideration that their conveying action is either none for the former or in the opposite direction for the latter. The drag flow through the reverse paddles section is backward. To get a forward flow, the pressure flow has to be greater than the drag flow. That is also true in other flow restricting elements. The 90° angle paddles have zero drag flow. As such the pressure drop across a 90° paddle section will be lower than that of a similar length reverse section with similar paddle thickness. The pressure drop will increase as the reverse paddles staggering angle decreases. The mechanical energy input is expected to increase with increasing pressure drop.

TABLE 2.1

Summary of Twin Lead Paddle Arrangements in a Corotating Extruder

	Narrow Thickness Paddles				Wide Thickness Paddles			
Paddle thickness (T, in D equiv.)	0.125	0.125	0.125	0.125	0.25	0.25	0.25	0.25
Paddle staggering angle (°)	30	45	60	90	30	45	60	90
Pitch (P, in D equiv.)	1.5	1.0	0.75	0.5	3.0	2.0	1.5	1.0
No. of paddles per pitch	12	8	6	4	12	8	6	4
Helix angle ϕ (°)	25.5	17.7	10.8	9.0	43.7	32.5	25.5	17.7

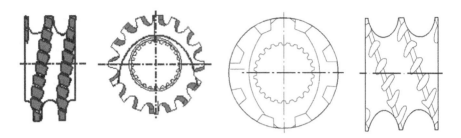

FIGURE 2.25 Example of cut flight mixing screw elements. (Courtesy of Coperion Inc., Ramsey, New Jersey.)

Cut Flight Screws

Standard screws are sometimes cut in a certain way to increase their leakage flow and improve distributive mixing (Figures 2.24 and 2.25). This can be done to screws with forward as well as reverse conveying direction. In forward conveying screws, the net drag flow is reduced, resulting in increased residence time distribution. In reverse conveying cut screws, the net drag flow is also reduced, resulting in decreased pressure drop, residence time, and energy input. Note that there is no self-wiping of the screws in the cut flight region. That could be a disadvantage with sticky and stagnating materials.

Reverse Screws

In other cases, solid reverse screws are used to create significant flow resistance within the extruder. Single or twin lead screws have been used. To avoid excessive restriction, the reverse screw section is typically short (up to 0.5 D long) with a medium helix angle (17.7°). Sometimes, reverse and forward elements (0.25 D length) are placed after each other to create a mixing section of up to 1.5 D long. In twin-screw extruders, a spacer has to be positioned between these successive elements to avoid interference with the screw elements on the other shaft. By design, reverse screw elements provide a significant degree of back mixing, thus widening the residence time distribution. Like other restricting elements, the pressure flow across reverse screw needs to be greater than their drag flow to achieve a net forward flow.

Throttle Valve

A throttle valve assembly is typically placed at the end of the extruder before the die assembly. It can provide additional control of the degree of extruder fill behind the die without having to change the screw design. This feature is particularly useful for multiproduct extrusion operations requiring different screw designs. In some cases, the throttle valve is designed to provide convenient start-up/shutdown material diverting capability (Figure 2.26).

INFLUENCE OF DESIGN AND OPERATING CONDITIONS ON THE EXTRUDER PERFORMANCE

The discussion in this section will aim at outlining the effect of major independent variables on the following response variables:

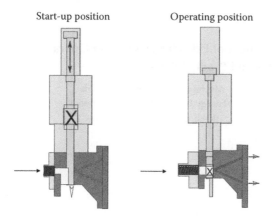

FIGURE 2.26 Combined start-up with butterfly throttle valve. (Courtesy of Coperion Inc., Ramsey, New Jersey.)

a. Average retention time
b. Product pressure at the die
c. Product temperature at the die
d. Specific mechanical energy input
e. Motor torque
f. Product density, assuming that the product is expanding. It is important to note that product expansion (as it emerges through the die) takes place in two directions, radial (sectional) and longitudinal (axial). For a similar expansion degree, these elements may not be the same, depending on the extent of extrudate swelling at the die.

The discussion will be general and intended for starved fed, multi-ingredient food extrusion cooking applications. The dependent variables response may not be linear and can vary in magnitude, depending on the application. The trend of the independent variables effect is summarized in Table 2.2.

Screw Design

Increasing the internal flow resistance within the extruder by placing flow restricting elements such as neutral and reverse kneading paddles/screws and discs will increase the melt filled section and retention time. This will increase the mechanical energy input requiring higher motor torque and resulting in higher product temperature. The melt viscosity at the die plate decreases, producing a lower die pressure. The increased product temperature gives a higher vapor expansion, resulting in lower density extrudate. The radial expansion may not increase due to lower viscosity, resulting in lower extrudate swell at the die.

Die Design

Increasing the die resistance to the melt flow by reducing the number of die outlets or decreasing the physical dimensions of the opening area would generate a higher

TABLE 2.2

Summary of the Effect of Extruder Design and Operating Conditions on the Extrusion Process and Product Response

Independent Variable	Mechanical Energy Input (kWh/kg)	Motor Torque (Nm)	Die Product Temp. (°C)	Die Product Pressure (psi)	Expanding Product Density (g/l)
Screw design, increasing restriction to the flow "stronger screw"	↑	↑	↑	↓	↓
Die design, increasing resistance to the flow	↑	↑	↑	↑	↓
Increasing barrel temp	↓	↓	↑	↓	↓
Increasing screw speed	↑	↓	↑	↓	↓
Increasing feed rate	↓	↑	↓	↑	↑
Increasing moisture content	↓	↓	↓	↓	↑
Increasing fat content	↓	↓	↓	↓	↑
Increasing sugar content	↓	↓	↓	↓	↑
Increasing high water absorbing fiber content	↑	↑	↑	↑	↑
Increasing low water absorbing fiber content	↓	↓	↓	↓	↑
Increasing high water absorbing protein content	↑	↑	↑	↑	↑
Increasing low water absorbing protein content	↓	↓	↓	↓	↑

pressure drop across the die. This would require a longer melt pumping section before the die, producing a longer retention time, and a higher rate of mechanical energy input. The latter will result in a higher motor torque and product temperature at the die. The resulting lower melt viscosity produces a lower product pressure increase than otherwise would be the case. The higher shear rate within the die can also impact the resulting pressure, depending on the non-Newtonian nature of the material. The product density decreases because of the higher melt temperature. The radial expansion is expected to increase due to greater extrudate swell.

BARREL TEMPERATURE

Increasing the barrel temperature increases the melt temperature, resulting in a decrease of the viscosity within the extruder. This reduces the mechanical energy input and motor torque. The die pressure generally decreases though it may not be significant. The product density is expected to decrease. The radial expansion may decrease due to lower extrudate swell at the die.

Barrel temperature can also influence the conveying efficiency of the extruder. That is because it may change the friction factor between the material and the barrel liner. Usually, the hotter the surface, the lower will be the friction factor, resulting in greater slippage and a higher degree of fill in the extruder. This influence can be more pronounced in single-screw than in twin-screw extruders. Therefore, increasing the barrel temperature may actually also result in a higher torque and a higher mechanical energy input.

Screw Speed

Increasing the screw speed increases the shear rate and mechanical energy input in all the melt filled sections. However, it also reduces the length of the melt pumping sections. It generally results in a higher mechanical energy input (but lower motor torque), greater melt temperature, lower die pressure, and higher product expansion.

With low conveying capability screw elements and paddles, the degree of fill may significantly rise with screw speed reduction below a certain threshold level. In this case, the reverse of the above result takes place, causing a step rise in the motor torque and mechanical energy input.

Screw speed has a more significant influence on the energy input in highly restricted screw configurations, such as those containing a high portion of reverse paddles/screw elements. The degree of fill in restricted, nonforward flow sections does not change regardless of the screw speed.

Feed Rate

Increasing the feed rate with an equivalent increase in the die openings generally reduces the average retention time in the extruder. This results in a lower specific mechanical energy input, lower product temperature, higher melt viscosity at the die, and higher extrudate density. The motor torque increases but not linearly with the feed rate.

The drop in the specific mechanical energy input could be explained by the fact that the total melt length does not linearly increase with feed rate. It increases only in the forward conveying sections. The restrictive sections fill length remains constant regardless of the feed rate. However, if a larger proportion of the forwarding elements become full (for example, small pitch screws getting choked), the rate of mechanical energy input increases with increasing feed rate above a certain threshold.

Moisture Content

Water has a significant effect on the viscosity of food materials. It also affects the temperature at which starch is gelatinized or melted. With decreasing melt viscosity, the specific mechanical energy input decreases, resulting in a lower motor torque, lower product temperature, and a higher product density. Generally, the effect of moisture content on the die pressure is higher than that of the resultant temperature decrease. Therefore, the die pressure generally decreases with increasing moisture content.

Fat Content

Liquid or solid fat affects the extruder performance in different ways depending on its type and concentration and the extruder design and operating conditions. If the added fat is well mixed with the feed, it acts as a viscosity reducing agent and affects the extruder in much the same way as water. Oils and fats, however, do not reduce the starch melting temperature. At relatively high oil/fat concentration, the extruder may not be able to generate adequate energy (through viscous heat dissipation) to cook the starch. The use of steam in such cases becomes necessary. At high fat concentration and high extrusion temperatures, some fat stripping may take place at the die exit with the flashed off steam.

If the added fat cannot be thoroughly mixed before or within the extruder, it may lubricate the screws and barrel surfaces, thus reducing their friction factor and extruder mixing performance. The increase in slippage on the barrel surface reduces the drag flow component and the pumping efficiency of the extruder. These conditions may also be accompanied with significant fluctuation in the degree of screw fill and pressure profile, causing severe extruder instability.

Sugar Content

Sugar is highly soluble in water and has a lower melting temperature than starch. Both of these factors contribute to reducing the material viscosity. Its effect on the cooker extruder performance in this respect is that of a viscosity reducing agent, but to a lower extent than that of water. At relatively low sugar concentrations, the extruder is generally able to raise the temperature through viscous heat dissipation of mechanical energy input. At very high concentrations, it becomes more difficult and barrel heating or the use of pregelatinized starch may be necessary.

At low sugar concentrations, the product density is not greatly affected. Generally, however, the density increases with increasing sugar concentration.

Fiber Concentration

The effect of fiber on the extruder performance is not consistent. This is likely impacted by the water absorption index (WAI) of the individual fibers under the prevailing extrusion conditions. High WAI fibers (such as psyllium) increase the material viscosity and the mechanical energy input. On the other hand, increasing wheat fiber (bran) concentration at the expense of wheat flour has a neutral to slightly reducing effect on the mechanical energy input. The die pressure generally decreases with increasing wheat bran concentration.

Unlike starch, fibers alone do not have the ability to form highly expanded porous structures. Thus, increasing their concentration at the expense of starch generally increases the product density. Fibers, like other filler ingredients, influence the expanded product and produce a smaller cell size structure. With some fibers, this is very pronounced even at very low concentrations when the impact on density is not yet significant.

Protein Concentration

The influence of proteins on the extruder performance is also inconsistent and likely to be dependent on their WAI and impact on material viscosity characteristics.

In this way, they perform in a similar manner to that of fibers. Generally, cereal and legume proteins are not as good as starch when it comes to forming a highly expanded porous structure. Therefore, increasing their concentration at the expense of starch generally increases the expanded product density.

Salt Concentration

Salts are included in the feed mix to impart certain taste and nutritional and functional attributes. Sodium chloride is the most common, but other salts are also incorporated. Due to their relatively low concentrations, their influence on the extruder performance is generally not significant. They do not expand and, like fibers, act as fillers diluting the starch concentration. However, certain salts such as sodium bicarbonate can influence the expansion process as well as the cell structure.

REFERENCES

Booy, M.L. 1980. Isothermal flow of viscous liquids in corotating twin screw devices. *Polym Eng Sci* 20: 1220.

Darnell, W.H. and Mol, E.A.J. 1956. Solids conveying in extruders. *SPE J* 12: 20.

Dekker, J. 1976. Verbesserte schneckenkonstruktion fur das extrudieren von polypropylen. *Kunstoffe* 66: 130.

Harman, D.V. and Harper, J.M. 1974. Modeling a forming food extruder. *J Food Sci* 39: 1099.

Jackopin, S. and Herman, H. 1978. An analysis of the conveying characteristics of twin screw extruder mechanism. *ANTEC* 24: 498.

Janssen, L.P.B.M. 1978. *Twin Screw Extrusion*. Amsterdam: Elsevier Scientific Publishing.

Jasberg, B.K., Mustakas, G.C. and Bagley, E.B. 1979. Extrusion of defatted soy flakes-model of a plug flow process. *J Rheology* 23(4): 437.

Klenk, P. 1968. Visuelle Untersuchungen zum Wandgleitverhalten hochpolymerer, Schmelzen, *Plastverabeiter* 21: 537.

Levine, L. 1989. Scaleup, experimentation, and data evaluation. In *Extrusion cooking*, eds. C. Mercier, P. Linko and J.M. Harper, 57–90. St. Paul, MN: American Association of Cereal Chemists.

Levine, L. and Miller, R.C. 2007. Extrusion processes. In *Handbook of food engineering*, 2nd edition, eds. D.R. Heldman and D.B. Lund, 799–846. Boca Raton: CRC Press.

Lindt, J.T. 1981. Pressure development in the melting zone of a single screw extruder. *Polym Eng Sci* 21: 1162.

Maddock, B. 1959. SPE ANTEC. *Tech. Papers* 15: 383–389.

Mailefer, C.H. 1959. Swiss Patent, 363: 149.

Rauwendaal, C. 1986. *Polymer Extrusion*. New York: Hanser Publishers.

Rauwendaal, C. 1998. Polymer mixing, A Self Study Guide, Carl Hanser Verlag, Munich.

Rowell, H.S. and Finlayson, D. 1928. Screw viscosity pumps. *Engineering* 126: 249.

Sablani, S.S. 2007. Gelatinization of starch. In *Food Properties Handbook*, 2nd Edition, ed. M. Shafiur Rahman, C., 292: CRC Press.

Tadmor, Z., Dukdevani, I. and Klein, I. 1967. Melting in plasticating extruders–Theory and experiments. *Polym Eng Sci* 7: 198.

Tanner, R.I. 1970. A theory of die-swell. *J Polym Sci* 8: 2067.

Yacu, W.A. 1985. Modeling a twin screw corotating extruder. *J Food Process Eng* 8: 1.

Yacu, W.A. 1992. Scale up of food extruders. In *Food extrusion science and technology*, ed. J.L. Kokini, C.-T. Ho and M.V. Karwe, 465–472. New York, N.Y.: Marcel Dekker Inc.

3 Raw Materials for Extrusion of Foods

Suvendu Bhattacharya

CONTENTS

INTRODUCTION

Extrusion of foods is a versatile approach of food process engineering combining several unit operations such as conveying, thermomechanical change and degradation, mixing, and shaping. There are major improvements in the

- Design of various extruders including the screw elements
- Control of the extruder and data capturing facility during extrusion
- Understanding of the extrusion behavior and product characteristics of food ingredients as evidenced from several hundred research publications and about 10 books

As a result, formulations can be altered to reduce the cost of raw materials and improve product attributes. In addition, because of the demand for newer foods, several preprocessing and postprocessing operations like moist heating, frying, drying, toasting, baking, flavoring, and coating have been added to make the products more attractive in terms of taste, texture, mouthfeel, and nutritional status; and also by creating different shapes and serving special purpose(s). It is thus obvious that many raw materials can be processed with the possibility of production of different products. Hence, it is desirable to discuss the raw materials with specific reference to products.

Based on the convenience of use, extruded products can be categorized into: (a) ready-to-eat (RTE), (b) half-product/intermediate (ready-to-process) products requiring baking/toasting/frying at the consumer end, and (c) raw material to be used for further processing and formulating other products.

A wide range of raw materials from various sources and different compositions can be fed to the same extruder for developing the products that can vary in cellular structure, shape, texture, and density (Figure 3.1). Generally, free flowing powders are fed into the extruder. The input mechanical and thermal energy pressurises/compresses the food powders to convert them into a viscoelastic fluid. Therefore, the characterization of raw material is important for food extrusion. These are composition (moisture, protein, and fat content), particle size, surface friction, hardness, and cohesiveness of particles. Apart from the main raw materials (Table 3.1), additives, though usually added at a low level, play a major role in the extrusion characteristics and the extrudate properties. These additives may be a plasticiser, lubricant, binder, nutrient/fortification agent, and aid for expansion agent, flavor, or simply a taste improver.

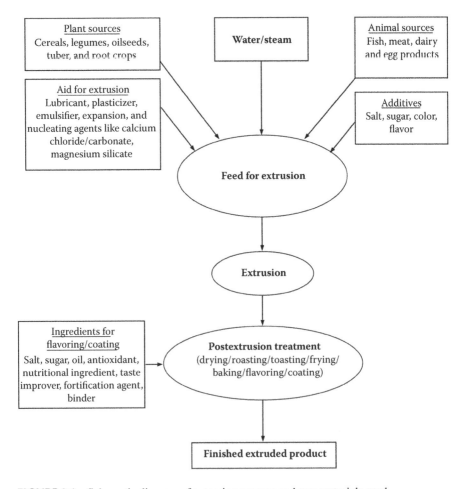

FIGURE 3.1   Schematic diagram of extrusion process and raw materials used.

TABLE 3.1
Extruded Products and Raw Materials Used for Extrusion

Product Group	Product Example	Raw Materials Used	Type of Extruder Used	Type of Product	Reference
Cereals and Starches					
Modified grain flours	Baby food, weaning food	Cereal grit/flour	TS	RTU	Desrumaux, Bouvier, and Burri (1998); Hagenimana, Ding, and Fang (2006); Altan, McCarthy, and Maskan (2009); Guha, Ali, and Bhattacharya (1997)
Modified starches	Raw materials for speciality foods/ convenience foods	Corn, potato, wheat, tapioca starch	SS/TS	RTU	Fletcher, Richmond, and Smith (1985); Cai and Diosady (1993); Della Valle et al. (1995); Ascheri, Bernal-Gomez, and Carvalho (1998)
Expanded product	Blended extruded product	Corn, soy, whey	SS	RTE	Aguilera and Kosikowski (1978)
Expanded product	Blended extruded product	Corn, soy, rice	SS	RTE	Molina, Braham, and Bressani (1983)
Expanded product	Blended extruded product	Rice, sweet potato	SS	RTE	Ascheri, Bernal-Gomez, and Carvalho (1998)
Expanded product	Blended extruded product	Corn, soy bean oil	TS	RTU	Konstance et al. (1998)
Restructured Products					
TVP	Meat substitute like soy chunk/nugget	Defatted soy flour	SS	RTC	Atkinson (1970); Harper (1981)
Muscle foods	Restructured fish/meat mince	Mince from fish/ meat/poultry and wheat/cassava	SS/TS	RTE/ RTC	Yu, Mitchell, and Abdullah (1981); Noguchi (1990); Bhattacharya, Das, and Bose (1993)
Breakfast Cereals					
Breakfast cereals	Corn/rice/wheat flakes, puffed products	Corn/rice/wheat	SS/TS	RTE	Altschul (1998); Eastman, Orthoefer, and Solorio (2001)

(continued)

TABLE 3.1 (Continued)
Extruded Products and Raw Materials Used for Extrusion

Product Group	Product Example	Raw Materials Used	Type of Extruder Used	Type of Product	Reference
		Legumes, Pulses, Oilseeds			
Soy flour	Trypsin inhibited soy flour	Defatted soy flour	SS	RTU	Van den Hout et al. (1998)
Legume product	Expanded extrudates	Pinto bean	SS	RTC	Balandrán-Quintana et al. (1998)
Almond product	Expanded extrudates	Almond and pregelatinized wheat flour, saccharose, and NaCl	TS	RTU	De Pilli et al. (2005)
		Snacks			
Expanded snacks	Extruded product	Rice/wheat	TS	RTE	Ding et al. (2005, 2006)
Flavored snacks	Blended extruded product	Cassava, pigeon pea	TS	RTE	Rampersad, Bodrie, Comissiong (2003)
Expanded snacks	Blended extruded product	Wheat flour, corn starch, brewer's spent grain, and red cabbage	TS	RTE	Stojceska et al. (2009)
High fiber snacks		Jatoba flour, cassava starch	SS	RTE	Chang et al. (1998)
Expanded snacks	Blended extruded product	Wheat-potato/corn–soy	TS	RTE high protein snack	Bhattacharya, Sudha, and Rahim (1999); Baskaran and Bhattacharya (2004)
Legume-cereal snacks	Blended extruded product	Rice–chickpea/green gram	TS	RTE	Bhattacharya and Prakash (1994); Bhattacharya (1997)
		Pasta Products			
Forming extrusion products	Blended cereal pasta	Wheat, egg	SS	RTC	Zardetto and Rosa (2009)
Forming extrusion products	Legume pasta	Yellow pea flour	TS	RTC	Wang et al. (1999)
		Miscellaneous			
Confections	Extruded confectionery	Chocolate	TS	RTE	Mulji et al. (2003)

TABLE 3.1 (Continued)
Extruded Products and Raw Materials Used for Extrusion

Product Group	Product Example	Raw Materials Used	Type of Extruder Used	Type of Product	Reference
Reactive extrusion	Specialty starch	Corn/potato starch, sodium tripolyphosphate/ trimetaphosphate/ sulphate	TS	Drug release agent	O'Brien et al. (2009)
Flavors	Base for flavor	Flavoring compounds	TS	RTC	Kollengode, Hanna, and Cuppett (1996)
Dairy products	Expanded dairy product	Whey protein concentrate, pregelatinized corn starch, CaCl$_2$, and NaCl	TS and super critical fluid	RTC	Manoi and Rizvi (2009)
Lipid oxidation	Lipid stability in extruded corn	Corn meal, butylated hydroxytoluene (BHT), cinnamic acid	SS	RTP	Camire and Dougherty (1998)
Gelatinization and liquefaction	Enzymatic hydrolysis of sago starch	Tapioca (sago) starch with α-amylase	TS	RTP	Govindasamy, Campanella, and Oates (1997)
Expanded extrudates	Cereal– vegetable	Barley flour and tomato pomace	TS	RTE	Altan, McCarthy, and Maskan (2008)
Flat bread	Blend of several ingredients	Wheat and rye flours, whole grain, soy protein, gluten, sodium caseinate, and milk powder	TS	RTE	Antila et al. (1983)
Sterilization	Extrusion sterilization	Herbs/spices	SS	RTU	Bayusik and Chen (1980)

TS Twin-screw extruder; *SS* Single-screw extruder; *RTU* ready-to-use; *RTE* ready-to-eat; *RTC* ready-to-cook; *RTP* ready-to-process

Classification of ingredients by their functional roles in extrusion cooking is available (Guy 1994, 2001). These groups are: (1) structure-forming materials, (2) dispersed-phase filling materials, (3) ingredients that act as plasticizers and lubricants, (4) soluble solids, (5) nucleating substances, (6) coloring materials, and (7) flavoring substances. The structure-forming materials provide the structure of an extruded product that is created by forming a melt fluid from biopolymers and blowing bubbles of water vapor to form a foam-like expanded structure. The dispersed-phase filling materials in cereal snack foods show a continuous phase of starch polymers and by fibrous materials such as cellulose or bran or proteins. These polymers will form separate phases within the continuous starch phase. The third group substances, plasticizers and lubricants, such as water, serve to reduce interactions by plasticizing the dry polymer forms, transforming them from solids to deformable plastic fluids. The addition of increasing amounts of water reduces the dissipation of mechanical energy. Oils and fats produce large effects on the processing of starch even at low levels of 1%–2%; higher levels may reduce the degradation of the starch polymer to such an extent that no expansion is obtained. The fourth group consists of soluble solids such as sugars or salts, which may be added to a recipe for flavoring or humectant properties. The nucleating substances are the fifth group that increase bubble nucleation in the hot melt fluid inside the extruder, and examples include powdered calcium carbonate and talc (magnesium silicate). The sixth group is the coloring substances, present in naturally occurring raw materials such as corn, or the permitted colors. Flavoring substances are usually added either during extrusion or in secondary operations like postextrusion treatment.

Considering convenience, it is desirable to mention important raw materials linking specific extruded products. Among all the ingredients, the most important are the moisture content and the added water. The moisture content of the feed ingredients is one of the principal techniques for controlling the extrusion process (temperature and rate) and extruded product characteristics. Specifically, moisture content affects product density, expansion, cooking, product rehydration, and starch gelatinization. Water may be supplied while mixing the feed prior to extrusion or directly added into the moving feed during extrusion in the form of steam or water. For example, during the manufacture of texturized vegetable protein (TVP) or texturized plant protein (TPP), the usual practice is the addition of water directly in the feed section of the single-screw extruder. Addition of steam serves two purposes—increasing the moisture content and supplying the thermal energy. The level of moisture decides the melt viscosity, which in turn dictates the density of the finished product apart from physiochemical changes. In the case of twin-screw extrusion cooking, the minimum acceptable moisture content to run the extruder steadily is approximately 10% while it is more than 13% for a single-screw machine; the torque and energy requirements are high at low moisture levels.

The moisture content of the extrusion feed may be grouped into three classes such as: (a) low moisture feed containing up to 25% moisture, (b) intermediate moisture contents between 25% and 50%, and (c) high moisture feed having 50% and above. During low moisture extrusion, the most commonly encountered process of food extrusion, a free flowing moist feed is subjected to extrusion cooking to get a cooked and low density product. Cereal, legume, and oilseed meals (like corn grit and defatted soy flour) are the common raw materials. The requirement of a drying step is usually mandatory if the extruded product has a moisture content above 5%. The product obtained in this case is either in the RTE or ready-to-process form. The latter product requires further processing such as baking, frying, roasting, or toasting prior to consumption. Examples include puffed corn balls and breakfast cereals, and products obtained with single-screw collet extruders. For the intermediate moisture feed, the feed contains much higher moisture levels such that the product coming out of the extruder initially expands but collapses later to yield a product with a higher density. The feed is conventionally made out of a common cereal such as wheat flour along with other additives. Postextrusion drying of extrudates is mandatory to ensure safe moisture content. The texture of the product is harder, compared with the first case. The most common example is pasta products, which are simply produced by employing the technique of forming extrusion, wherein the target is to obtain a shaped product and the rise in temperature and the extent of cooking are marginal.

High moisture extrusion can be accomplished by employing a twin-screw extruder along with special die and cooling facilities (Noguchi 1990). Defatted soybean flour at a moisture content of 60% can be continuously extruded and the resultant products have been reported to be soft but elastic. Wet extrusion seems to be a promising area to make use of low-cost fish and meat by-products including surimi. Texturization of minced fish has been reported, and the product possesses an attractive white appearance and aligned stringy structure. The possible additives include isolated soy protein, casein, stearyl monoglyceride, and egg white powder. The texturized products that can also be obtained in this manner are crabs, shrimps, and beef analogs. The extrudates need to be consumed immediately or refrigerated or frozen for use in the near future.

POSTEXTRUSION TREATMENT AND RAW MATERIALS

Postextrusion treatment of extrudates includes several processing steps such as baking, frying, roasting, toasting, flavoring, and coating. It serves many purposes such as to ensure a safe moisture content, converting ready-to-cook/serve extrudates into RTE products, further expansion to make the finished product soft in texture and attractive to eat, development of roasted flavor and surface crispness, and avoiding bland taste. One limitation of the extrusion cooked product is the development of bland taste, which results due to the escape of superheated steam that takes away inherent flavoring compounds present in the extrusion feed. The common ingredients (Table 3.2) that are used for such purposes are salt; sugar; spices; color; oleoresin; oil; protein isolates; honey; maltodextrin; edible gum; corn syrup; liquid glucose; organic acid, e.g., citric acid; antioxidants; wax; fruit and vegetable powders; minerals and vitamins; different flavors, e.g., cheese, fruit, fish, spice, vanilla, rock salt, and monosodium glutamate. The list may be expanded by several innovative items such as tamarind, autolysed yeast, molasses, smoke flavor, vinegar, and sour cream (Seighman 2001). Salt is added after the cheese/oil mixture and sprayed onto the surface of the product to aid its adhesion. The addition of 1.5% to 2% salt is usually adequate to give the proper taste, though the use of higher levels of salts (3%) is possible as some consumers prefer high salt content in snack foods.

The number of published articles on postextrusion treatment is limited, though the importance of this step for commercial success is undeniable. Many extruded products fail to attract consumers due to improper postextrusion treatment. Agarwal and Bhattacharya (2007) used soybean isolate for coating extruded corn balls after conducting a simulation study. This study indicates an alternative method to increase the protein content or the incorporation of heat-sensitive nutrients.

TABLE 3.2

Common Postextrusion Treatments and Raw Materials Used for Selected Products

Name of Product	Type of Treatment	Machinery Employed	Raw Materials Used during Postextrusion Treatment
TVP	Drying	Continuous belt dryer	Nil
Low-density cereal/legume/	Drying	Continuous belt/rotary inclined dryers	Nil
tuber and root crop based	Roasting	Air/contact inclined surface roaster	Nil
snack	Flavoring	Batch inclined drum/enrober/ continuous rotary drum with spray/powder dispensing system	Salt, oil, sugar, flavors, binders, antioxidants, color, minerals, and vitamins
Breakfast cereals	Roasting/ toasting	Air/contact surface roaster/toaster	Nil
	Coating	Batch or continuous drum/enrober	Sugar, malt powder, flavor, liquid glucose, oil
Soy granules	Size reduction and drying	Granulators/disintegrator/rotary slicer and continuous dryer	Flavor, salt

PROTEIN RICH EXTRUDED FOODS AND NUTRITIOUS FOODS FOR MASS FEEDING

Diets consumed by a majority of the world's population are based mainly on cereals and contain a small amount of pulses, vegetables, milk, and other animal foods. Consequently, protein-calorie malnutrition, anemia, and diseases due to vitamin deficiencies are widely prevalent. Protein rich and protective foods such as milk, eggs, meat, and fish are not available in adequate quantities and are sometimes beyond the reach of low-income groups.

The residual cake (45%–50% protein) that is obtained as a by-product from major oilseeds like groundnut, sesame, mustard, rapeseed, and soybean is a potential source of good quality edible protein, and efforts have been provided for the development of processes to obtain edible quality oilseed meals free from toxic factors in these oilseeds. RTE/RTC-extruded protein rich foods by proper blending of oilseed meals, pulses, and cereals are possible. These protein enriched snack foods have a good potential for use in nutritional intervention program for preschool and school going children, social welfare programs, and to meet natural calamities, e.g., floods, drought, cyclones, epidemics, and famine conditions.

Nutritionally balanced foods are necessary for maintaining good health in the case of normal humans, but nutritious foods enriched with protective nutrients are essential to improve the health status of vulnerable segments of the population, e.g., children, mothers, and geriatrics. Nutritious foods for nutritional intervention programs have been implemented throughout the world, and extrusion cooking technology has been frequently and successfully applied for this purpose. The criteria for selecting a particular raw material or usually a blend depend on low cost, easy availability, and protein content. The finished product is expected to contain a minimum of 16% protein, an energy content of 340 kcal per 100 g of food, less than 6% moisture, and the shelf-life should be at least 6 months when packaged in conventional low-cost flexible packaging materials like low density polyethylene (LDPE) or polypropylene (PP). It is also fortified with minerals (containing calcium, iron, and others) and vitamin premixes (containing a significant quantity of B-vitamins) for which the recommended daily allowance (RDA) becomes the guideline. The product may be directly consumed as an RTE snack or is powdered and mixed with milk or water and sugar for feeding. A blend of cereals and pulses at an appropriate proportion is usually extruded as cereals are a rich source of carbohydrates, whereas grain legumes are a good source of dietary proteins. Cereal proteins are deficient in lysine, an essential amino acid, whereas legume proteins are rich in lysine. In most of the cases, corn–soy–skim milk (CSM), corn–soy blend (CSB), and wheat–soy blend (WSB) are used for nutritional assistance programs (Harper 1981). In CSM, the blends consist of 64 parts of corn meal, 24 parts of defatted soy flour, and 5 parts each of skim milk powder and refined soybean oil. In the case of CSB, corn meal and soy flour are 64% and 22%, respectively, while in WSB, the wheat and soy flour are 73% and 20%, respectively; mineral and vitamin premix are also added.

BAKERY AND PASTA PRODUCTS

The extrusion process was originally started for the manufacture of pasta products, wherein a manually or electrically operated screw press was employed to obtain a shaped product. It is a key operation in the process as it determines both the rate of production and the quality of the end product. Now the technology has been adopted for the manufacture of bakery products, e.g., biscuits, cookies, and cakes. The forming extrusion process, wherein pressure development is low, allows continuous kneading and shaping of the dough. The cookie is normally made from relatively soft dough that, on extrusion from a die aperture, is cut into pieces of an appropriate size by means of a reciprocating wire. These dough pieces are deposited on the oven band.

Pasta products, which include macaroni, spaghetti, and vermicelli, are made from a basic mixture of flour or semolina and water, to which some other ingredients may be added, and are formed into some convenient shapes, e.g., elbows, tubes, shells, alphabets, numerals, stars, wheels, rings, and rice; they are either immediately cooked and eaten, or dried for consumption at a later time.

Most noodles contain egg. The principal ingredient required for quality macaroni is durum wheat so that the resulting macaroni product is mechanically strong, has a smooth surface and a bright yellow color. Enriched macaroni or noodles are made by adding vitamins and minerals (Harper 1981).

Though wheat has been the most common principal raw material, other cereals such as rice and corn as well as legumes and their blends are also used. Though quality pasta is made from durum wheat, it is not uncommon to use flours from different varieties of semi durum and soft wheat, corn, and even other cereals like rice, particularly in Asia. A great variety of raw materials are used to make special types of pasta including rice, corn, oats, potato starch, and mixtures of cereal and legume. Corn and potato are the other raw materials that hold promise in the formulation and development of newer pasta products.

LEGUME PRODUCTS

Individual legume flours are extruded, though legume–cereal blends are more common. The added advantages of using legume with cereal are a higher protein content, and often, an attractive taste and flavor. Several pulses have been subjected to extrusion cooking, e.g., green gram and Bengal gram (chickpea) (Bhattacharya 1997; Bhattacharya and Prakash 1994), kidney bean, cowpea (Van der Poel, Stolp, and Van Zuilichem 1992), lentil, faba bean (Lombardi Boccia, Di Lullo, and Carnovale 1991), pinto bean, and pigeon pea (Rampersad, Badrie, and Comissiong 2003). These studies indicate alternative raw material like legumes to focus on health benefits from the point of high protein content and balanced essential amino acid contents.

EXTRUSION OF OILSEEDS

High-pressure high-temperature extrusion through small openings converts the raw oilseeds into cooked and partially inflated particles known as collets. These collets are a better material than expeller pressed cakes for solvent extraction. High pressure and temperature developed in the extruder provide rapid cooking for oilseeds and an advantage of inactivating some of the antinutritional factors present in the oilseeds. They also help in inactivating some of the enzymes such as urease in soybean, and lipase in rice bran. A serious problem in the utilization of oilseed protein is that some of them contain antinutritional and toxic factors that need to be either inactivated or rendered innocuous by suitable processing.

Protein inhibitors exert their antinutritional effect by causing pancreatic hypertrophy/hyperplasia, resulting in inhibition of growth. The toxicity of lectins could be attributed to their ability to bind to specific receptor sites in the intestine. Most of the antinutritional factors are heat labile. Optimization of the processes to inactivate these factors in individual oilseeds is necessary.

TEXTURIZED VEGETABLE PROTEIN

Two types of textured meat-like products are manufactured by employing the extrusion–texturization process. The meat extender is produced by employing a single-screw extruder operating at high temperatures and pressures, followed by drying. After hot water hydration, this TVP is used to extend/replace meat for use as pizza toppings, meat sausages, and fabricated food formulations. The second type of product is meat analogs that can be used in place of meat. Both these products exhibit fiber formation due to extrusion cooking of defatted soybean flour and consequent alignment when passing through the restriction or die. Soybean is commercially available as flakes, grits, and flour of different sizes. The proximate composition of defatted soy flour indicates a protein ($N \times 6.25$) content of 45%–50% and less than 1% fat. A high nitrogen solubility index (NSI) is desirable, and commercially available samples have an NSI between 50 and 70. The defatted soy flour samples are usually toasted such that the antinutritional factors like trypsin inhibitors (TI) and hemagglutinate are reduced to a value of 10 to 15 units. Atkinson (1970) suggested the addition of a small quantity of cereal flour (approximately 10%) to obtain a faster rehydrating product. The other ingredients that can be added are permitted colors, flavors, minerals, emulsifiers, and cross-linking enhancers like elemental sulfur and calcium chloride (Harper 1981). Other defatted oilseed flours, such as cottonseed proteins, have been attempted, but such products have not seen commercial success.

SNACKS

The market for extrusion-puffed snacks is huge and may approach U.S. $1 billion. Extruded snacks come in a variety of shapes other than the common collets or curls, including tubes, wheels, rings, hats, mushroom sticks, and scoops. Application of flavors has expanded from the basic cheese or oil to include barbecue, pizza, vinegar, onion, and garlic. Though initially expanded snacks were based on corn, they are now produced from potato and wheat and rice flour along with mixtures of modified and unmodified starches. For example, pretzels, one of the important traditional snack foods, is manufactured by employing a forming single-screw extruder wherein the major raw materials used are wheat flour, yeast, salt, water, malt, and hydrogenated soybean oil (Groff 2001).

The evolution of snack foods has been described with certain types of snacks being associated with a specific generation of products (Harper 1981). The first generation of snacks considered are the conventional potato chips and baked crackers. Second generation snacks encompass the puffed collets. A low moisture extrusion (<15%), where the major source of energy comes from the dissipation of the mechanical energy input to the extruder, is used to produce highly expanded collets from corn grits, other cereals, and/or starch. Once extruded, the collets are dried or baked to less than 5% moisture content and enrobed with flavors and oil. Higher moisture containing collets can be fried directly after extrusion cooking. The third generation snacks can involve extrusion to produce a variety of shapes and textures that are not possible with collet extrusion technology. Mixtures of cereals, starch (modified and unmodified), vegetable oil, and emulsifiers can be first cooked in an extruder and then formed in a second extruder. If pregelatinized or precooked ingredients are used, the cooking step can be eliminated and the product is formed directly in a forming extruder. The shelf-stable pellets are called semi- or half-products, and these may be shipped for puffing outside the plant. These snacks can also be puffed/cooked prior to consumption in domestic sectors using a simple baking or microwave oven. Another form of third generation snack involves dual or coextrusion of dissimilar materials to form a single piece. The concept of simultaneous extrusion of two materials forming a single product has led to some interesting new product concepts. At present, there is a good demand for such products in the international market.

BREAKFAST CEREALS

Commercially available breakfast cereals are predominantly corn and wheat based foods though other grains such as rice, oat, sorghum, or a combination of grains like multigrain breakfast cereals are also produced through extrusion cooking. The specific advantages of such combinations include the possibility of different textures, micronutrient content, and lowering the cost of extrudate feed. Corn flakes, sweet in taste and/or flavored with chocolate, are the most popular item among consumers. Breakfast cereals made out of whole grain containing a reasonable amount of fiber are gaining popularity among consumers.

Extrusion processing of breakfast cereals has been discussed by Eastman, Orthoefer, and Solorio (2001), covering details of extrusion processes and the extruders employed for the manufacture of RTE breakfast cereals and the ingredients used. Altschul (1998) described a method for the manufacture of an edible, full-dimensional toy figure that is constructed of prepared breakfast cereal for consumption with a liquid such as milk.

Ingredients affect the characteristics of the finished product in various ways. The addition of oils/fats/lipids, fibers, protein, and starch containing high amylose increases the density of the extrudates while sucrose and a small amount of finely ground calcium silicate and calcium carbonate reduce the density.

CHARACTERISTICS OF RAW MATERIALS

A high fraction of all human food energy is in the form of starchy foods that is mostly derived from cereal grains, legumes, and starchy root crops. In the raw or native state, starch exists in the form of granules and requires thermal treatment in the presence of water to make it edible and digestible. Extrusion cooking technology is ideally suitable for this purpose where the time requirement is low and mechanical degradation also occurs to make this process more effective and useful. Starch provides the much needed structure, texture, and mouthfeel of the extrudates. During extrusion cooking, the uncooked starchy fractions undergo mainly gelatinization, dextrinization, and shearing of polymer chains though the chance of retrogradation cannot be ruled out. Corn and sorghum starches are similar and most commonly used because of their low cost and easy availability. Very small granules characterize rice starch. Wheat starch forms gels having lower viscosity and a tender structure. The waxy starches are composed of a higher proportion of amylopectin, and specific varieties of corn, sorghum, and rice are used for this purpose. Waxy starches are used to reduce the water uptake in cereals and baked foods so that they remain crunchy for some time in milk or water. Conversely, high amylose starches from special varieties of corn are available with up to 80% amylose. These starches behave in a manner opposite to that of waxy starches (Harper 1981). Starches from root and tuber crops like potato and tapioca (cassava) also find applications in extruded products. Their applications, particularly in snack foods, are due to their characteristic flavor and good swelling power in the presence of water.

Water is an important factor for the process of gelatinization of starches. Based on stoichiometry, water molecules bonded to each available hydroxyl group and starch, a minimum of 25% is required. Low levels of water are sufficient to interact with the starch in extrusion ingredients to plasticize the mass and form dough. It is worth mentioning here that most of the extrusion cooking process for the manufacture of expanded cereals and/or legume products is practiced with moisture content less than 25%, and use of 13%–17% is the most common practice. Apparently, the gelatinization temperature is highest for starches from tuber crops, followed by root starches while the cereal starches possess the lowest magnitude. Amylose has been reported to provide lightness, surface and texture regularities, elasticity, and a sticky characteristic to the extruders. Products high in amylopectin are harder and less expanded.

Corn meal is marginally smaller than grits and contains less than 1% oil, is low in ash and fiber content, possesses a long shelf-life, and is bright in color without black specks (Rooney and Suhendro 2001). Corn meal is used extensively to produce popular extruded products such as corn puffs, curls, and other expanded products; they are further fried or baked, flavored or coated. Specifications of corn meal for extrusion vary depending on the type of product desired and the type of extruder used. Single-screw extruders including commonly used collet extruders for the manufacture of corn puffs require a feed that does not contain any flour or fine particles. On the other hand, twin-screw extruders can accommodate more fine particles in the feed. The products obtained from the dry milling of corn have been cited by Rooney and Suhendro (2001). Particle size distribution affects the ease of cooking, expansion of the product, and the relative crunch of the finished product, and an increase in protein content tends to decrease the expansion of the extruders.

Meals and grits from sorghum are easily extruded to achieve white or light colored bland tasting products that can easily be colored and flavored. Corn has a strong flavor that can significantly interact with seasonings used for snack foods, but sorghum and rice do not have this problem. The expansion capabilities of sorghum grits are equivalent to those of corn grits and meal. In addition, sorghum is usually lower in price than corn. However, the proper variety of sorghum has to be selected as some varieties contain bitter and off-tasting components. The presence of aflatoxin in corn must be controlled and should be below the safety limits.

The application of degermed cereal and pulse grits and flours is more common than that of starches. These grits/flours are predominantly starch. They also contain some protein, ash, and a very small amount of fat. Among the cereals and legumes, the most common ingredients are debranned and degermed corn grit and defatted soybean flour. The specific advantage of using corn grit is its low cost, easy availability, and the easy production of expanded products. The protein content is between 6% and 9%, while ash and fat contents are less than 1%. The behavior of sorghum grits is to that of corn grits with a marginally lower water absorption index. Rice grits are a preferred commodity for extruded snacks in areas where rice is a staple food. A good expansion along with a bland taste and attractive white color are the features of rice extrudates. Wheat, with a higher protein content than corn and rice, is less preferred for expanded snack preparations while being suitable for extruded breakfast cereals and pasta products. The other cereals that are sometimes used include barley, oats, and triticale. Two root crops such as potato and tapioca are often used for developing RTE/RTC snack products. Extrusion of potato flakes indicates a higher expansion when the temperature of extrusion is increased. Extrudates from tapioca are cost-effective and provide a light color.

Among the other ingredients used for food extrusion, emulsifiers are the compounds that have hydrophilic and hydrophobic ends on their molecules. When added to water/oil mixtures, they align themselves at the interface of the water and oil and reduce the natural interfacial tension that exists there. Emulsifiers are normally added at very low levels (<1%). They can affect the texture, cell size, and density of an extruded product and act as a lubricant for the dough. Lecithin (a natural phospholipid and a by-product of soy oil refining) and hydroxylated lecithin show extensive emulsification capabilities. Emulsifiers are known to form complexes with amylose in starch and consequently retard crystallization and retrogradation. Complexes are also formed between amylose and fat during extrusion. If excess fat (>5%) exists, the complexes cannot accommodate all the fat, meaning a reduction in the dough viscosity and expansion of the extruded product. If a higher percentage of fat is used, discontinuous extrusion strands may appear in addition to poor expansion. The different food emulsifiers are glyceryl monostearate (GMS), propylene glycol monostearate (PGMS), sorbitan monostearate, polyoxythylene sorbitan monostearate (Polysorbate 60), sodium stearoyl-2-lactylate (SSL), and dioctyl sodium sulfosuccinate (DSS).

Fats and oils added to extrusion feed tend to weaken the resulting dough, reduce the strength of the product, and increase the plasticity as it leaves the extruder. They can coat starch granules, therefore restricting the amount of water they absorb. The sequence in which fat is added to the product and its distribution significantly affects the extrusion cooking process.

The addition of acid or basic salts to feed materials can alter the characteristics of an extruded product, particularly if significant quantities of protein are present; they are known as pH modifiers. Changes in the rehydration rate, flavor, density, texture, and toughness of the product accompany pH modification. Calcium chloride and calcium carbonate are the commonly used acid salts. Dilute or weak acids such as lactic, citric, or acetic acid or weak bases such as ammonium hydroxide are sometimes used. Care needs to be taken for pH modification of ingredients that may corrode the barrel and screw of the extruder and increase the maintenance costs.

The addition of sugar to the extrusion formulations can alter the taste and texture of products, but the resulting material may be very sticky and difficult to handle. The addition of reducing sugars can have a detrimental effect on available lysine and browning can occur.

The hydrocolloids alter the thickening or gelling properties. They are useful in developing high moisture dough-like systems to produce certain types of snacks or fabricated foods. The alginates, guar, and locust bean gums do not thicken until they react with calcium ions, giving them unique properties that are useful in formulating high-moisture foods. Hydrocolloids also find use as thickeners in coating material formulations to develop flavored/coated extruded products.

CONCLUSIONS

Almost all the food ingredients are used for extrusion cooking. Novel ingredients, cutting-edge extrusion technology, and innovative processing methods need to be combined to yield new types of extruded products such as snacks and breakfast cereals with ever-widening appeal to health-conscious consumers, to shoppers seeking a crispier breakfast cereals, and to gastronomes who favor shrimp-flavored snacks. The challenge lies in developing breakfast cereals of uniform size and shape that can retain their integrity and crispness for a longer duration, i.e., 5 minutes in warm milk. With the current emphasis on health and wellness, whole grain foods will remain a top trend in new product development. Low-cost snacks and breakfast cereals with improved formulation for health benefits, e.g., low fat content, high fiber content, and lower energy content, appear to be the emphasis on extruded products in the near future.

REFERENCES

Agarwal, K. and Bhattacharya, S. 2007. Use of soybean isolate in coated foods: Simulation study employing glass balls and substantiation with extruded products. *Int J Food Sci Tech* 42: 708–713.

Aguilera, J.M. and Kosikowski, F.V. 1978. Extrusion and roll-cooking of corn-soy-whey mixtures. *J Food Sci* 43(1): 225–227.

Altan, A., McCarthy, K.L. and Maskan, M. 2008. Evaluation of snack foods from barley–tomato pomace blends by extrusion processing. *J Food Eng* 84: 231–242.

Altan, A., McCarthy, K.L. and Maskan, M. 2009. Effect of extrusion cooking on functional properties and in vitro starch digestibility of barley-based extrudates from fruit and vegetable by-products. *J Food Sci* 74(2): E77–E86.

Altschul, R.L. 1998. Edible toy figures constructed of breakfast cereal. US Patent 5804235.

Antila, J., Seiler, K., Seibel, W. and Linko, P. 1983. Production of flat bread by extrusion cooking using different wheat/rye ratios, protein enrichment and grain with poor baking ability. *J Food Eng* 2(3): 189–210.

Ascheri, J.L.R., Bernal-Gomez, M.E. and Carvalho, C.W.P. 1998. Production of snacks from mixtures of rice and sweet potato flours by thermoplastic extrusion. I. Chemical characterization, expansion index and apparent density. *Alimentaria* 293: 71–77.

Atkinson, W.T. 1970. Meat-like protein food product. US Patent 3488770.

Balandrán-Quintana, R.R., Barbosa-Cánovas, G.V., Zazueta-Morales, J.J., Anzaldúa-Morales, A. and Quintero-Ramos, A. 1998. Functional and nutritional properties of extruded whole pinto bean meal (*Phaseolus vulgaris* L.). *J Food Sci* 63(1): 113–116.

Baskaran, V. and Bhattacharya, S. 2004. Nutritional status of the protein of corn-soy based extruded products evaluated by rat bioassay. *Plant Foods Hum Nutr* 59: 101–104.

Bayusik, M.J. and Chen, P.H. 1980. Method of sterilizing spices. US Patent 4210678.

Bhattacharya, S. 1997. Twin-screw extrusion of rice-green gram blend: Extrusion and extrudate characteristics. *J Food Eng* 32: 83–99.

Bhattacharya, S. and Prakash, M. 1994. Extrusion of blends of rice and chick pea flours: A response surface analysis. *J Food Eng* 21: 315–330.

Bhattacharya, S., Das, H. and Bose, A.N. 1993. Effect of extrusion process variables on texture of blends of minced fish and wheat flour. *J Food Eng* 19: 215–235.

Bhattacharya, S., Sudha, M.L. and Rahim, A. 1999. Pasting characteristics of an extruded blend of potato and wheat flours. *J Food Eng* 40: 107–111.

Cai, W. and Diosady, L.L. 1993. Model for gelatinization of wheat starch in a twin-screw extruder. *J Food Sci* 58(4): 872–875.

Camire, M.E. and Dougherty, M.P. 1998. Added phenolic compounds enhance lipid stability in extruded corn. *J Food Sci* 63(3): 516–518.

Chang, Y.K., Silva, M.R., Gutkoski, L.C., Sebio, L. and Da Silva, M.A.A.P. 1998. Development of extruded snacks using jatoba (*Hymenaea stigonocarpa Mart*) flour and cassava starch blends. *J Sci Food Agric* 78: 59–66.

De Pilli, T., Severini, C. Baiano, A., Derossi, A., Arhaliass, A. and Legrand, J. 2005. Effects of operating conditions on oil loss and properties of products obtained by corotating twin-screw extrusion of fatty meal: Preliminary study. *J Food Eng* 70: 109–116.

Della Valle, G., Boché, Y., Colonna, P. and Vergnes, B. 1995. The extrusion behaviour of potato starch. *Carbohydrate Polymers* 28(3): 255–264.

Desrumaux, A., Bouvier, J.M. and Burri, J. 1998. Corn grits particle size and distribution effects on the characteristics of expanded extrudates. *J Food Sci* 63(5): 857–863.

Ding, Q-B., Ainsworth, P., Plunkett, A., Tucker, G. and Marson, H. 2006. The effect of extrusion conditions on the functional and physical properties of wheat-based expanded snacks. *J Food Eng* 73(2): 142–148.

Ding, Q-B., Ainsworth, P., Plunkett, A., Tucker, G. and Marson, H. 2005. The effect of extrusion conditions on the physicochemical properties and sensory characteristics of rice-based expanded snacks. *J Food Eng* 66(3): 283–289.

Eastman, J., Orthoefer, F. and Solorio, S. 2001. Using extrusion to create breakfast cereal products. *Cereal Foods World* 46(10): 468, 470–471.

Fletcher, S.I., Richmond, P. and Smith, A.C. 1985. An experimental study of twin-screw extrusion-cooking of maize grits. *J Food Eng* 4(4): 291–312.

Govindasamy, S., Campanella, O.H. and Oates, C.G. 1997. Enzymatic hydrolysis of sago starch in a twin-screw extruder. *J Food Eng* 32(4): 403–426.

Groff, E.T. 2001. Perfect pretzel production. In *Snack foods processing*, eds. E.W. Lusas and L.W. Rooney, 369–384. Lancaster, Pennsylvania: Technomic Publications.

Guha, M., Ali, S.Z. and Bhattacharya, S. 1997. Twin-screw extrusion of rice flour without a die: effect of barrel temperature and screw speed on extrusion and extrudate characteristics. *J Food Eng* 32: 251–267.

Guy, R.C.E. 1994. Raw materials. In *The technology of extrusion cooking*, ed. N.D. Frame, 52–72. London: Blackie Academic & Professional.

Guy, R.C.E. 2001. Raw materials for extrusion cooking. In *Extrusion cooking: Technologies and applications*, ed. R.C.E. Guy, 5–28. Cambridge, England: Woodhead Publishing Limited.

Hagenimana, A., Ding, X. and Fang, T. 2006. Evaluation of rice flour modified by extrusion cooking. *J Cereal Sci* 43(1): 38–46.

Harper, J.M. 1981. *Extrusion of Foods*. Vol 2, Boca Raton, Florida: CRC Press.

Kollengode, A.N.R., Hanna, M.A. and Cuppett, S. 1996. Volatiles retention as influenced by method of addition during extrusion cooking. *J Food Sci* 61(5): 985–990.

Konstance, R.P., Onwulata, C.I., Smith, P.W., Lu, D., Tunick, M.H., Strange, E.D. and Holsinger, V.H. 1998. Nutrient-based corn and soy products by twin-screw extrusion. *J Food Sci* 63(5): 864–868.

Lombardi Boccia, G., Di Lullo, G. and Carnovale, E. 1991. In-vitro iron dialysability from legumes: Influence of phytate and extrusion cooking. *J Sci Food Agric* 55(4): 599–605.

Manoi, K. and Rizvi, S.S.H. 2009. Physicochemical changes in whey protein concentrate texturized by reactive supercritical fluid extrusion. *J Food Eng* 95(4): 627–635.

Molina, M.R., Braham, J.E. and Bressani, R. (1983). Some characteristics of whole corn: whole soybean (70:30) and rice: whole soybean (70:30) mixtures processed by simple extrusion cooking. *J Food Sci* 48(2): 434–437.

Mulji, N.C., Miquel, M.E., Hall, L.D. and Mackley, M.R. (2003). Microstructure and mechanical property changes in cold-extruded chocolate. *Food Bioproducts Process* 81(2): 97–105.

Noguchi, A. 1990. Recent research and industrial achievements in extrusion cooking in Japan. High temperature/short time (HTST) processing, guarantee for high quality food with long shelf-life. In *Processing and quality of foods*, eds. P. Zeuthen, J.C. Cheftel, C. Eriksson, T.R. Gormley, P. Linko and K. Paulus, 203–214. London: Elsevier Applied Science.

O'Brien, S., Wanga, Y.-J., Vervaet, C. and Remon, J.P. 2009. Starch phosphates prepared by reactive extrusion as a sustained release agent. *Carbohydrate Polymers* 76: 557–566.

Rampersad, R., Badrie, N. and Comissiong, E. 2003. Physicochemical and sensory characteristics of flavored snacks from extruded cassava/pigeon pea flour. *J Food Sci* 68(1): 363–367.

Rooney, L.W. and Suhendro, E.L. 2001. Food quality of corn. In *Snack foods processing*, eds. E.W. Lusas and L.W. Rooney, 39–71. Lancaster, Pennsylvania: Technomic Publications.

Seighman, B. 2001. Snack food seasonings. In *Snack foods processing*, eds. E.W. Lusas and L.W. Rooney, 495–516. Lancaster, Pennsylvania: Technomic Publications.

Stojceska, V., Ainsworth, P., Plunkett, A. and Ibanoglu, S. 2009. The effect of extrusion cooking using different water feed rates on the quality of ready-to-eat snacks made from food by-products. *Food Chem* 114: 226–232.

Van den Hout, R., Jonkers, J., Van Vliet, T. and Van Zuilichem, D.J. 1998. Influence of extrusion shear forces on the inactivation of trypsin inhibitors in soy flour. *Food Bioproducts Process* 76(3): 155–161.

Van der Poel, A.T.F.B., Stolp, W. and Van Zuilichem, D.J. 1992. Twin screw extrusion of two pea varieties: Effects of temperature and moisture level on antinutritional factors and protein dispersability. *J Sci Food Agric* 58: 83–87.

Wang, N., Bhirud, P.R., Sosulski, F.W. and Tyler, R.T. 1999. Pasta-like product from pea flour by twin-screw extrusion. *J Food Sci* 64(4): 671–678.

Yu, S.Y., Mitchell, J.R. and Abdullah, A. 1981. Production and acceptability testing of fish crackers (*keropok*) prepared by the extrusion method. *J Food Tech (Int J Food Sci Tech)* 16: 51–58.

Zardetto, S. and Rosa, M.D. 2009. Effect of extrusion process on properties of cooked, fresh egg pasta. *J Food Eng* 92: 70–77.

4 Nutritional Changes during Extrusion Cooking

Mary Ellen Camire

CONTENTS

INTRODUCTION

Today, a wide array of extruded foods and food ingredients are available, but there is no formal system to track how much of these materials are consumed. Thus, the role of extruded foodstuffs in human health cannot be accurately estimated. Extruded snack foods account for a large share of extruded foods consumed in the U.S., although many consumers perceive snack foods to be unhealthy (Dinkins 2000). Ready-to-eat extruded breakfast cereals have been targeted for nutritional enhancements including fortification and the addition of dietary fiber-rich and other health-promoting ingredients. Extrusion offers the means to convert whole grains to palatable and convenient products. Whole grain consumption in the U.S. is less than one serving per day, far below the recommended three servings (Wells and Buzby 2008). The ability to reduce toxins and microbes while increasing the digestibility makes extrusion well-suited to the production of nutritious foods for at-risk populations, particularly

TABLE 4.1

Factors that Influence Nutritional Properties During Extrusion

Primary	Secondary
Extruder model	Pressure
Feed composition, including moisture	Specific mechanical energy
Feed particle size	Mass (product) temperature
Feed rate	
Barrel temperature profile	
Screw configuration	
Screw speed	
Die geometry, size and number	

those in the developing economies. International research projects have focused on simple single-screw extruders because these devices are relatively inexpensive and easy to maintain. Friction from the rotation of the screw may be sufficient to thoroughly cook the food, thereby decreasing the reliance on expensive and/or scarce fuel sources. Extrusion facilitates the development of nutritious food blends made from indigenous crops such as beans, millet, and cassava as well as from donated products such as dried milk and flour. Extruded pellets can be ground into flours or meals that can be mixed with milk or water to create gruels for infants or weak older children and adults. Advantages and limitations of extrusion for nutrition aid projects have been reviewed (Harper and Jansen 1985).

Retention of nutritional quality during extrusion is dependent on a multitude of factors, particularly those that affect the temperature of the food mass within the extruder barrel (Table 4.1). Several review articles have addressed chemical and nutritional changes in extruded foods (Björck and Asp 1983; Camire 1998; Camire, Camire, and Krumhar 1990; Cheftel 1986; de la Gueriviere, Mercier, and Baudet 1985). So, this chapter will emphasize recent research while providing an overview of trends previously reported in the literature.

MACRONUTRIENTS

CARBOHYDRATES

This section will focus on starch and sugars in extruded foods. Major physico-chemical reactions involving these compounds include gelatinization, hydrolysis, Maillard reactions, and caramelization. Control of carbohydrate reactions during extrusion is critical for product nutritional and sensory quality since these molecules account for a major portion of many extruded foods.

Starch digestibility directly influences how much energy may be derived from food. Humans and other monogastric animal species cannot easily digest ungelatinized starch. Some people, such as infants, require easily digestible food for maximum energy, while obese persons may benefit from consuming a satiating food that contains starch that is resistant to digestion. Gelatinization occurs at much lower moisture levels (12%–22%) during extrusion than in other food operations. The rate of gelatinization tends to increase with increasing mass temperature, shear, and pressure. Other food compounds, particularly lipids, sucrose, dietary fiber (DF), and salts, also affect gelatinization (Jin, Hsieh, and Huff 1994). Improved digestibility results even if complete gelatinization does not occur (Wang, Casulli, and Bouvier 1993).

Passage through the extruder barrel can shear amylopectin branches, but both amylose and amylopectin may undergo reductions in molecular weight. Larger corn amylopectin molecules underwent the greatest molecular weight reduction (Politz, Timpa, and Wasserman 1994). Starch degradation can be modulated to some extent by changes in the screw configuration (Gautam and Choudhoury 1999). The starch depolymerization that may occur during extrusion can be exploited to increase the digestibility of foods, especially foods used to wean infants. Uji is an east African porridge-type food made from fermented grains. Extruding a corn–millet blend, with or without prior fermentation that could be used for uji, resulted in total starch reductions of 8%–12% and increases in *in vitro* starch digestibility from 20 mg/maltose/g starch to over 220 mg/maltose/g starch (Onyango et al. 2005).

Consumption of rapidly digested starch can cause a rapid rise in blood sugar and insulin levels after meals that can lead to insulin insensitivity and Type 2, or adult-onset, diabetes. The rise in blood glucose after eating is often measured as the glycemic index (GI), with glucose or white bread used as an arbitrary control with a value of 100 (Brouns et al. 2005). Relatively few studies have examined the impact of extrusion processing on the GI in human volunteers. Extrusion in a single-screw extruder did not affect starch hydrolysis and predicted the GI of amaranth seeds, while other processes such as popping and roasting did (Capriles et al. 2008). Healthy adults exhibited no differences in postprandial glucose, insulin, or satiety responses after consuming meals containing bread made with chickpea flour or with extruded chickpea flour (Johnson, Thomas, and Hall 2005). However, foods such as maize that are higher in starch content may provoke different responses. *In vitro* barley starch digestibility increased with extrusion, but the addition of fruit and vegetable by-products and processing at high barrel temperatures reduced starch digestibility (Altan, McCarthy, and Maskan 2009a).

Extrusion has been evaluated as a means to produce digestion-resistant starch (RS). RS is of increasing interest to food product developers, not only due to its potential for reducing caloric value, but also for digestive health and other benefits (Kendall et al. 2004). Theander and Westerlund (1987) reported transglycosidation in extruded wheat flour that was theorized to have resulted from novel linkages between free amylopectin fragments and other molecules. While Politz, Timpa, and Wasserman (1994) did not report differences in 2,3-glucose linkages in extruded wheat flour, a patent was granted for producing as much as 30% RS by reacting high amylose starch with pullulanase prior to extrusion (Chiu, Henley, and Altieri 1994). Increased amylose content is another factor involved with RS production during extrusion. Slower screw speed (30 rpm) produced higher levels of RS in extruded maize, mango, and banana starches; both fruit starches had higher amylose and RS content (González-Soto et al. 2006). Whole grain breakfast cereals made from B63 wheat flour containing nearly 23% amylose (0.78 amylose/amylopectin ratio) had

3.1%–4.3% RS after extrusion, compared with other flours that contained 0.7%–1.5% RS (Chanvrier et al. 2007). RS was also higher in the blends of the B63 high-amylose wheat with maize. Extrusion temperature had no effect on RS creation. High-amylose maize extruded through a capillary rheometer contained less amylose and RS than did the raw starch, but different results may occur with a single or twin-screw extruder (Htoon et al. 2009).

The reaction with acids also favors RS formation during extrusion, but the amount of acid necessary to significantly increase RS may lead to substantial losses in sensory acceptability. RS and DF values increased when 7.5% citric acid was mixed with cornmeal, and 30% high-amylose cornstarch with 5% or 7.5% citric acid resulted in values of 12% more RS than in extruded cornmeal alone, compared with slightly more than 2% in 100% cornmeal (Unlu and Faller 1998). Hwang, Kim, and Kim (1998) demonstrated that oligosaccharides and polydextrose could be formed from glucose-citric acid mixtures extruded at different barrel temperatures. More oligosaccharides and polydextrose were produced at higher temperatures; for example, a 93.7% yield occurred at 200°C. The extrusion of acid-modified corn starch under varying shear conditions did not significantly increase RS, but subsequent heating of the extruded starches for 1–3 days resulted in levels as high as 15% (Hasjim and Jane 2009).

DIETARY FIBER

Although the legal definition of DF has been contentious, most scientists acknowledge that indigestible components of plants are important for human health. The selection of the analytical procedure for quantifying DF can influence interpretation of extrusion effects on these compounds. For example, measurement of total DF before and after extrusion will not provide information regarding the relative molecular size or solubility of DF molecules. Older methods for DF quantification cannot recover some types of RS and lower molecular weight nondigestible oligosaccharides. McCleary (2007) has developed a procedure for measuring these compounds with total DF; the assay has received first approval from AOAC International as an approved method. Trends for the estimation of DF changes during extrusion are also challenging because DF molecules interact with other food components and starch may be converted to RS.

Extrusion cooking can be used to modify DF-rich materials for use as nutritious food ingredients or as biofuels. Orange pulp residue from juice processing is rich in pectin and other forms of soluble DF (SDF). Extrusion of dried orange pulp over a range of screw speeds, barrel temperatures, and feed moisture contents indicated that longer residence times and higher temperatures promoted the formation of SDF (Larrea, Chang, and Martínez Bustos 2005). In the same study, the highest values for soluble pectin were found under conditions of high barrel temperature and low moisture. While these trends are interesting, the study used a laboratory single-screw extruder with a barrel 380 mm long and 19 mm in diameter; scale-up and applicability of twin-screw extruders is not clear. Oat bran extruded on a twin-screw extruder (screw speed = 600 rpm, 180°C, 40% moisture) had greater levels of SDF with higher molecular weight than native bran, but the number of $\beta 1 \rightarrow 4$ linkages were lowest after extrusion compared with steam-heated and highly milled brans (Zhang et al. 2009). The food matrix plays an important role in DF modulation as well. Extrusion increased the insolubility but not the SDF in durum wheat bran (Esposito et al. 2005). Formation of RS and the production of Maillard reaction products were proposed by the researchers as mechanisms for the apparent insoluble DF increase.

LIPIDS

Lipids are a concentrated form of energy providing 9 kcal/g. Overconsumption of dietary lipid is associated with serious health problems such as heart disease, cancer, and obesity. It is challenging to extrude foods containing more than 10% lipids because high levels of lipids reduce shear within the extruder barrel. The production of high-fat expanded products is very difficult. However, some snack foods may be fried after extrusion to remove moisture and increase crunchiness and flavor.

Extrusion has been used to aid oil extraction since oil is freed during the cooking and shearing operations (Nelson et al. 1987). Free oil may actually leak from the die. This problem can be resolved to some extent by mixing a high-starch material with the fatty ingredient. Amylose–lipid complexes form during extrusion, but the complexes can be resistant to some lipid extraction techniques. Lipid recovery is improved when extruded foods are first digested with acid or amylase and then extracted with an organic solvent. Total fat was not significantly changed in extruded whole wheat, but only half of the ether-extractable lipids were recovered unlike extruded wheat bran, which was low in starch and thus had higher amounts of extractable lipids (Wang, Casulli, and Bouvier 1993). Cornmeal extruded at 50°C–60°C or 85°C–90°C contained over 75% bound lipids, but extrusion at 120°C–125°C only bound 70% of the lipids (Guzman, Lee, and Chichester 1992).

Despite interest in the health benefits of omega-3 fatty acids, very few studies have been published on the stability of these highly unsaturated lipids. Both docosahexaenoic (DHA) and eicosapentaenoic (EPA) acids were retained in chum salmon muscle extruded with 10% wheat flour (Suzuki et al. 1988). Jerky-style snacks made from salmon meat exhibited no losses of DHA and EPA after extrusion; oat fiber increased lipid retention compared with high-amylose corn starch and tapioca starch (Kong et al. 2008). Another nutritional issue is the safety of *trans* fatty acids but extruded corn and soy contained only 1.5% *trans* fatty acids (Maga 1978).

Lipid oxidation reduces nutritional and sensory quality in foods and feeds. Many factors contribute to lipid oxidation in extruded foods (Artz, Rao, and Sauer 1992). Screw wear results in higher concentrations of pro-oxidant minerals. Iron and peroxide values were higher in extruded rice and dhal, compared with similar products that were just dried (Semwal, Sharma, and Arya 1994). Increased surface area in expanded products also increases the exposure of lipids to oxygen. On the other hand, lipolytic enzymes and other enzymes that promote oxidation may be inactivated during extrusion, and starch–lipid complexes formed in the barrel may be more resistant to oxidation. Packaging under nitrogen or vacuum in opaque containers may further protect the extruded foods.

PROTEINS

Reviews on extrusion of proteins have been written by Areas (1992) and Camire (1991). Extrusion improves protein digestibility by denaturation, which exposes enzyme-accessible sites. Upon denaturation, most proteins including enzymes, enzyme inhibitors, and allergens lose activity. The extent of denaturation is often estimated by changes in protein solubility in water or aqueous solutions. Such changes are greater under high shear extrusion conditions (Della Valle, Quillien, and Gueguen 1994); mass temperature and moisture also influence protein solubility. Wheat protein solubility was reduced even at the relatively low process temperatures being used in pasta making (Ummadi, Chenoweth, and Ng 1995).

High barrel temperatures and low moisture promote Maillard reactions during extrusion (Ilo and Berghofer 2003). Reducing sugars, including those formed by shearing of starch and sucrose, can react with lysine, thereby lowering the protein nutritional value. Since lysine is the limiting essential amino acid in cereals, additional losses of this nutrient from extruded foods could slow the growth in children and young animals. Blends of cornmeal, full-fat soy flakes, and soy isolates or concentrates had good retention of lysine and could be mixed with water or milk to create porridges or gruels, which could be used as weaning foods (Konstance et al. 1998). The amount and type of sugars present in the food matrix influence amino acid stability during extrusion. Fructose was more protective of essential amino acids (EAA) than was galactose in model systems of wheat flour mixed with either egg or milk proteins (Singh, Wakeling, and Gamlath 2007). Milk protein blends processed at 125°C with 30% protein and 8% galactose retained only 40% of lysine but about 80% of the other EAA. Processors of special use foods such as weaning foods should evaluate all the ingredients for their potential effects on protein quality.

Unlike starches and lipids, protein is the only macronutrient where increased digestibility is always preferred, since protein is essential for the growth and development of lean tissues. High protein digestibility is especially important for weaning foods and other foods consumed by children. Fermentation of a corn–millet blend followed by extrusion significantly increased the *in vitro* digestibility of both soluble and insoluble proteins, while extrusion alone increased only the digestibility of insoluble proteins, which were much higher than the previously–fermented samples (Onyango et al. 2005). Losses of over 10% of the original lysine in the blend occurred in the fermented–extruded blend and the blend extruded with 1.0% added citric acid, presumably due to increased Maillard reactions. Racemization of EAA to their D-epimers can also restrict digestibility. Csapó et al. (2008) reported that racemization increased with increasing barrel temperature in extruded full-fat soy.

Protein digestibility is affected in part by the presence of trypsin inhibitors. Extrusion reduced trypsin inhibitor activity (TIA) by 88%–91% in breadfruit–corn–soy blends (Nwabueze 2007). TIA in full-fat soybeans declined from 28.4 to 1.9 mg/g dry matter (DM) in samples extruded at the highest temperature (160°C) (Clarke and Wiseman 2007). Adding bean flour to corn starch to increase snack protein content can also add TIA. Anton, Fulcher, and Arntfield (2009) reported that navy beans had higher levels of TIA than did small red beans and that extrusion significantly reduced the levels of antinutrients.

Soy protein has been identified as a cholesterol-lowering food ingredient. Extrusion texturization of soy isolate did not reduce its effects on rat serum cholesterol, excretion of cholesterol and other steroids in feces, or protein nutrition compared with nonextruded soy (Fukui et al. 1993). Feeding a high-moisture extruded soy isolate to mice for 90 days revealed no apparent health differences, compared with animals fed unextruded soy (MacDonald, Pryzbyszewski, and Hsieh 2009).

MICRONUTRIENTS

VITAMINS

Vitamins are essential for human health because they are enzyme cofactors that cannot be produced by humans. Vitamins differ in their chemical structure and are classified as water-soluble or lipid-soluble. The diverse nature of these compounds is reflected in their varying stability under extrusion cooking conditions. Killeit (1994); and Riaz, Asif, and Ali (2009) have summarized factors affecting vitamin retention in extruded foods.

Vitamin A and the Carotenoids

Vitamin A is essential for proper immune system functions, and deficiency of this vitamin is a leading cause of blindness in many nations with developing economies. Vitamin A and related carotenoids are not stable in the presence of oxygen and heat, thus they are particularly vulnerable during extrusion. Beta-carotene is an antioxidant that is a vitamin A precursor. Beta-carotene is added to foods to make them more orange in color, but it is unstable when heated. Fortification of an extruded rice substitute called Ultra Rice® provides a simple means to add micronutrients to the local rice. Concerns about potential losses of vitamin A from this product led to an investigation of numerous antioxidant systems. Combinations of hydrophilic and hydrophobic antioxidants with chelating agents and moisture sequestrants maintained as much as 85% of the added vitamin A after 6 months of storage (Li et al. 2009). Vitamin A stability generally declined as postextrusion storage temperature increased.

Other Lipid-Soluble Vitamins

Vitamins D and K are stable during food processing, and little is known about the effects of extrusion on these vitamins. Vitamin E and related tocopherols are present in many foods naturally and may be added as antioxidants. Rice bran tocopherol decreased in samples extruded at higher temperatures during storage (Shin et al. 1997).

B Vitamins

Refined grains in the U.S. and many other nations must be enriched with thiamin, riboflavin, niacin, and folic acid to prevent deficiencies of these vitamins. Retention of these added nutrients, as well as native vitamins, during extrusion has been studied for several decades. Thiamin, also known as vitamin B_1, is the most heat-labile vitamin. Breakdown of thiamine during wheat flour extrusion is a first-order reaction (Guzman-Tello and Cheftel 1987) with potential thiamine losses varying from

5% to 100% (Killeit 1994). The high temperatures required to puff products can have a detrimental effect on thiamine content. The extrusion of wheat flour at lower temperatures combined with puffing produced by carbon dioxide aided in thiamine retention; increasing feed moisture and barrel temperature decreased thiamin levels while the reduction of residence time by increasing screw speed appeared to protect the vitamins (Schmid, Dolan, and Ng 2005). Based on a series of experiments using a short-barrel, single-screw extruder, Athar et al. (2006) concluded that riboflavin and niacin were more stable than thiamin and pyridoxine, and that retention of specific vitamins varies with the composition of the material being extruded.

Ascorbic Acid

Ascorbic acid (vitamin C) is susceptible to heat destruction and oxidation. Surface application of ascorbic acid postextrusion is the typical solution to potential losses within the extruder barrel. Ascorbic acid added to cassava starch was retained by at least 50% (Sriburi and Hill 2000). Blueberry concentrate appeared to protect 1% added vitamin C in an extruded breakfast cereal, compared with a product containing corn, sucrose, and ascorbic acid (Chaovanalikit et al. 2003).

MINERALS

Many mineral nutrients are heat-stable, thus fortification prior to extrusion is possible. While the minerals may not be destroyed during processing, their bioavailability could be reduced due to their inclusion in a matrix of macromolecules, particularly DF and phytate. Changing screw speed had no effect on phytate retention in wheat, rice, and oat bran, but insoluble fiber decreased in rice and oat bran after extrusion (Gualberto et al. 1997). When a high-fiber cereal product was fed to seven persons with ileostomies, mineral availability was reduced even though DF and phytate values were not affected by extrusion (Sandberg et al. 1986; Kivistö et al. 1986). Extrusion reduced phytate levels in wheat flour (Fairweather-Tait et al. 1989). Inactivation of phytases during extrusion in these studies may partly explain these findings. Although phytic acid was lower under all processing conditions, total phytate was not affected. Several studies reported that phytate in legumes was also not affected by extrusion (Lombardi-Boccia, Di Lullo, and Carnovale 1991; Ummadi, Chenoweth, and Ng 1995). Phytate was not reduced by the extrusion of corn–millet blends (Onyango et al. 2005).

Several studies have assessed the effect of extrusion on *in vitro* and *in vivo* mineral availability. The solubility of iron under conditions similar to digestion and subsequent ability to dialyze across a membrane is used to assess bioavailability. While extrusion slightly increased iron availability in corn snacks (Hazell and Johnson 1989), high-shear extrusion reduced dialyzable iron, compared with low-shear extrusion of navy beans, lentils, chickpeas, and cowpeas (Ummadi, Chenoweth, and Ng 1995). Extrusion did not reduce the zinc bioavailability of 85:15 blends of semolina and soy protein concentrate (Kang 1996). Extrusion barrel temperature and moisture content had little impact on the *in vitro* iron and zinc dialyzability of bean flour (Drago et al. 2007). Iron-depleted Indian children fed extruded rice containing micronized ground ferric pyrophosphate for 7 months had significantly higher iron reserves than did schoolmates fed control rice (Moretti et al. 2006). Another child nutrition study in the Philippines found that extruded rice containing micronized dispersible ferric pyrophosphate was as effective as extruded rice containing ferrous sulfate in combating iron deficiency anemia (Angeles-Agdeppa et al. 2008).

High-fiber foods may abrade the interior of the extruder barrel and screws, resulting in increased mineral content. Potato peels extruded under higher temperature had as much as 38% more total iron after extrusion (Camire, Zhao, and Violette 1993). Extruded corn, which is fairly low in fiber, showed no difference in total, elemental, or soluble iron, even in the presence of antioxidant additives (Camire and Dougherty 1998). Iron content in extruded potato flakes increased with barrel temperature (Maga and Sizer 1978). Screw wear iron had high bioavailability in rats fed extruded corn and potato (Fairweather-Tait et al. 1987). Extrusion did not reduce the utilization of iron and zinc from wheat bran and wheat in adult human volunteers (Faireather-Tait et al. 1989).

PHYTOCHEMICALS

PHENOLIC AND FLAVONOID COMPOUNDS

Plants contain numerous phenolic compounds including phenolic acids, anthocyanins, flavonolols, catechins, and tannins. The role of these compounds and their metabolites is not clearly understood, but their function as antioxidants may offer protection against cancer and cardiovascular disease (Crozier, Jaganath, and Clifford 2009). Analysis of these chemicals is challenging since they may be bound to nonstarch polysaccharides and are not easily released during extraction procedures. Characterization of phenolic compounds and their breakdown products in extruded foods by mass spectrometry has been reported in very few research studies. Such characterization is essential in order to understand how extrusion alters these compounds and their potential health effects.

Phenolic acids are among the most common and smallest phenolic compounds. Typically they are measured by a spectrometric procedure that reports total phenolics. The majority of published research papers have reported significant decreases in total phenolics after extrusion (Camire 1998). Ferulic acid, the predominant phenolic acid in grains, underwent only minor losses during twin-screw extrusion of germinated brown rice (Ohtsubo et al. 2005). White, Howard, and Prior (2010) extruded mixtures of corn starch with cranberry pomace, the residue of cranberry juice processing. Total anthocyanins decreased after extrusion, and barrel temperatures of 170°C and 190°C resulted in greater losses. Total flavonols, especially quercetin 3-rhamnoside, increased, however, by 30%–34%. Procyanidins are the compounds believed to be responsible for the antibacterial properties of cranberries. Procyanidins with a degree of polymerization (DP) of 1–2 were higher in extruded blends, but oligomers with a DP of 4–9 decreased. This shift suggests that the larger moieties were depolymerized. Similar trends in procyanidin changes were reported for extruded sorghum (Awika et al. 2003) and blueberry pomace–white sorghum blends (Khanal et al. 2009a). Extruding sorghum increased the bioavailability of catechins to pigs by 50% (Gu et al. 2008). All of the aforementioned studies were done with small extruders; effects of scale-up are not yet known.

Anthocyanins provide attractive red, blue, and purple colors in foods. Total anthocyanins in grape pomace were reduced by 18%–53% by extrusion (Khanal, Howard, and Prior 2009b). Ascorbic acid fortification of corn breakfast cereals containing blueberry by-products experienced significant losses of total anthocyanins, compared with those without added vitamin C, but all extruded samples underwent losses of the pigments (Chaovanalikit et al. 2003).

Total antioxidant activity in foods varies according to food composition, prior processing, and handling as well the analytical procedure used to measure activity. Nutrients such as ascorbic acid, beta-carotene, tocopherols, and selenium serve as antioxidants, and numerous phenolic and flavonoid compounds have antioxidant activity. Although antioxidant activity measured by the 2-2-diphenyl-1-picrylhydrazyl (DPPH) method was not correlated with phenolic content in extruded barley, both traits declined with extrusion (Altan, McCarthy, and Maskan 2009b). Antioxidant activity dropped by 60%–68% and phenolics decreased by 46%–60%, but no clear trend emerged for the effects of barrel temperature or screw speed on the retention of either factor. Antioxidant activity and total phenols also decreased after extrusion in the blends of corn starch with navy or small red beans (Anton, Fulcher, and Arntfield 2009). Small but significant increases in both total antioxidant capacity and total phenols were reported for extruded blends of red cabbage with corn starch or with wheat flour (Stojceska et al. 2009).

Tannins form indigestible complexes with proteins. Increasing feed moisture and screw speed reduced the tannin levels in breadfruit–corn–soy blends (Nwabueze 2007). Extrusion reduced tannins in a corn–millet blend by 58%; 67% losses occurred in fermented and then extruded samples (Onyango et al. 2005). The addition of lactic or citric acid also reduced tannins to a lesser extent.

OTHER PHYTOCHEMICALS

Brown rice extruded after germination for 72 hours had poor retention of oryzanol (Ohtsubo et al. 2005). Blends of wheat and ginseng powder (*Panax ginseng* C.A. Meyer) were twin-screw extruded, by combining a variety of feed moistures, screw speeds and final barrel temperatures (Chang and Ng 2009). Extrusion enhanced the extractability of several ginsenosides and caused the formation of novel compounds. The Rg3 ginsenoside has been studied for its health benefits; the phytochemical was recovered in extruded samples from only three of the experimental treatments suggesting that extrusion conditions must be optimized for the retention of specific compounds.

Soy isoflavones have estrogenic activity, but their role in human health is still being elucidated. Okara, a tofu by-product, when mixed with wheat flour showed reductions postextrusion in the aglycone genistein (Rinaldi, Ng, and Bennick 2000). Daidzin and genistin glucosides increased; total isoflavone values were 40% lower in the okara samples extruded at higher temperature.

Increasing barrel temperature caused decarboxylation of isoflavones, causing higher proportions of acetyl derivatives in the blends of 20% soy protein concentrate with cornmeal (Mahungu et al. 1999). Extruded corn–soy blends were less effective in preventing proliferation of breast cancer cells *in vitro* even though the levels of the biologically active aglycones were not affected by extrusion (Singletary et al. 2000).

SUMMARY

Extrusion cooking can lead to both improvements and decreases in nutritional quality. Additional research is needed to understand the effects of extrusion processing parameters on the changes at the molecular level. The impact of extrusion on phytochemicals and other healthful food components is an evolving area that is expected to expand significantly in the next few years.

REFERENCES

Altan, A., McCarthy, K.L., and Maskan, M. 2009a. Effect of extrusion cooking on functional properties and *in vitro* starch digestibility of barley-based extrudates from fruit and vegetable by-products. *J Food Sci* 74: E77–E86.

Altan, A., McCarthy, K.L., and Maskan, M. 2009b. Effect of extrusion process on antioxidant activity, total phenolics and β-glucan content of extrudates developed from barley-fruit and vegetable by-products. *Int J Food Sci Tech* 44: 1263–1271.

Angeles-Agdeppa, I., Capanzana, M.V., Barba, C.V., Florentino, R.F., and Takanashi, K. 2008. Efficacy of iron-fortified rice in reducing anemia among schoolchildren in the Philippines. *Int J Vitam Nutr Res* 78: 74–86.

Anton, A.A., Fulcher, R.G., and Arntfield, S.D. 2009. Physical and nutritional impact of fortification of corn starch-based extruded snacks with common bean (*Phaseolus vulgaris* L.) flour: Effects of bean addition and extrusion cooking. *Food Chem* 113: 989–996.

Areas, J.A.G. 1992. Extrusion of food proteins. *Crit Rev Food Sci Nutr* 32: 365–392.

Artz, W.E., Rao, S.K., and Sauer, R.M. 1992. Lipid oxidation in extruded products during storage as affected by extrusion temperature and selected antioxidants. In *food extrusion science and technology*, eds. J.L. Kokini, C.-T. Ho, and M.V. Karwe, 449–461, New York: Marcel Dekker, Inc.

Athar, N., Hardacre, A., Taylor, G., Clark, S., Harding, R., and McLaughlin, J. 2006. Vitamin retention in extruded food products. *J Food Compos Anal* 19: 379–383.

Awika, J.M., Dykes, L., Gu, L., Rooney, L.W., and Prior, R.L. 2003. Processing of sorghum (*Sorghum bicolor*) and sorghum products alters procyanidin oligomer and polymer distribution and content. *J Agric Food Chem* 51: 5516–5521.

Björck, I., and Asp, N.-G. 1983. The effects of extrusion cooking on nutritional value-a literature review. *J Food Eng* 2: 281–308.

Brouns, F., Bjorck, I., Frayn, K.N., Gibbs, A.L., Lang, V., Slama, G., and Wolever, T.M. 2005. Glycaemic index methodology. *Nutr Res Rev* 18: 145–171.

Camire, M.E. 1991. Protein functionality modification by extrusion cooking. *J Am Oil Chem Soc* 68: 200–205.

Camire, M.E. 1998. Chemical changes during extrusion cooking. In *Process-induced chemical changes in food,* eds. F. Shahidi, C.-T. Ho, and H. van Chuyen, 109–121. New York: Plenum Press.

Camire, M.E., Camire, A., and Krumhar, K. 1990. Chemical and nutritional changes in foods during extrusion. *Crit Rev Food Sci Nutr* 29: 35–57.

Camire, M.E. and Dougherty, M.P. 1998. Added phenolic compounds enhance lipid stability in extruded corn. *J Food Sci* 63: 516–518.

Camire, M.E., Zhao, J., and Violette, D.A. 1993. In vitro binding of bile acids by extruded potato peels. *J Agric Food Chem* 41: 2391–2394.

Capriles, V.D., Coelho, K.D., Guerra-Matias, A.C., and Arêas, J.A. 2008. Effects of processing methods on amaranth starch digestibility and predicted glycemic index. *J Food Sci* 73:H160–164.

Chang, Y.H., and Ng, P.K. 2009. Effects of extrusion process variables on extractable ginsenosides in wheat-ginseng extrudates. *J Agric Food Chem* 57: 2356–2362.

Chanvrier, H., Appleqvist, I.A.M., Bird, A.R., Gilbert, E., Htoon, A., Li, Z., Lillford, P.J., Lopez-Rubio, A., Morell, M.K., and Topping, D.L. 2007. Processing of novel elevated amylose wheats: Functional properties and starch digestibility of extruded products. *J Agric Food Chem* 55: 10248–10257.

Chaovanalikit, A., Dougherty, M.P., Camire, M.E., and Briggs, J. 2003. Ascorbic acid fortification reduces anthocyanins in extruded blueberry-corn cereals. *J Food Sci* 68:2136–2140.

Cheftel, J.C. 1986. Nutritional effects of extrusion cooking. *Food Chem* 20: 263–283.

Chiu, C.W., Henley, M., and Altieri, P. 1994. Process for making amylase resistant starch from high amylose starch. United States Patent. Patent No: 5,281,276. Date of Patent: Jan. 25, 1994.

Clarke, E., and Wiseman, J. 2007. Effects of extrusion conditions on trypsin inhibitor activity of full fat soybeans and subsequent effects on their nutritional value for young broilers. *Brit Poultry Sci* 48: 703–712.

Crozier, A., Jaganath, I.B., and Clifford, M.N. 2009. Dietary phenolics: Chemistry, bioavailability and effects on health. *Nat Prod Rep* 26: 1001–1043.

Csapó, J., Varga-Visi, E., Lóki, K., Albert, C., and Salamon, S. 2008. The influence of extrusion on loss and racemization of amino acids. *Amino Acids* 34: 2872–92

de la Gueriviere, J.F., Mercier, C., and Baudet, L. 1985. Incidences de la cuisson-extrusion sur certains parametres nutritionnels de produits alimentaires notamment céréaliers. *Cah Nutr Diet* 20: 201–210.

Della Valle, G., Quillien, L., and Gueguen, J. 1994. Relationships between processing conditions and starch and protein modifications during extrusion-cooking of pea flour. *J Sci Food Agric* 64: 509–517.

Dinkins, J. 2000. Beliefs and attitudes of Americans towards their diet. *Nutr Insight* 19: 1–2.

Drago, S.R., Velasco-González, O.H., Torres, R.L., González, R.J., and Valencia, M.E. 2007. Effect of the extrusion on functional properties and mineral dialyzability from *Phaseolus vulgaris* bean flour. *Plant Foods Hum Nutr* 62: 43–48.

Esposito, F., Arlotti, G., Bonifati, A.M., Napolitano, A., Vitale, D., and Fogliano, V. 2005. Antioxidant activity and dietary fibre in durum wheat bran by-products. *Food Res Int* 38: 167–1173.

Fairweather-Tait, S.J., Portwood, D.E., Symss, L.L., Eagles, J., and Minski, M.J. 1989. Iron and zinc absorption in human subjects from a mixed meal of extruded and non-extruded wheat bran flour. *Am J Clin Nutr* 49: 151–155.

Fairweather-Tait, S.J., Symss, L.L., Smith, A.C., and Johnson, I.T. 1987. The effect of extrusion cooking on iron absorption from maize and potato. *J Sci Food Agric* 39: 341–348.

Fukui, K., Aoyama, T., Hashimoto, Y., and Yamamoto, T. 1993. Effect of extrusion of soy protein isolate on plasma cholesterol level and nutritive value of protein in growing male rats. *J Jap Soc Nutr Food Sci* 46: 211–216.

Gautam, A., and Choudhoury, G.S. 1999. Screw configuration effects on starch breakdown during twin screw extrusion of rice flour. *J Food Process Preserv* 23: 355–375.

González-Soto, R.A., Sánchez-Hernández, L., Solorza-Feria, J., Núñez-Santiago, C., Flores-Huicochea, E., and Bello-Perez, L.A. 2006. Resistant starch production from non-conventional starch sources by extrusion. *Food Sci Tech Int* 12: 5–11.

Gu, L., House, S.E., Rooney, L.W., and Prior, R.L. 2008. Sorghum extrusion increases bioavailability of catechins in weanling pigs. *J Agric Food Chem* 56: 1283–1288.

Gualberto, D.G., Bergman, C.J., Kazemzadeh, M., and Weber, C.W. 1997. Effect of extrusion processing on the soluble and insoluble fiber, and phytic acid contents of cereals bran. *Plant Foods Human Nutr* 51: 187–198.

Guzman L.B., Lee, T.-C., and Chichester, C.O. 1992. Lipid binding during extrusion cooking. In *Food extrusion science and technology*, eds. by J.L. Kokini, C.-T. Ho, and M.V. Karwe. New York: Marcel Dekker, Inc.

Guzman-Tello, R., and Cheftel, J.C. 1987. Thiamine destruction during extrusion cooking as an indicator of the intensity of thermal processing. *Int J Food Sci Tech* 22: 549–562.

Harper, J.M., and Jansen, G.R. 1985. Production of nutritious precooked foods in developing countries by low-cost extrusion technology. *Food Reviews Int* 1: 27–97.

Hasjim, J., and Jane, J.-L. 2009. Production of resistant starch by extrusion cooking of acid-modified normal-maize starch. *J Food Sci* 74: C556–C562.

Hazell, T., and Johnson, I.T. 1989. Influence of food processing on iron availability in vitro from extruded maize-based snack food. *J Sci Food Agric* 46: 365–374.

Htoon, A., Shrestha, A.K., Flanagan, B.M., Lopez-Rubio, A., Bird, A.R., Gilbert, E.P., and Gidley, M.J. 2009. Effects of processing high amylose maize starches under controlled conditions on structural organization and amylase digestibility. *Carb Polymers* 75: 236–245.

Hwang, J.-K., Kim, C.-J., and Kim, C.-T. 1998. Production of glucooligosaccharides and poly-dextrose by extrusion reactor. *Starch* 50: 104–107.

Ilo, S., and Berghofer, E. 2003. Kinetics of lysine and other amino acids loss during extrusion cooking of maize grits. *J Food Sci* 68: 496–502.

Jin, Z., Hsieh, F., and Huff, H.E. 1994. Extrusion cooking of corn meal with soy fiber, salt, and sugar. *Cereal Chem* 71: 227–234.

Johnson, S.K., Thomas, S.J., and Hall, R.S. 2005. Palatability and glucose, insulin and satiety responses of chickpea flour and extruded chickpea flour bread eaten as part of a breakfast. *Eur J Clin Nutr* 59: 169–176.

Kang, S.-Y. 1996. Zinc bioavailability in a semolina/soy protein mixture was not affected by extrusion processing, M.S. Thesis, East Lansing, MI, Michigan State University.

Kendall, C.W., Emam, A., Augustin, L.S., and Jenkins, D.J. 2004. Resistant starches and health. *J AOAC Int* 87: 769–774.

Khanal, R., Howard, L.R., Brownmiller, C.R., and Prior, R.L. 2009a. Influence of extrusion processing on procyanidin composition and total anthocyanin contents of blueberry pomace. *J Food Sci* 74: H52–H58.

Khanal, R.C., Howard, L.R., and Prior, R.L. 2009b. Procyanidin content of grape seed and pomace, and total anthocyanin content of grape pomace as affected by extrusion processing. *J Food Sci* 74: H174–H182.

Killeit, U. 1994. Vitamin retention in extrusion cooking. *Food Chem* 49: 149–155.

Kivistö, B., Andersson, H., Cederblad, G., Sandberg, A.-S., and Sandstrom, B. 1986. Extrusion cooking of a high-fiber cereal product. 2. Effects on apparent absorption of zinc, iron, calcium, magnesium and phosphorus in humans. *British J Nutr* 55: 255–260.

Kong, J., Dougherty, M.P., Perkins, L.B., and Camire, M.E. 2008. Composition and consumer acceptability of a novel extrusion-cooked salmon snack. *J Food Sci* 73: S118–S123.

Konstance, R.P., Onwulata, C.I., Smith, P.W., Lu, D., Tunick, M.H., Strange, E.D., and Holsinger, V.H. 1998. Nutrient-based corn and soy products by twin-screw extrusion. *J Food Sci* 63: 864–868.

Larrea, M.A., Chang, Y.K., and Martínez Bustos, F. 2005. Effect of some operational extrusion parameters on the constituents of orange pulp. *Food Chem* 89: 301–308.

Li, Y.O., Lam, J., Dioady, L.L., and Jankowski, S. 2009. Antioxidant system for the preservation of vitamin A in Ultra Rice. *Food Nutr Bull* 30: 82–89.

Lombardi-Boccia, G., Di Lullo, G., and Carnovale, E. 1991. In-vitro iron dialysability from legumes: influence of phytate and extrusion cooking. *J Sci Food Agric* 55: 599–605.

MacDonald, R.S., Pryzbyszewski, J., and Hsieh, F.H. 2009. Soy protein isolate extruded with high moisture retains high nutritional quality. *J Agric Food Chem* 57: 3550–3555.

Maga, J.A., 1978. Cis-trans fatty acid ratios as influenced by product and temperature of extrusion cooking. *Lebensm-Wiss u–Technol* 11: 183–184.

Maga, J.A., and Sizer, C.E. 1978. Ascorbic acid and thiamin retention during extrusion of potato flakes. *Lebensm-Wiss u–Technol* 11: 192–194.

Mahungu, S.M., Diaz-Mercado, S., Li, J., Schwenk, M., Singletary, K., and Faller, J. 1999. Stability of isoflavones during extrusion processing of corn/soy mixture. *J Agric Food Chem* 47: 279–284.

McCleary, B.V. 2007. An integrated procedure for the measurement of total dietary fibre (including resistant starch), non-digestible oligosaccharides and available carbohydrates. *Anal Bioanal Chem* 389: 291–308.

Moretti, D., Zimmermann, M.B., Muthayya, S., Thankachan, P., Lee, T.C., Kurpad, A.V., and Hurrell, R.F. 2006. Extruded rice fortified with micronized ground ferric pyrophosphate reduces iron deficiency in Indian schoolchildren: A double-blind randomized controlled trial. *Am J Clin Nutr* 84: 822–829.

Nelson, A.I., Wijeratne, W.B., Yeh, S.W., Wei, T.M., and Wei, L.S. 1987. Dry extrusion as an aid to mechanical expelling of oil from soybeans. *J Am Oil Chem Soc* 64: 1341–1347.

Nwabueze, T.U. 2007. Effect of process variables on trypsin inhibitor activity (TIA), phytic acid and tannin content of extruded African breadfruit-corn-soy mixtures: A response surface analysis. *LWT* 40: 21–29.

Onyango, C., Noetzold, H., Ziems, A., Hofmann, T., Bley, T., and Henle, T. 2005. Digestibility and antinutrient properties of acidified and extruded maize-finger millet blend in the production of uji. *LWT* 38: 697–707.

Ohtsubo, K., Suzuki, K., Yasui, Y., and Kasumi, T. 2005. Bio-functional components in the processed pre-germinated brown rice by a twin-screw extruder. *J Food Comp Anal* 18: 303–316.

Politz, M.L., Timpa, J.D., and Wasserman, B.P. 1994. Quantitative measurement of extrusion-induced starch fragmentation products in maize flour using nonaqueous automatic gel-permeation chromatography. *Cereal Chem* 71: 532–536.

Riaz, M.N., Asif, M., and Ali, R. 2009. Stability of vitamins during extrusion. *Crit Rev Food Sci Nutr* 49: 361–368.

Rinaldi,V.E.A., Ng, P.K.W., and Bennick, M.R. 2000. Effects of extrusion on dietary fiber and isoflavone contents of wheat extrudates enriched with wet okara. *Cereal Chem* 77: 237–240.

Sandberg, A.S., Andersson, H., Kivistö, B., and Sandstrom, B. 1986. Extrusion cooking of a high-fibre cereal product. 1. Effects on digestibility and absorption of protein, fat, starch, dietary fibre and phytate in the small intestine. *Br J Nutr* 55: 245–254.

Schmid, A.H., Dolan, K.D., and Ng, P.K.W. 2005. Effect of extruding wheat flour at lower temperatures on physical attributes of extrudates and on thiamin loss when using carbon dioxide gas as a puffing agent. *Cereal Chem* 82: 305–313.

Semwal, A.D., Sharma, G.K., and Arya, S.S. 1994. Factors influencing lipid autooxidation in dehydrated precooked rice and bengalgram dhal. *J Food Sci Technol* 31: 293–297.

Shin, T.S., Godber, J.S., Martin, D.E., and Wells, J.H. 1997. Hydrolytic stability and changes in E vitamers and oryzanol of extruded rice bran during storage. *J Food Sci* 62: 704–708.

Singh, S., Wakeling, L., and Gamlath, S. 2007. Retention of essential amino acids during extrusion of protein and reducing sugars. *J Agric Food Chem* 55: 8779–8786.

Singletary, K., Faller, J., Li, J.Y., and Mahungu, S. 2000. Effect of extrusion on isoflavone content and antiproliferative bioactivity of soy/corn mixtures. *J Agric Food Chem* 48: 3566–3571.

Sriburi, P., and Hill, S.E. 2000. Extrusion of cassava starch with either variations in ascorbic acid concentration or pH. *Int J Food Sci Tech* 35: 141–154.

Stojceska, V., Ainsworth, P., Plunkett, A., and Ibanoğlu, S. 2009. The effect of extrusion cooking using different water feed rates on the quality of ready-to-eat snacks made from food by-products. *Food Chem* 114: 226–232.

Suzuki, H., Chung, B.S., Isobe, S., Hayakawa, S., and Wada, S. 1988. Changes in ω(omega)-3 polyunsaturated fatty acids in the chum salmon muscle during spawning migration and extrusion cooking. *J Food Sci* 53: 1659–1661.

Theander, O., and Westerlund, E. 1987. Studies on chemical modifications in heat-processed starch and wheat flour. *Starch/Stärke* 39: 88–93.

Ummadi, P., Chenoweth, W.L., and Ng, P.K.W. 1995. Changes in solubility and distribution of semolina proteins due to extrusion processing. *Cereal Chem* 72: 564–567.

Unlu, E., and Faller, J.F. 1998. Formation of resistant starch by a twin-screw extruder. *Cereal Chem* 75: 346–350.

Wang, S., Casulli, J., and Bouvier, J.M. 1993. Effect of dough ingredients on apparent viscosity and properties of extrudates in twin-screw extrusion cooking. *Int J Food Sci Tech* 28: 465–479.

Wells, H.F., and Buzby, J.C. 2008. Dietary Assessment of Major Trends in U.S. Food Consumption, 1970–2005, Economic Information Bulletin No. 33. Economic Research Service, U.S. Dept. of Agriculture.

White, B.L., Howard, L.R., and Prior, R.L. 2010. Polyphenolic composition and antioxidant capacity of extruded cranberry pomace. *J Agric Food Chem* Published on-line 12/18/2009. DOI: 10.1021/jf902838b.

Zhang, M., Liang, Y., Pei, Y., Gao, W., and Zhang, Z. 2009. Effect of process on physicochemical properties of oat bran soluble dietary fiber. *J Food Sci* 74: C628–C636.

5 Rheological Properties of Materials during the Extrusion Process

Kathryn L. McCarthy, Daniel J. Rauch, and John M. Krochta

CONTENTS

INTRODUCTION

Because of increased concerns about the effects of waste products on the environment, edible and biodegradable films have become the subjects of great interest. While there has been much work done on the creation of edible and biodegradable films as a batch process, the manufacture of these films must be economical on an industrial scale to be commercially viable. Extrusion is explored as a means to produce these films on a large scale.

Work has been done to utilize whey protein as a component of these films. Whey is a liquid by-product of the cheese-making process. Each kilogram of cheese produced requires over 10 L of milk and creates a waste stream of 9 L of liquid whey. The high biological oxygen demand (BOD) and chemical oxygen demand (COD) of whey make it a burden to dispose of properly (O'Shea 2003). Many researchers have worked to find the uses of whey and its components. One use of whey protein is the formation of edible films; whey protein films can be made from an aqueous solution, with glycerol (GLY) or another plasticizer added to improve film mechanical properties (Sothornvit and Krochta 2001). The resulting films have moderately good tensile properties and good oxygen-, aroma-, and oil-barrier properties (Han and Krochta 2001; Hong and Krochta 2006), but are poor moisture barriers (Perez-Gago, Nadaud, and Krochta 1999). While the solution-casting method used on the laboratory-scale is appropriate for studying the properties of films that would be formed as coatings on foods from aqueous solution, it does not lend itself well to the industrial-sized production of films that would be used as stand-alone food wraps, layers between food components, or sealed to form food pouches. In the laboratory-scale production of films, aqueous solutions of whey protein are heated to denature and cross-link the protein (Perez-Gago and Krochta 1999). The water-evaporation step of the solution-cast film production is both time and energy consuming.

Extrusion processing is a method that avoids these pitfalls of solution-casting. Guided by the thermal transitions found for whey protein–glycerol mixtures by differential scanning calorimetry (DSC), extruder temperature profiles were identified, which result in the softening, melting, and flowing of the glycerol-plasticized whey protein to form flexible, transparent films (Hernandez-Izquierdo et al. 2008). As part of that study, it was demonstrated that these sheets can subsequently be pressed and stretched into thin transparent films that have good heat-sealing capabilities and closely resemble their solution-cast counterparts (Hernandez-Izquierdo and Krochta 2009). While that work demonstrated the possibility of making whey protein sheets by extrusion, additional investigation has provided an understanding of the relationship between extruder operating conditions and the resultant extrudate properties (Rauch 2008).

This chapter focuses on a methodology to take a successful product (i.e., solution-cast whey protein films) and create a similar extruded product. In particular, in-line viscosity measurements and extrudate rheological properties were examined to evaluate the similarities and differences between the extruded films and their solution-cast counterparts.

EXPERIMENTAL

MATERIALS AND METHODS

A laboratory-scale 30 mm corotating twin-screw extruder (Model MPC/V-30 APV, Staffordshire, England) with a System9000 torque rheometer (Haake Buchler, Paramus, NJ, U.S.A.) that provided computer control and data acquisition was used to produce whey protein films. Instantized whey protein isolate powder (WPI) (Davisco Foods, Le Sueur, MN, U.S.A.) at 96.8% protein (db) and 6.48% moisture

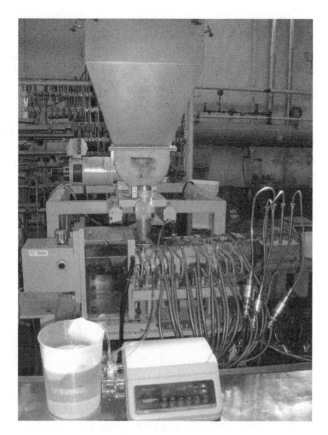

FIGURE 5.1 APV 30 mm corotating twin-screw extruder during operation.

(wb) was fed into the extruder with a twin-screw volumetric feeder (Type T-20, K-Tron Corp., Pitman, NJ, U.S.A.). A vegetable glycerol (Starwest Botanicals, Rancho Cordova, CA, U.S.A.) and water mixture (70/30 w/w) was pumped into a downstream port with a peristaltic pump (Masterflex 7524-00, Cole-Parmer, Vernon Hills, IL, U.S.A.).

Figure 5.1 illustrates the experimental setup. The extruder barrel had a clamshell design consisting of three independent temperature zones controlled by electrical heating and compressed air cooling. A fourth and a fifth temperature zone provided electrical heating to the transition piece between the extruder and die and to the slit die. Set temperatures for the five temperatures zones were 40°C–60°C–130°C–145°C–120°C for all trials. The slit die (Haake Buchler, Paramus, NJ, U.S.A.) had the dimensions of 1.47 mm × 20 mm × 150 mm and was equipped with three flush mount pressure transducers (Dynisco, Type PT422A, Sharon, MA, U.S.A.). The pressure transducers were positioned at 44.45 mm, 82.55, and 120.65 mm from the entrance of the slit die. The height and width of the die were 1.47 mm and 20 mm, respectively. Data were acquired every 6 s for extruder torque, die pressure, and melt temperatures.

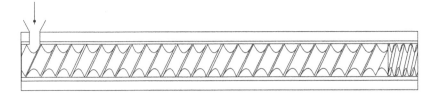

FIGURE 5.2   Schematic of the screw configuration of the APV 30 mm corotating twin-screw extruder (side view).

Figure 5.2 illustrates the screw configuration used for all trials: 12 *L/D* of twin lead feed screws and 1 *L/D* single lead discharge screw. The configuration was chosen to primarily convey the melt and minimize shear damage.

EXPERIMENTAL DESIGN

The experimental design was a circumscribed central composite design (CCD) with screw speed and flow rate as the two experimental factors. The design required five levels for each factor and augmented an existing 2×2 factorial design. For a CCD, star points are at some distance α from the center, which is $\sqrt{2}$ for a two factor experimental design. Mass flow rates for the combined WPI and glycerol–water mixture were determined to yield 52% glycerol (db) in the extrudate. Table 5.1 gives the experimental design for the extrusion trials with coded and actual variable levels. Runs within a block were randomized. The experimental responses were specific mechanical energy (SME), residence time (RT), and apparent viscosity (η) in the die, and tensile properties of the extrudate. The trials given in Table 5.1 were replicated.

TABLE 5.1
Experimental Design for Extrusion Experiment with Coded and Actual Variable Levels

Block	Coded Level		Actual Level	
	X_1	X_2	Flow Rate (g/min)	Screw Speed (rpm)
1	1	−1	25	100
1	1	1	25	150
1	0	0	20	125
1	−1	−1	15	100
1	−1	1	15	150
2	1.414	0	27	125
2	0	−1.414	20	90
2	0	0	20	125
2	0	1.404	20	160
2	−1.414	0	13	125

RESPONSES

Operating data (e.g., torque, pressure readings at the die) and extrudate ribbons were collected under steady state conditions as characterized by torque values that varied less than 10%. Under steady state conditions, operating data were acquired over a time interval of 30 min to determine the average values for SME and apparent viscosity.

Residence Time Distribution and Mean Residence Time

The residence time distribution (RTD) and the mean residence time (MRT) were determined based on monitoring the color of the extrudate in response to a pulse input of the dye. Color readings were acquired using a GretagMacbeth Color-Eye 7000A Colorimeter (GretagMacbeth, New Windsor, NY, U.S.A.) and creating a calibration curve with varying amounts of Erythrosin B Red 140 Dye (Sigma-Aldrich, St. Louis, MO, U.S.A.) in WPI. Measurement results were displayed in Hunter values of L^* (lightness), a^* (red–green), and b^* (yellow–blue). Due to the red color of the dye, a^* values were the most sensitive to changes in the dye level and the resulting expression for Δa^* was used (Equation 5.1).

$$\Delta a^* = a^*_{dyed} - a^*_{undyed} \tag{5.1}$$

The resulting regression expression for the calibration curve was

$$\Delta a^* = 337.62c + 0.476; \ R^2 = 0.997 \tag{5.2}$$

over $0 \leq c \leq 5 \times 10^{-2}$ mg dye/g WPI, where c is the mass fraction of dye.

The RTD for each trial was determined by evaluating the color response of a pulse input of the dyed WPI; the proportion of dye to WPI was 1:199. The mass of the dyed WPI pulse input was 10% of the WPI feed rate for 1 s. For example, if the WPI was flowing at 10.3 g/min (i.e., 0.17 g/s), the pulse input was 0.017 g, added to the feed port during a 1 s time interval along with the 10.3 g/min WPI. At the time of the pulse input, a stop watch was started and marks were made on the extrudate piece every 5 s for feed rates ≥ 20 g/min and every 10 s for feed rates <20 g/min. Collection of the extruded sheet was terminated when no more red dye exited the extruder (i.e., 10–13 min after pulse input). The extrudate strip was then analyzed using the color meter and Δa^* values were determined. The average Δa^* was based on 20 color readings. All values of Δa^* fell in the range modeled by Equation 5.2, with maximum Δa^* values below 17. The color readings were converted to mass fractions of the dye, c (Equation 5.2). For the RTD of material in the extruder, the following expressions were used:

$$E(t) = \frac{c}{\int_0^\infty c \, dt} \cong \frac{c_i}{\sum_{i=0}^\infty c_i \Delta t} \tag{5.3}$$

for the exit age distribution and

$$\bar{t} = \int_0^t tE(t)dt \cong \frac{\sum_{i=0}^{t} t_i c_i \Delta t}{\sum_{i=0}^{\infty} c_i \Delta t} \tag{5.4}$$

for the MRT ($\bar{t}$).

Specific Mechanical Energy

SME, the energy imparted to the material in the extruder per unit mass, was calculated by Equation 5.5.

$$SME = \frac{T * N}{\dot{m}} \tag{5.5}$$

with torque (T) in units of $N \cdot m$, screw speed (N) in radians per second, and mass flow rate ($\dot{m}$) in kilograms per second.

Apparent Viscosity

The force balance, which equates pressure forces to viscous forces during viscometric flow, provides the relationship between the shear stress at the wall (σ_w), and shear rate at the wall ($\dot{\gamma}_w$),

$$\sigma_w = \frac{-(\Delta P)}{2L} h \tag{5.6}$$

where ΔP is the pressure drop over the length L and h is the height of the slit die (Steffe 1996). For a Newtonian fluid, shear rate is a function of volumetric flow rate through the slit die,

$$\dot{\gamma}_{w,Newt} = \frac{6Q}{h^2 w} \tag{5.7}$$

where Q is the volumetric flow rate and w is the width of the slit die. If the constitutive equation of the fluid being tested is not known, the general solution comparable to the Rabinowitsch–Mooney solution for pipe flow is (Steffe 1996)

$$\dot{\gamma}_w = \left(\frac{2n'+1}{3n'}\right)\frac{6Q}{h^2 w} \tag{5.8}$$

where the correction factor n' is defined as

$$n' = \frac{d\ln(\sigma_w)}{d\ln(6Q/wh^2)} \tag{5.9}$$

For a Newtonian fluid, n' is equal to one; for a power law fluid, n' is equal to the flow behavior index, n. Using the expressions above, the apparent viscosity η is determined by

$$\eta = \frac{\sigma_w}{\dot{\gamma}_w} \tag{5.10}$$

Tensile Properties

Tensile properties were measured using an Instron Universal Testing Machine, model 1122 (Instron Corp., Canton, MA, U.S.A.), at a crosshead speed of 50 mm/min, following the procedure outlined in the ASTM method D882-97 (ASTM 1998). Ten bone-shaped pieces of extruded sheet were tested per trial. The samples were placed in controlled humidity chambers and conditioned at 50% ± 5% relative humidity (RH), achieved using saturated solutions of magnesium nitrate (Fisher Scientific, Pittsburgh, PA, U.S.A.), for a minimum of 48 hours prior to testing. During the subsequent tensile tests, all efforts were made to maintain 50% RH conditions. Immediately prior to testing, sheet thickness was measured at five positions with an electronic digital caliper (Model S 225, Fred V. Fowler Co., Newton, MA, U.S.A.). Tensile strength (pulling force per initial sheet cross-sectional area at break, MPa), Young's modulus (stiffness as determined by the ratio of stress to strain in the linear region of the stress–strain curve, MPa), and elongation at break (degree to which the sheets stretched before breaking, %) were measured for each sample.

RESULTS AND DISCUSSION

QUALITATIVE DESCRIPTION OF EXTRUDATE

The goal of this work was to use extrusion as a processing technology to prepare food-safe films that can be used for biodegradable packaging material. Extruder conditions were identified in which the extrudate films were visually similar under all operating conditions. Figure 5.3 illustrates representative samples of the WPI/glycerol/water extrudate. Extrudate made at the lower feed rate conditions are the bottom strips in the photo, with the flow rate increasing toward the top of the photo. Under all conditions given in Table 5.1, the extrudate was translucent, yellow in color, and smooth. Color measurements were performed on the extrudate using the GretagMacbeth Color-Eye 7000A Colorimeter in $L^*a^*b^*$ notation and reported in Rauch (2008). There were significant differences in a^* and b^* values as a function of mass flow rate, but the differences were not perceptible

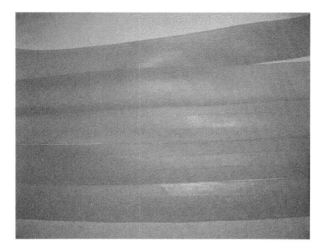

FIGURE 5.3   Representative ribbons of WPI/glycerol/water extrudate at increasing feed rates.

visually. Though the slit die temperature was over 100°C, the WPI extrudates were not expanded products. The average thickness was from 1.27 mm (die dimension) to 1.59 mm. Statistically, the change in dimension was not a function of screw speed or mass flow rate. Normally, a thinner die would be used to make films for wraps, separation layers for food components, and food pouches. Alternatively, thick films of the nature produced in this study can be heated and compressed to thin films. When that was done, the films appeared colorless and transparent (Rauch 2008).

The approach to measure in-line viscosity was to acquire pressure measurements at the slit die at known volumetric flow rates. Extrusion conditions were straightforward in terms of producing different products while maintaining a given temperature profile along the extruder length, using one screw configuration, and ensuring constant mass fraction of glycerol to WPI (db). Typical in-line viscosity measurements of shear viscosity (Equations 5.6 through 5.10) are made on a material that is unchanging through the experimental trials. Even under the best of circumstances, this is not the case for in-line viscosity measurements during extrusion. Therefore, apparent viscosity values of the extrudate melt in the slit die reflect different time/ temperature treatment under different operating conditions at the same shear rate through the die (e.g., flow rate). The second experimental factor in this work, e.g., screw speed, yielded different SME input at the same flow rate. In turn, the rheological properties of extrudate measured off-line also vary due to different operating conditions.

RESIDENCE TIME DISTRIBUTION

The RTD in the extruder describes the time history of the material as it flows from the feed inlet to the die. The RTD is frequently characterized by the exit age

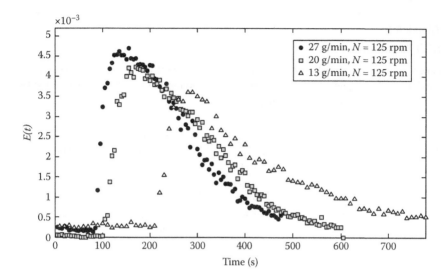

FIGURE 5.4 RTD as a function of flow rate at a screw speed of $N = 125$ rpm.

distribution. The MRT ($\bar{t}$) is the first moment of the exit age distribution and is given in Equation 5.4. In addition, the minimum RT, t_o, represents the RT of the most rapid "particle." To characterize the spread around the MRT, the variance was calculated

$$\sigma^2 = \int_0^\infty \left(t - \bar{t}\right)^2 E(t)dt \tag{5.11}$$

and is the second moment centered on the mean. Figure 5.4 illustrates exit age distributions for a material flowing at three flow rates at a screw speed of 125 rpm. Typical of the results reported in other extrusion studies (as cited by Poulesquen and Vergnes 2003) the RTD is shifted toward the lower time when the feed rate increases, with a decrease in the broadness of the distribution. As the flow rate increases, the minimum RT also decreases. For these curves, the minimum residence time were 220, 105, and 90 s as the flow rate increased from 13 to 27 g/min. Likewise, the variance decreased from 24,234 to 8,745 as the flow rate increased from 13 to 27 g/min. Table 5.2 gives the minimum RT, MRT, and variance for all combinations of mass flow rate and screw speed (Table 5.1). For each condition, the average is given and plus/minus one standard deviation. Initially, an analysis of variance (ANOVA) was performed on the 2×2 factorial design (as given by the operating conditions in Block 1, Table 5.1). Only the mass flow rate was a significant factor; means were not the same as the mass flow rate changed from 15 to 25 g/min.

A regression was performed on the entire CCD design, and again the screw speed term did not significantly contribute to the regression equations for MRT, minimum RT, or variance. For the MRT, the regression equation was

TABLE 5.2

RTD Parameters

Flow Rate (g/min)	Screw Speed (rpm)	t_0 (s) (average ± 1 stdev)	MRT (s) (average ± 1 stdev)	Variance (average ± 1 stdev)
13	125	203 ± 17	387 ± 8	21611 ± 1948
15	100	166 ± 16	358 ± 24	17258 ± 2350
	150	170 ± 12	370 ± 4	17455 ± 1583
20	90	129 ± 13	302 ± 15	13415 ± 1381
	125	111 ± 11	271 ± 12	12215 ± 1124
	160	115 ± 15	286 ± 15	14174 ± 3120
25	100	83 ± 13	242 ± 16	10324 ± 737
	150	88 ± 19	245 ± 28	12341 ± 801
27	125	85 ± 7	219 ± 3	8955 ± 402

$$\bar{t} = 296.4 - 58.3X_1 \quad R^2 = 0.868 \tag{5.12}$$

where X_1 is the coded variable for flow rate. This regression equation format allows ease of visualization as the flow rate varies from the values of the star points: −1.414 to 1.414. For instance, the MRT decreased from an average value of 294.4 s as the flow rate increased from 0 to 1.414 (coded). Physically, axial mixing is improved by a reduction in feed rate; uniformity of heat treatment is improved by increased feed rate.

Visually, the degree of fill in the extruder was examined at the end of several trials. As shown in Figure 5.2, the screw configuration consisted entirely of the conveying screw and one L/D discharge screw. Material did not build up within the extruder barrel during operation, except at the discharge screws. Consequently, increasing the screw speed had less impact than if the extruder had a higher degree of fill. In contrast to the conveying screws, the transition piece and the slit die operated at 100% fill. Therefore, due to the low degree of fill and the lack of dependence of the RT on screw speed, it was possible to more easily compare extrudates prepared at different screw speeds.

SPECIFIC MECHANICAL ENERGY

Both feed rate (X_1) and screw speed (X_2) had significant effects on SME. The regression expression was

$$SME = 825.7 - 124.6X_1 + 96.8X_2 \quad R^2 = 0.571 \tag{5.13}$$

Values of SME, as defined in Equation 5.5, increased as screw speed increased and decreased as mass flow rate increased. When torque was evaluated separately, the trend was reversed. Torque values decreased as screw speed increased and increased as flow rate increased. However, the effect of torque did not counterbalance the over-all effect of screw speed and mass flow rate.

APPARENT VISCOSITY

Modeling apparent viscosity has been approached in a number of ways for the extrusion of protein dough, most notably by Morgan, Steffe, and Ofoli (1989) and then extended to the extrusion of starch doughs by Mackey and Ofoli (1990), using analogies from polymerization kinetics and polymer rheology. These researchers developed a comprehensive model that predicts the effects of shear rate, temperature, moisture, and temperature–time history on the apparent viscosity. Experimental data from laboratories and from literature were used to evaluate the model parameters. The focus of the work was to develop a model to predict product quality, aid in scale-up, and use in process control (Morgan, Steffe, and Ofoli 1989; Mackey and Ofoli 1990). However, the approach requires extensive experimental designs. For less extensive studies, the purpose of modeling the apparent viscosity is to characterize the material by relating it to the experimental factors and/or experimental responses.

With that target, the data required to calculate the in-line apparent viscosity for the WPI melts were the pressure drops over the die length between pressure transducers and volumetric flow rate. For each experimental trial, pressure measurements were acquired over 30 consecutive minutes after steady state conditions were achieved. Figure 5.5 illustrates representative pressure profiles in the slit die. Curves are shown for a screw speed of 125 rpm and mass flow rates of 13, 20, and 27 g/min. The average pressure readings for each transducer were plotted; the linear regression for pressure yielded R^2 values equal to or greater than 0.994. The slope (Figure 5.5) was equal

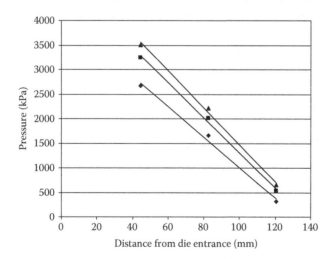

FIGURE 5.5 Representative pressure profiles in the slit die, shown for a screw speed of $N = 125$ rpm at flow rates of 13 g/min (diamonds), 20 g/min (squares), and 27 g/min (triangles). Values of the slope are −30.99, −35.52, and −37.45 kPa/mm, respectively. Coefficient of determination values for the linear regressions are $R^2 \geq 0.994$.

to $-\Delta P/L$ in Equation 5.6. The wall shear stress was calculated with Equation 5.6. For the shear rate, the volumetric flow rate was the product of the mass flow rate and the density of the extrudate within the extruder. This density was approximated by the composition of the feed component at 1.234 g/cm³. In comparison, the measured density of the extrudate product was a little higher at 1.282 g/cm³, due to the slightly lower moisture content.

The flow through the die was not expected to follow Newtonian behavior, therefore the apparent shear rate was corrected with n' (Equation 5.9) to yield the wall shear rate (Equation 5.8). Typically, the correction factor could be expected to be a function of extruder screw speed because different extruder conditions would inherently yield extrudates with different physical properties due to the time/temperature treatment. The correction factor at various screw speeds was evaluated, but the physical properties did not change sufficiently to justify multiple values of n' (see Figure 5.3). The value of $n' = 0.235$ was used to correct all shear rate values. The apparent viscosity (Equation 5.10) was plotted against the corrected shear rate on an ln–ln plot (Figure 5.6). Values of the apparent viscosity ranged from 4.6 to 8.6 Pa · s. The extrudate was characterized as a power law fluid and the correction factor n' was identical to the flow behavior index, n. The constitutive expression for a power law fluid is

$$\eta = K\dot{\gamma}_w^{n-1} \tag{5.14}$$

where K is the consistency index. The regression yielded a flow behavior index of $n = 0.237$ and a consistency index of 3,535 Pa · s^n ($R^2 = 0.909$). Similar results were found for soy protein plastics that were evaluated as a renewable, biodegradable alternative to fuel-based plastic resins. The flow behavior of the soy protein formulations was described by the power law model with flow behavior indices between 0.18 and 0.44 (Ralston and Osswald 2008).

For the WPI, the apparent viscosity values were a function of the experimental factors; changes in both flow rate and screw speed affected the apparent visccosity at $p \leq 0.05$. The linear regression, using coded variables, yielded

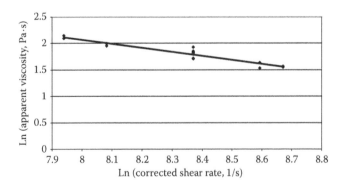

FIGURE 5.6 The apparent viscosity was characterized as a power law fluid. The regression yielded a flow behavior index of $n = 0.237$ and a consistency index of 3535 Pa · s^n ($R^2 = 0.909$).

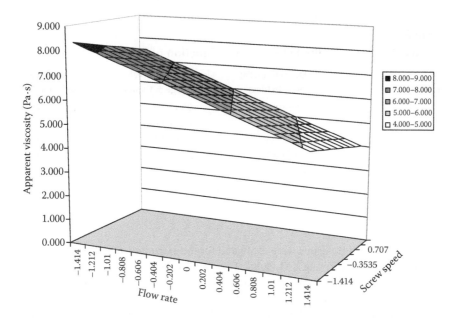

FIGURE 5.7 Response surface for in-line apparent viscosity as a function of screw speed and mass flow rate.

$$\eta = 6.19 - 1.23X_1 - 0.31X_2 \quad R^2 = 0.940 \tag{5.15}$$

The apparent viscosity decreased with both increasing flow rate (X_1) and increasing screw speed (X_2) as illustrated in the response surface (Figure 5.7).

To assess the relationship between apparent viscosity and other experimental responses, apparent viscosity was plotted against SME and against MRT. The best correlation was with MRT, with a coefficient of determination of $R^2 = 0.878$ (Figure 5.8). Apparent viscosity increased as MRT increased, which occurred at the lower flow rates (Equation 5.15). In other words, the apparent viscosity increased linearly with the increase in time/temperature treatment within the extruder.

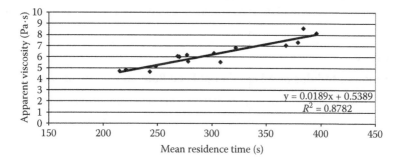

FIGURE 5.8 Correlation between apparent viscosity and MRT.

TABLE 5.3

Tensile Properties of the Extrudate as a Function of Feed Rate

Flow Rate (g/min)	Tensile Strength (MPa) (average ± 1 stdev)	Elongation at Break (%) (average ± 1 stdev)	Modulus (MPa) (average ± 1 stdev)
13	2.31 ± 0.15	90.1 ± 8.9	7.28 ± 0.67
15	2.35 ± 0.38	90.3 ± 11.8	8.33 ± 2.13
20	2.68 ± 0.26	98.6 ± 10.4	9.50 ± 1.85
25	2.80 ± 0.18	102.3 ± 7.7	10.22 ± 2.50
27	2.93 ± 0.12	97.5 ± 8.7	10.71 ± 2.42

TENSILE PROPERTIES

Three rheological properties were evaluated for the extrudate: tensile strength, elongation at break, and Young's Modulus. For each of these properties, different feed rates affected the mean values at a level of significance of $p < 0.05$. Table 5.3 gives the average values and plus/minus one standard deviation of these extrudate properties. Tensile strength and elongation at break increased with increasing feed rate. Tensile strength is strongly influenced by the cross-linking of heat-denatured whey proteins (Perez-Gago and Krochta 2001). Thus, extrudate could possibly become stronger with more exposure of the material to the heated barrel. However, the opposite effect occurred; film strength was greater at lower RTs. A possible cause may be that the WPI was overexposed to heat and shear at the higher RTs, with a resulting reduction in whey protein molecular weight. Therefore, lower RTs, which provided adequate cross-linking, produced stronger films.

Young's Modulus is a measure of flexibility, lower values being more flexible material. For the WPI–glycerol/water mixture, values of Young's Modulus varied with flow rate and screw speed at levels of significance less than 0.05. Representative values for two flow rates and two screw speeds are given in Table 5.4. The sheets became more flexible as shear and RT increased.

To assess the relationship between the extrudate rheological properties and melt viscosity, the extrudate properties were plotted against the in-line apparent viscosity. The best correlation was between tensile strength and apparent viscosity, with a coefficient of determination of $R^2 = 0.366$ (Figure 5.9). Loosely speaking, higher

TABLE 5.4

Young's Modulus as a Function of Feed Rate and Screw Speed

Flow Rate (g/min)	Screw Speed (RPM)	Modulus (MPa) (average ± 1 stdev)
15	100	9.77 ± 1.95
	150	6.89 ± 1.08
25	100	11.52 ± 2.32
	150	8.92 ± 1.97

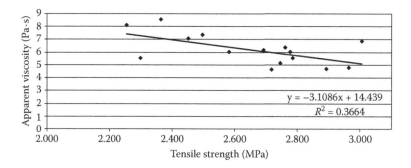

FIGURE 5.9   Correlation between apparent viscosity and tensile strength.

apparent viscosities yielded lower tensile strengths. In part, this may be due to the longer RT (Figure 5.8) and possible overexposure of the WPI to the heat and shear environment.

COMPARISONS BETWEEN SOLUTION-CAST FILMS AND EXTRUDATES

The intent of this work was to take a product (i.e., solution-cast whey protein films) and create a similar extruded product in order to develop a continuous process to replace a current batch process. Therefore, it is important to compare the physical properties of the films made by the two methods. Solution-cast films with 55% glycerol (db) were made according to the McHugh and Krochta method (1994); the amount of glycerol level was just slightly higher than that of the extruded films at 52% db. Aqueous solutions of 5% (w/w) WPI for the cast films were prepared and heated at 90°C for 30 min in a water bath and then cooled to room temperature (~23°C). Glycerol with a weight equal to the that of the whey protein isolate was added as a plasticizer. The films were cast by pipetting 28.8 g solution onto rimmed, smooth, circular Teflon discs. Films were dried for around 18 h at room temperature and were peeled from the casting surface. Tensile properties were tested in the same manner as described in the Experimental section. These cast films had a tensile strength of 1.4 MPa, Young's modulus of 26 MPa, and elongation at break of 68% (Hernandez-Izquierdo et al. 2008). By comparison to solution-cast films, the extrudates were considerably more flexible as characterized by lower Young's Modulus values and had greater toughness, which is the product of tensile strength and elongation at break. The extruded WPI films had an average toughness value of 251, compared to the solution-cast films at 95.

Scanning electron microscopy (SEM) images were taken using a Phillips XL30 microscope (Philips Electronics North America, New York, NY, U.S.A.). Images of the top surfaces of the solution-cast sheet and extruded sheet are illustrated in Figure 5.10. The surface of the extruded film was quite different from that of the solution-cast film, with a much higher degree of roughness due to the processing environment within the extruder. Extruded sheets had more and larger pores as compared with solution-cast films.

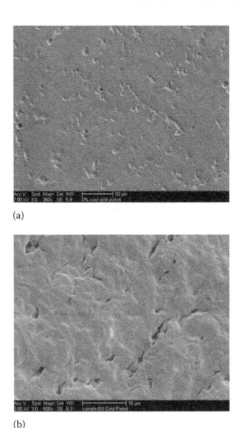

(a)

(b)

FIGURE 5.10   Scanning electron micrographs of (a) a solution-cast film and (b) an extruded film.

SUMMARY

Edible and biodegradable polymers are being developed as a biodegradable and renewable replacement for synthetic polymers. Extrusion has been shown to be a viable technology to produce these edible/biodegradable films on an industrial scale. This chapter focused on the use of whey protein as a potential component in these films due to its functional properties. In-line viscosity measurements were acquired to characterize the melt; apparent viscosity measurements ranged from 4.6 to 8.6 Pa · s. The melt was characterized as a power law fluid with a flow behavior index of 0.237. The apparent viscosity was correlated with other experimental responses to provide insight into the extrusion process. In particular, the apparent viscosity was correlated to the MRT, which was an indication of time–temperature treatment within the extruder. This time–temperature treatment resulted in differing degrees of cross-linking of heat-denatured whey proteins, which resulted in a range of tensile properties of the extrudate films.

ACKNOWLEDGMENT

The studies reported here were supported by the California Dairy Research Foundation.

REFERENCES

American Society for Testing and Materials [ASTM]. 1998. Standard test method for tensile properties of thin plastic sheeting. D882-97. West Conshohocken, PA: ASTM.

Han, J.H., and Krochta, J.M. 2001. Physical properties and oil absorption of whey-protein-coated paper. *J Food Sci* 66(2): 294–299.

Hernandez-Izquierdo, V.M., and J.M. Krochta. 2009. Thermal transitions and heat sealing of glycerol-plasticized whey protein films. *Packag Technol Sci* 22: 255–260.

Hernandez-Izquierdo, V.M., Reid, D.S., McHugh, T.H., Berrios, J.D.J., and Krochta, J.M. 2008. Thermal transitions and extrusion of glycerol-plasticized whey protein mixtures. *J Food Sci* 73(4): E169–E175.

Hong, S.I., and Krochta, J.M. 2006. Oxygen barrier performance of whey-protein-coated plastic films as affected by temperature, relative humidity, base film and protein type. *J Food Eng* 77(3): 739–745.

Mackey, K.L., and Ofoli, R.Y. 1990. Rheological modeling of corn starch doughs at low to intermediate moisture. *J Food Sci* 55(2): 417–423.

McHugh, T.H., and Krochta, J.M. 1994. Dispersed phase particle size effects on water vapor permeability of edible emulsion films. *J Food Process Preserv* 18:173–188.

Morgan, R.G., Steffe, J.F., and Ofoli, R.Y. 1989. A generalized viscosity model for extrusion of protein doughs. *J Food Process Eng* 11: 55–78.

O'Shea, J. 2003. *The Alcohol Textbook.* Murtagh and Assoc., Virginia.

Perez-Gago, M.B., and Krochta, J.M. 1999. Water vapor permeability of whey protein emulsion films as affected by pH. *J Food Sci* 64(4): 695–698.

Perez-Gago, M.B., and Krochta, J.M. 2001. Denaturation time and temperature effects on solubility, tensile properties, and oxygen permeability of whey protein edible films. *J Food Sci* 66(5): 705–710.

Perez-Gago, M.B., Nadaud, P., and Krochta, J.M. 1999. Water vapor permeability, solubility, and tensile properties of heat-denatured versus native whey protein films. *J Food Sci* 64(6): 1034–1037.

Poulesquen, A., and Vergnes, B. 2003. A study of residence time distribution in co-rotating twin-screw extruders. Part I: Theoretical Modeling. *Polymer Eng Sci* 43(12): 1841–1848.

Ralston, B.E., and Osswald, T.A. 2008. Viscosity of soy protein plastics determined by screw-driven capillary rheometry. *J Polymers Envt* 16: 169–176.

Rauch, D.J. 2008. Screw speed, feed rate, and composition effects on extruder operating conditions and extruded whey protein films. MS thesis. University of California, Davis, CA, USA, p.118.

Sothornvit, R., and Krochta, J.M. 2001. Plasticizer effect on mechanical properties of beta-lactoglobulin films. *J Food Eng* 50(3): 149–155.

Steffe, J.F. 1996. *Rheological Methods in Food Process Engineering, 2nd edition.* East Lansing, MI: Freeman Press.

6 Development of Extruded Foods by Utilizing Food Industry By-Products

Aylin Altan and Medeni Maskan

CONTENTS

INTRODUCTION

The food industry produces large amounts of waste, in particular fruit and vegetable waste resulting from the production, preparation, and consumption of food. According to the AWARENET (2004) report, 3 billion tons of waste generated each year in Europe and up to 222 million tons of food chain waste from all food processing sectors, which are mainly meat, fish, milk, fruits and vegetables, and wine, are produced annually across these key food sectors. Among the key sectors, fruit and vegetable processing creates huge amounts of waste, approximately 150 million tons per year (AWARENET 2004).

The recovery of food waste is rapidly expanding around the world. In the past, the ways of dealing with fruit and vegetable waste were often dumping or using without treatment for animal feed or as fertilizers because traditional methods of food preparation resulted in relatively small amounts of locally produced domestic waste, which were easier to deal with by disposing or using as animal feed. However, at present, technological advances in processing techniques and diversity in the marketplace and consumer consumption have increased over time, and this has brought increasing demand for their conversion into useful products for preventing pollution of the environment as well as for economic motives and the utilization of valuable constituents from food processing waste. Besides a major disposal problem for the industry concerned, food processing waste has the potential to be converted into useful products and utilized as supplements for other industries (e.g., cosmetics and pharmaceutics), or for the functional ingredient for consumers as a source of functional compounds. Legislation is also strongly forcing industries to find new end-uses for these by-products. Considering this, and valuable substances of fruit and vegetable by-products in particular antioxidants, dietary fibers, minerals, and essential fatty acids that are beneficial for health, there is a need to put into practice processes and technologies that help the conversion of by-products into useful products.

Several researchers have been working on the development of multifunctional ingredients from fruit and vegetable residues and its application in different food products, e.g., pie fillings, crackers, cookies, and bread. For example, apple pomace has been suggested in bakery products, pie fillings, and cookies as a potential food ingredient (Wang and Thomas 1989; Carson, Collins, and Penfield 1994) and used as a source of dietary fiber in wheat bread (Masoodi and Chauhan 1998). One viable method for the utilization of fruit and vegetable by-products into useful products is extrusion processing due to its versatility, high productivity, relative low cost, energy efficiency, and lack of effluents.

This chapter deals with the development of extruded foods using food industry by-products by extrusion technology and also discusses how extrusion process parameters affect the physical, functional, textural, sensorial, and nutritional characteristics of these products.

KEY REASONS FOR BY-PRODUCTS RECOVERY AND EXAMPLES OF FOOD INDUSTRY BY-PRODUCTS

Fruit and vegetable processing by-products pose increasing disposal issues and severe loss of valuable substances, e.g., dietary fiber, lycopene, antioxidants, pectin, essential fatty acids, antimicrobials, and minerals. These functional components have many health-promoting benefits. For example, dietary fiber in by-products acts as a bulking agent, normalizing intestinal motility and preventing diverticular disease. Some types may also be important in reducing colonic cancer, in lowering serum cholesterol levels, and in preventing hyperglycemia in diabetic patients (Larrea, Chang, and Martinez-Bustos 2005a; Thebaudin et al. 1997). Pectic substances of dietary fiber have a hypocholestrolemic effect because they form complex with bile acids and prevent their reabsorption in the small intestines. The occurrence of prostate cancer in men and breast cancer in women, as well as the risk of cardiovascular disease, may be reduced by lycopene intake (Kris-Etherton et al. 2002). Antioxidants in fruit and vegetable by-products reduce oxidative stress through the inhibition of lipid peroxidation, a factor that is currently linked to a host of diseases such as cancer and heart disease. Fruit and vegetable by-products, potato peels such as are rich in dietary fiber (Arora, Zhao, and Camire 1993; Camire et al. 1997; Laufenberg, Kunz, and Nystroem 2003) and tomato, grape, and apple pomaces are probably the most important sources of dietary fiber and antioxidants (Bobek, Ozdin, and Hromadova 1998; Moure et al. 2001). Apple pomace has been shown to be a good source of polyphenols, which are predominantly localized in the peels. Citrus seeds and peels have been found to possess high antioxidant activity (Schieber, Stintzing, and Carle 2001). Polyphenols extracted from apple pomace could be successfully separated by column chromatography, which would result in the isolation of epicatechin, caffeic acid, three dihydrochalcone glycosides, and five different quercetin glycosides (Lu and Foo 1997). Results show that apple pomace contains a high level of polyphenols that could be commercially exploited. Research has been carried out to evaluate some functional properties of fiber concentrates from apple and citrus fruit residues in order to use them as potential fiber sources in the enrichment of foods (Figuerola et al. 2005). All the fiber concentrates had a high content of dietary fiber (between 44.2 and 89.2 g/100 g dry matter), with a high proportion of insoluble dietary fiber, and the characteristics found in the concentrates suggested many potential applications such as volume replacement and thickening or texturizing in the development of foods reduced in calories and rich in dietary fiber.

The amounts of particular dietary fiber fractions has been studied in samples containing apple, black currant, chokeberry, pear, cherry, and carrot pomaces (Nawirska and Kwaśniewska 2005). In each pomace sample, pectins occurred in the smallest amounts and the content of lignin was very high for black currant and cherry pomaces or comparatively high for pear, chokeberry, apple, and carrot pomaces, respectively. Comparison of the contents of hemicellulose and pectin in the investigated study showed that the highest contents of these species were in chokeberry pomace (41%) and the lowest in apple pomace (36%). It could therefore be anticipated that the chokeberry and apple pomaces, would be equally good sorbents for heavy metals (Nawirska and Kwaśniewska 2005). Peschel and his group (2006) investigated 11 fruit and vegetable by-products and two minor crops for their extraction yield, total phenolic content, and antioxidant activity to determine the industrial exploitation potential. Extracts with the highest activity, economic justification, and phenolic content were obtained from apple (48.6 mg Gallic acid equivalents [GAE]/g dry extract), pear (60.7 mg GAE/g), tomato (61.0 mg GAE/g), golden rod (251 mg GAE/g), and artichoke (514.2 mg GAE/g). Apple, golden rod, and artichoke byproducts have been extracted on pilot plant scale and their antioxidant activity has been confirmed by the determination of their free radical scavenging activity and the inhibition of simulated linoleic acid peroxidation. The authors demonstrated the possibility of recovering high amounts of phenolics with antioxidant properties from fruit and vegetable residuals not only for food but also for cosmetic applications.

Grape pomace is the press residue remaining when grapes are processed for wine-making. The pomace consists of pressed skins, disrupted cells from the grape pulp, seeds, and stems. Among fruits, grapes constitute one of the major sources of phenolic compounds and their pomace is particularly rich in phenols (Meyer, Jepsen, and Sørensen 1998). Anthocyanins, catechins, flavonol glycosides, phenolic acids, alcohols, and stilbenes are the principal phenolic constituents of grape pomace (Schieber, Stintzing, and Carle 2001). Numerous studies have demonstrated the antioxidant and health promoting effects of phenolic compounds present in grapes and wine, particularly in relation to cardiovascular diseases (Scalbert et al. 2005). The evidence is clearly mounting that grape polyphenols are absorbed by the body and increase the total antioxidative capacity of blood plasma or decrease the peroxidation of low density lipoprotein (LDL) (Shrikhande 2000). Grape pomace therefore represents a potentially valuable source of phenolic antioxidants that may have technological applications as functional food ingredients and possible nutritional benefits. Grape pomace is also characterized by a high content of dietary fiber and associated polyphenols (Valiente et al. 1995) and could be used as a potential ingredient for dietary fiber-rich supplements (Martín-Carrón, Saura-Calixto, and Goñi 2000).

Grape seeds are rich sources of polyphenolics that have been shown to act as strong antioxidants and exert health promoting effects (Jayaprakasha, Singh, and Sakariah 2001). Meyer et al. (1997) studied phenolic extracts from 14 different types of fresh grapes in terms of the inhibition of human LDL oxidation *in vitro*. The relative antioxidant activity increased with grape seed crushing and with longer extraction times. This seed extraction provided greater concentration of flavan-3-ol flavonoid compounds of antioxidation potential in inhibiting LDL oxidation. Natural flavonoids can donate hydrogen to and/or react with superoxide anions, hydroxyl radicals, and lipid peroxyl radicals, all of which can cause lipid peroxidation *in vitro*, leading to LDL oxidation implicated in the development of atherosclerosis (Shrikhande 2000). Antioxidant-rich fractions of grape pomace have been extracted using ethyl acetate, methanol, and water (Chidambara Murthy, Singh, and Jayaprakasha 2002). As the methanol extract also showed high antioxidant activity, it might be directly correlated to the high phenolic content of the methanol extract of grape pomace. The data obtained in this study revealed that the grape pomace extracts are free radical scavengers and primary antioxidants, which react with free radicals. The ability of the grape pomace extract to quench hydroxyl radicals seems to directly relate to the prevention of propagation of the process of lipid peroxidation, and the extract seems to be a good scavenger of active oxygen species, thus reducing the rate of the chain reaction (Chidambara Murthy, Singh, and Jayaprakasha 2002).

Tomato is one of the most popular vegetables, used as a salad; in food preparations; and as juice, soup, puree, ketchup, and paste. During tomato processing a by-product, known as tomato pomace, is generated. This by-product represents, at most, 4% of the fruit weight (Del Valle, Camara, and Torija 2006). Tomato pomace consists of the dried and crushed skins and seeds of the fruit. The chemical composition of tomato pomace collected at different steps during tomato processing for paste has been studied in order to assess the quality of this by-product (Del Valle, Camara, and Torija 2006). The average value of tomato pomace composition (on a dry weight basis) was 59.03% fiber, 25.73% total sugars, 19.27% protein, 7.55% pectins, 5.85% total fat, and 3.92% mineral content, with no great differences between samples collected at different steps during the processing. Tomato pomace could be used as a potential source of fiber, protein, or fat (most of which is polyunsaturated). Research has been carried out to determine the fatty acid composition and physico-chemical characteristics of the oils extracted from industrial tomato seed waste from hot and cold break treatments (Cantarelli, Regitano-d'arce, and Palma 1993). Results indicated that the oil yields of tomato seeds were 19.0% and 14.5% for cold and hot break treatments. The total saturated and unsaturated fatty acid compositions were 29.4% and 70.6% for cold break seed oil and 31.3% and 68.6% for hot break seed oil. In both treatments palmitic acid was the major saturated fatty acid, followed by stearic acid. Linoleic acid was the major unsaturated fatty acid, followed by oleic acid. Both oleic and linoleic acids added up to over 60% of the total fatty acids, being higher in cold break seed oils. Both treatments produced high nutritional oil quality. The skin, another important component of pomace, was utilized for extracting the red pigment using organic solvents (Tonucci et al. 1995). Lycopene is an excellent natural food color and also serves as a micronutrient with important health benefits (Kaur et al. 2005). Baysal, Ersus, and Starmans (2000) clearly stated that a large quantity of carotenoids is lost as waste in tomato processing. Supercritical CO_2 extraction of lycopene and β-carotene from tomato paste waste resulted in recoveries of up to 50% when ethanol was added by 5%. Lycopene is a major carotenoid in human serum tissues and in the diet. It is unique among the carotenoids in that it has one major food source—tomatoes. Epidemiologic studies suggest that a diet rich in lycopene is related to a decreased risk of certain diseases, particularly cancers of the digestive tract, prostate, and pancreas as well as cardiovascular disease and HIV infection. The chemoprotective effect of lycopene is thought to be due to its role as an antioxidant (Johnson 2000).

Toor and Savage (2005) determined the major antioxidants and antioxidant activity in different fractions (skin, seeds and pulp) of three tomato cultivars in New Zealand. The skin fractions of all cultivars had higher levels of total phenolics, total flavonoids, lycopene, ascorbic acid, and antioxidant activity compared to their pulp and seed fractions. The skin and seeds of all cultivars on average contributed 53% to the total phenolics, 52% to the total flavonoids, 48% to the total lycopene, 43% to the total ascorbic acid, and 52% to the total antioxidant activity present in tomatoes. Rao and Agarwal (2000) reviewed the role of antioxidant lycopene in cancer and heart disease. Epidemiological studies and a small number of animal and experimental studies have provided evidence in support of its protective role in heart disease and cancer. Dietary intake of tomatoes and tomato products containing lycopene has been shown to be associated with a decreased risk of chronic diseases such as cancer and cardiovascular diseases.

The number of examples for fruit and vegetable by-products can be increased, but this is not the primary subject for the current chapter. However, from the literature review given it is obvious that fruit and vegetable by-products contain numerous valuable components. Recently, there is an increasing attempt to incorporate these into food products by extrusion technology.

APPLICATIONS OF THE EXTRUSION PROCESS TO THE DEVELOPMENT OF EXTRUDED FOODS USING FOOD INDUSTRY BY-PRODUCTS

The application of extrusion technology is one of the most economic processes, being used increasingly in the food industry for the development of new products such as snacks, baby foods, breakfast cereals, pasta products, texturized protein food stuffs, and modified starch from cereals. During extrusion cooking, the raw materials undergo many chemical and structural transformations, such as starch gelatinization, protein denaturation, complex formation between amylose and lipids, and degradation reactions of vitamins and pigments. Therefore, chemical and structural transformations in foods during extrusion cooking determine the quality of the extruded products (Bhattacharya and Prakash 1994; Yeh and Jaw 1998; Ilo, Liu, and Berghofer 1999). Applications of extrusion technology to evaluate food industry by-products are relatively recent. Table 6.1 shows a summary of applications for the extrusion process by utilizing food industry by-products. The extrusion process has been used to modify the functional properties of the fiber from potato peels and orange pulp (Camire et al. 1997; Larrea, Chang, and Martinez-Bustos 2005a). Camire et al. (1997) studied the effects of extrusion on the dietary fiber of potato peels obtained from the abrasion peeling method used by chip manufacturers and the steam peeling procedure used for the production of dehydrated potatoes to identify differences in dietary fiber composition between these types of peels. The total dietary fiber generally increased in steam peeled samples but the total dietary fiber of abrasion peeled samples did not change by the extrusion process. The results showed that the extrusion process increased soluble non-starch polysaccharides in both types of peels. Orange pulp has been extruded using a Brabender laboratory single-screw extruder to modify the properties of the fiber components (Larrea, Chang, and Martinez-Bustos 2005a). The composition of orange pulp was 9.79% proteins, 2.43% lipids, 2.66% ash, 9.72% total carbohydrates, and 74.87% total dietary fiber, consisting of 54.81% insoluble dietary fiber and 20.06% soluble dietary fiber. The extrusion process caused a decrease of 39.06% in insoluble dietary fiber and an increase of 80% in soluble dietary fiber. A higher value of soluble pectin was obtained under severe extrusion conditions caused by the highest barrel temperature and the lowest moisture content.

Press residue from black currant juice production has been investigated for the development of a nutritious breakfast cereal using an extrusion process at high temperatures (173°C–184°C) (Tahvonen et al. 1998). The composition of air dried residue was reported as fat content 11.2%, total fiber content 62.4%, and soluble dietary fiber content 2.1%, respectively. In the formulation, 20%–40% of dried black currant ingredients, 20%–40% mixed oat flour and bran, and 20%–30% potato starch were used with 10%–15% optimal moisture content for the flour mixture in the extruder. Products containing more than 40% of black currant press residue were hard and did not expand well. Thirty percent of black currant residue was successfully incorporated into a mix containing 30% oat flour and oat bran, 30% potato starch, 7.5% sugar, 1.5% malt extract, and 1% salt.

Onion waste mainly consists of brown skin and two fleshy scale leaves, which are a major by-product after the industrial peeling of onions. Extrusion of the white outer

TABLE 6.1
Summary of Work Published to Date on the Application of the Extrusion Process by Utilizing Food Industry By-Products

By-Product	Raw Material	Extruder Type	Extruder Condition	Reference
Potato peel	Potato peel	A Werner Pfleiderer ZSK-30 co-rotating twin-screw extruder	BT: 110°C, 150°C. MC: 30%, 35% (db). SS: 300 rpm	Camire et al. 1997
Orange pulp	Orange pulp	A Brabender single-screw extruder	BT: 83°C–167°C. MC: 22%–38%. SS: 126–194 rpm	Larrea, Chang, and Martinez-Bustos 2005a
Black currant press residue	Mixed oat flour and bran, potato starch	A Berstorff ZE-25 × 33D pilot extruder	BT: 173°C–184°C. MC: 11%–12%. SS: 280 rpm	Tahvonen et al. 1998
Onion peel	White outer fleshy scale leaves	An APV Baker co-rotating twin-screw extruder	BT: 100°C, 120°C, 140°C, 160°C. MC: 40%, 60%, 80%. SS: 100, 250 rpm	Ng et al. 1999
Corn bran	Corn meal	A Cerealtec single-screw extruder	BT: 150°C, 170°C, 190°C. MC: 16%, 19%, 22%. SS: 200 rpm	Mendonça, Grossmann, and Verhé 2000
Sweet whey solids, whey protein concentrate	Corn meal, rice flour, potato flour	A Werner Pfleiderer ZSK-30 co-rotating twin-screw extruder	BT: 125°C SS: 300 rpm	Onwulata, Konstance et al. 2001
Crab processing by-product (crab legs, meat and shell)	Corn meal, potato flake	A Werner Pfleiderer ZSK-30 co-rotating twin-screw extruder	BT: 157°C. MC: 21%–23%. SS: 150, 250 rpm	Murphy et al. 2003
Crab processing by-product (crab legs)	Corn meal	A Werner Pfleiderer ZSK-30 co-rotating twin-screw extruder	BT: 150°C. MC: 25%, 30%. SS: 200, 250, 350 rpm	Obatolu et al. 2005
Shrimp industry by-product (shrimp head protein and chitosan)	Cassava starch, corn starch rice and wheat flour	A Brabender single-screw extruder	BT: 150°C–170°C. MC: 15%. SS: 115, 220 rpm	Gibert and Rakshit 2005
Brewer's spent grain	Maize flour, corn starch, whole oat flour	A Werner Pfleiderer Continua 37 co-rotating twin-screw extruder	BT: 110°C. MC: 11.8%. SS: 100, 200, 300 rpm	Ainsworth et al. 2007

(continued)

TABLE 6.1 (Continued)
Summary of Work Published to Date on the Application of the Extrusion Process by Utilizing Food Industry By-Products

By-Product	Raw Material	Extruder Type	Extruder Condition	Reference
Brewer's spent grain	Wheat flour, corn starch	A Werner Pfleiderer Continua 37 co-rotating twin-screw extruder	BT: 120°C. MC: 14% (db). SS: 150–350 rpm	Stojceska Ainsworth, Plunkett, and Ibanoglu 2008
Tomato pomace	Barley flour	An APV co-rotating twin-screw extruder	BT: 140°C–160°C. MC: 21%–22%. SS: 150–200 rpm	Altan, McCarthy, and Maskan 2008a
Defatted hazelnut flour, durum clear flour, mixture of orange peel, grape seed and tomato pomace	Rice grits	A single-screw extruder	BT: 150°C–175°C. MC: 12%–18%. SS: 200–280 rpm	Yağcı and Göğüş 2008
Cauliflower trimmings (florets, stem, leaves)	Wheat flour, corn starch, oat flour	A Werner Pfleiderer Continua 37 co-rotating twin-screw extruder	BT: 120°C. MC: 9%–11%. SS: 25–350 rpm	Stojceska, Ainsworth, Plunkett, Ibanoglu, and Ibanoglu 2008
Grape pomace	Barley flour	An APV Baker co-rotating twin-screw extruder	BT: 140°C–160°C. MC: 21.7%. SS: 150–200 rpm	Altan, McCarthy, and Maskan 2008b
Grape pomace, grape seed	Decorticated white sorghum flour	A twin-screw Haake PolyLab system extruder	BT: 160°C, 170°C, 180°C, 190°C. MC: 45%. SS: 200 rpm	Khanal, Howard, and Prior 2009
Blueberry pomace	Decorticated white sorghum flour	A twin-screw Haake PolyLab system extruder	BT: 160°C, 180°C, 200°C. MC: 45%. SS: 150, 200 rpm	Khanal et al. 2009

BR Barrel temperature; *MC* Moisture content; *SS* Screw speed

fleshy scale leaves derived from onion waste through a co-rotating twin-screw extruder has been studied to modify cell wall polymers of onion waste (Ng et al. 1999). Extrusion cooking resulted in an increase in the solubility of pectic polymers and hemicelluloses of the cell wall material. Modified onion cell walls, which are especially rich in pectic polymers, could be a useful component of fiber enriched foodstuffs.

Corn bran was utilized through the production of expanded snacks with high fiber using the extrusion process (Mendonça, Grossmann, and Verhé 2000). Blends of corn bran (18%, 25%, and 32%) and corn meal were extruded using a single-screw extruder. This study showed that up to 32% of corn bran could be added with a lack of sensorial and physio-chemical characteristics. However, the best combination of variables to produce a snack with good sensory acceptability and high fiber content was 25% corn bran, 16% moisture, 190°C temperature of extrusion, and 0.4% glycerol monostearate. The fiber content of the selected snack was 16%, and, therefore, 100 g of this product would provide 64% of the recommended daily intake.

Whey is the principal by-product from dairy processing such as cheese and casein production. The utilization of whey is variable across the world. Even though whey products are utilized for many purposes, the overall usage of whey products is 50% and 60% of the total whey produced for the United States and Europe (Durham and Hourigan 2007). Therefore, Onwulata et al. (2001) aimed to incorporate whey products (sweet whey solids and whey protein concentrate) into extruded products to increase the utilization of whey products and improve the nutrient density of snacks by increasing the protein content. Corn meal, rice, or potato flour with whey products, sweet whey solids, or whey protein concentrates are extruded under low shear, high shear, and the combination of high shear and low moisture extrusion processing conditions to make snack products. Up to 25% of whey protein substitution could produce extrudates with good quality in terms of expansion and breaking strength of products.

Extrusion cooking has been used to add value to crustacean processing by-products. Crustacean processing by-products are traditionally disposed of in landfills or at sea. However, the shells of crustaceans are partly composed of calcium carbonate and they also contain significant amounts of protein and chitin (Goldhor and Regenstein 2007). Therefore, research has been focused on how to evaluate crab processing and shellfish by-products to create value added products with increasing protein, chitin, and calcium content using the extrusion process. Murphy et al. (2003) investigated the use of wet and dry crab processing by-products in the development of a seafood-flavored calcium rich expanded snacks. The blends of crab processing by-products (0%–40%) and corn meal and potato flakes were processed in a twin-screw extruder. The crab processing type (wet, dry) and level of incorporation had significant effects on the calcium content, expansion ratio, bulk density, and pH of the extrudates. A mildly flavored expanded snack could be produced from crab processing by-products with combinations of corn meal and potato flakes that results in a calcium rich product with a negligible microbial count after three months of room temperature storage. Later, Obatolu et al. (2005) investigated processing variables on the nutritional and physical quality of extrudates from a blend of corn meal and crab processing by-products. Corn meal was mixed with 10% ground crab leg and processed through a twin-screw extruder using moisture contents of 25% and 30% and screw speeds of 200, 250, and 300 rpm. The snack product extruded at 25% moisture content and 300 rpm had better physical properties compared to other treatments. The ground crab product was successfully incorporated in its wet form into corn meal. A cassava snack (chips and extrudate) was formulated using shrimp head protein and chitosan as by-products of the shrimp industry (Gibert and Rakshit 2005). The ratios of shrimp head protein and chitosan in the formulations were 0.5%–10%

and 0.5%–16%. The results suggested that chips and the extruded product confirmed their good potential for by-product valorization.

Brewer's spent grain is the main by-product of the brewing industry. Due to its high protein and fiber content and also its low cost, it has received attention for developing value-added food products. Ainsworth et al. (2007) developed a snack food from brewer's spent grain with a combination of chickpea flour, maize flour, oat flour, and corn starch. The ratios of brewer's spent grain into blends were changed from 10% to 30%, and replaced by maize flour. The addition of brewer's spent grain improved the nutritional value of chickpea based snacks in terms of protein and crude fiber. An increase in phytic acid and resistant starch content in the samples was also observed with the addition of brewer's spent grain in the blends, up to 30% based on the maize flour. Later, the same research group incorporated brewer's spent grain into a combination of wheat flour and corn starch to develop a ready-to-eat extruded snack (Stojceska, Ainsworth, Plunkett, and Ibanoglu 2008). Blends of brewer's spent grain (10%–30%) were processed in a twin-screw extruder at a constant feed rate of 25 kg/h, temperatures of 80°C–120°C, and a screw speed of 150–350 rpm. Ainsworth et al. (2007) also investigated the effects of brewer's spent grain, corn starch, and screw speed on selected physical and nutritional properties of the snacks. The dried brewer's spent grain had a composition of 20.30% protein, 53.39% fiber, 8.32% fat, and 10.76% carbohydrate content. The incorporation of up to 30% brewer's spent grain caused a wide variation in the fiber content of the extrudates, ranging from 1.1% to 14.4%. Adding up to 30% brewer's spent grain in the formulations produced acceptable physio-chemical characteristics, but the authors considered it better to add 20% of brewer's spent grain to obtain snack foods with properties similar to that of commercially available ones.

The by-product of tomato processing, known as tomato pomace, consists of the dried and crushed skins and seeds of the fruit. It has been reported that skin is an important component of pomace and is considered to be a source of lycopene. Fiber is the major compound of tomato by-product, up to 50% dry weight basis (Del Valle, Camara, and Torija 2006). Thus tomato by-product can be considered as a potential source of fiber for human food formulations. In this way, the use of tomato processing by-products could provide valuable substances and at the same time reduce the waste disposal problem. Research has been directed toward the use of tomato pomace in new formulations of snack foods. Altan, McCarthy, and Maskan (2008a) investigated the possibilities of snack foods from tomato pomace and barley flour blends. The blends were extruded in a co-rotating twin-screw extruder with a die temperature of 140°C–160°C and 150–200 rpm screw speed. As a result of sensory analysis, the incorporation of tomato pomace at 2% and 10% into barley flour had a higher preference in terms of color, texture, taste, and overall acceptability. The results showed that tomato pomace could be extruded with barley flour into an acceptable and nutritional snack. Rice grits were combined with food waste (3%–7%), partially defatted hazelnut flour (5%–15%), and durum clear flour (8%–20%) to develop an extruded snack food (Yağcı and Göğüş 2008). The food waste contained orange peels (80%), grape seeds (10%), and tomato pomace (10%). The physical and functional characteristics of products were affected mostly from changes in the partially defatted hazelnut flour content and less affected by fruit waste addition. At low defatted hazelnut flour contents, well expanded snack products were obtained with acceptable sensory properties.

Cauliflower by-products mainly consist of leaves and lesser amounts of stems. A recent study showed that cauliflower by-products are sources of phenolic compounds, mainly flavonol derivatives (Llorach et al. 2003). The authors proposed that 16 g (db) of cauliflower by-products could provide the same antioxidant capacity as one cup of tea of normal strength (1%–2%) or one glass of red wine. Cauliflower by-products are inexpensive and also have antioxidant and anticarcinogenic properties. Therefore, cauliflower by-products can be used to add value to foods. Research has been carried out to use cauliflower by-products to produce ready-to-eat snack foods (Stojceska, Ainsworth, Plunkett, Ibanoglu, and Ibanoglu 2008). Cauliflower trimmings were added at levels of 5%–20% into a formulation mix and samples were processed in a twin-screw extruder at temperatures of 80°C–120°C, screw speeds of 250–350 rpm, and feed rates of 20–15 kg/h. Extrusion cooking increased the total phenolics and antioxidant capacities but decreased protein *in vitro* digestibility and fiber content in the extruded products. The incorporation of up to 10% cauliflower by-products produced acceptable ready-to-eat expanded products.

To date, a number of studies have been conducted to determine or characterize the phenolic compounds of grape pomace from red and white grape varieties (Valiente et al. 1995; Meyer et al. 1997; Chidambara Murthy, Singh, and Jayaprakasha 2002; Kammerer et al. 2004). However, there has been limited research available about the direct incorporation of grape pomace for developing new foods. Recently, Altan, McCarthy, and Maskan (2008b) used grape pomace, from Thompson seedless grapes, with a combination of barley flour to develop nutritious extruded foods. The results showed that the incorporation of up to 10% grape pomace produced sensorially acceptable foods. Extrusion processing variables resulted in a reduction in phenolic contents of extrudates (Altan, McCarthy, and Maskan 2009a). Later, Khanal, Howard, and Prior (2009) studied the extrusion of grape pomace (variety Sunbelt) and grape seed (variety Riesling) with decorticated white sorghum to determine the effect of extrusion conditions on the procyanidin composition of grape seeds and grape pomace as well as the total anthocyanin content of grape pomace. High temperatures (160°C–180°C for grape seed and 160°C–190°C for grape pomace) were used to extrude mixtures of grape seed as well as pomace with decorticated white sorghum flour at a ratio of 30:70 and a moisture content of 45%. An extrusion temperature of 170°C and a screw speed of 200 rpm resulted in the highest increase in monomer contents, with a 120% increase in grape pomace and an 80% increase in grape seed over the unextruded control. Extrusion processing caused the reduction in total anthocyanin content of grape pomace by 18% to 53%. Khanal et al. (2009) also worked on the extrusion of blueberry pomace under temperatures of 160°C and 180°C and screw speeds of 150 and 200 rpm with decorticated white sorghum flour at the same ratio and moisture content of their previous work. A considerable increase in monomeric, dimeric, and trimeric procyanidins was found after the extrusion process at both temperatures and screw speeds. The highest monomer content was obtained at 180°C and 150 rpm screw speed and was 84% higher than the nonextruded control. On the other hand, the extrusion process reduced the total anthocyanin contents by 33% to 42%, and therefore additional treatments would be needed to retain the pigments.

EFFECTS OF EXTRUSION PROCESS ON PROPERTIES
OF EXTRUDED FOODS FROM FOOD BY-PRODUCTS

PHYSICAL PROPERTIES

Expansion

Expansion is an important property of extruded products being developed as snack foods. The expansion and texture formation of extrudates are complex even for products based on a single component. They depend on the viscoelastic properties of the melt, the mechanism of bubble nucleation and growth, as well as the plastisizing properties of water in the transition from fluid (melt) to viscoelastic and subsequently to a glassy state, which are all important for the expansion and final texture of extrudates. Even though the pressure drop at the die is not directly responsible for expansion, it causes bubble nucleation. This pressure difference between the vapor pressure of the water inside the nucleated bubbles and the pressure of the melt drives expansion (Arhaliass, Bouvier, and Legrand 2003). The multicomponent and multiphase structure of the melt would modify each of these processes, ultimately affecting the expansion and texture of the products (Zasypkin and Lee 1998).

The effect of by-product addition on extrudate expansion varied according to the level of incorporation and the source of the by-product. Onwulata, Smith et al. (2001) found that substituting whey protein concentrate at 25% in corn meal resulted in extrudates with comparable expansion with corn alone. Increasing the crab processing by-product content, either wet or dry, up to 10% did not reduce the expansion ratio, but a further increase in crab processing by-product decreased the expansion ratio of the extrudates (Murphy et al. 2003). Onwulata, Smith et al. (2001) proposed that the protein surrounded the available starch and thus limited its expansion. The radial expansion of corn based snacks decreased when the corn bran ratio was increased from 18% to 32% (Mendonça, Grossmann, and Verhé 2000). The deleterious effect of corn bran in radial expansion was explained by the fact that fiber causes rupturing of the walls of air cells as well as the external surface of extrudates, thereby preventing full expansion of the gas bubbles. Ainsworth et al. (2007) observed a decrease in expansion at increased brewer's spent grain levels in the formulations. According to the authors, the decrease in expansion resulted from a reduction of the starch amount in the formulations. Different studies of the same group showed that the incorporation of brewer's spent grain at levels of 20% and 30% into wheat flour and corn starch mix produced extrudates with reduced expansions (Stojceska, Ainsworth, Plunkett, and Ibanoglu 2008). Similar results were reported for the extrusion of rice grits in combination with durum clear flour, partially defatted hazelnut flour, and fruit waste (Yağcı and Göğüş 2009). The reduction in radial expansion (Figure 6.1) was interpreted as a dilution of starch and an increase in lipid content by increasing the amount of hazelnut flour. It has been reported that starch conversion decreases as the lipid levels increase by preventing a severe mechanical breakdown of the starch granules by shear stress and preventing water from being absorbed by starch. Therefore, reduced starch conversion or gelatinization causes a decrease in expansion (Moraru and Kokini 2003). A decrease in the expansion of barley based extrudates at a high level of tomato pomace (10%) was also reported

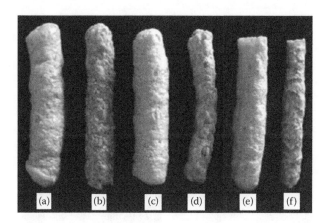

FIGURE 6.1 Photographs of snack samples from different combinations (a) DCF = 20%/PDHF = 7%/FW = 3%, (b) DCF = 12%/PDHF = 15%/FW = 3%, (c) DCF = 20%/PDHF = 5%/FW = 5%, (d) DCF = 12.8%/PDHF = 12.2%/FW = 5%, (e) DCF = 18%/PDHF = 5%/FW =7%, and (f) DCF = 8%/PDHF = 15%/FW = 7%. DCF: durum clear flour; PDHF: partially defatted hazelnut flour; FW: fruit waste. (Reprinted from Yağcı, S. and Göğüş, F. 2009. Development of extruded snack from food by-products: A response surface analysis. *J Food Process Eng* 32: 565–586. With permission.)

in the study of Altan, McCarthy, and Maskan (2008a). In addition to the possible reasons stated before, fibers have the ability to bind some of the moisture present in the matrix, thus reducing its availability for expansion (Moraru and Kokini 2003).

The incorporation of jatobá flour into cassava starch increased fiber content while that of starch decreased, interfering with the expansion of the products (Chang et al. 1998). The decrease in expansion due to increasing amounts of jatobá flour was attributed to the dilution of the starch content and an increase in the fiber content in the blend, resulting in an increase in the mass viscosity and restricting expansion ability. In the study of Yanniotis, Petraki, and Soumpasi (2007), the incorporation of wheat fiber into corn starch reduced the cell size (Figure 6.2), probably due to the premature rupture of gas cells, which reduced the overall expansion of the extrudates.

Stojceska, Ainsworth, Plunkett, and Ibanoglu (2008) found a correlation ($r^2 = 0.94$) between the mean cell area and the sectional expansion index. It shows that an increasing level of brewer's spent grain decreased the expansion of the product, resulting in a structure containing more small cells. The increasing bran level in rice flour decreased the radial expansion or product diameter, whereas the axial expansion increased with 10% rice bran followed by a reduction at 20% and 30% rice bran levels (Grenus, Hsieh, and Huff 1993). The long and stiff fiber molecules at a small concentration align themselves in the extruder in the direction of flow, reinforcing the expanding matrix and increasing its mechanical resistance in the longitudinal direction (Moraru and Kokini 2003). It has been reported that above a critical concentration, the fiber molecules disrupt the continuous structure of the melt, impeding its elastic deformation during extrusion. Hsieh et al. (1991) found that increasing sugar beet fiber content in corn meal increased the axial expansion but decreased its radial expansion. A similar phenomenon has been observed for the incorporation of wheat

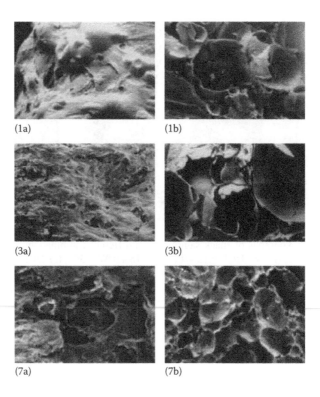

(1a)   (1b)

(3a)   (3b)

(7a)   (7b)

FIGURE 6.2 Micrographs of extrudates (magnification × 10): (1a–b) 100% corn starch, (3a–b) 90% corn starch and 10% pectin, (7a–b) 90% corn starch and 10% wheat fiber. Micrographs for extrudate surface (1a, 3a, 7a) and extrudate cross section (1b, 3b, 7b). (Reprinted from Yanniotis, S., Petraki, A. and Soumpasi, E. 2007. Effect of pectin and wheat fibers on quality attributes of extruded cornstarch. *J Food Eng* 80: 594–599. With permission.)

fiber or oat fiber content in corn meal by Hsieh et al. (1989). The degree of expansion is dependent on the size, number, and distribution of air cells inside the extrudates. With appropriate dough plasticity and elasticity, well expanded air pockets without ruptures can be formed during extrusion. In contrast, if the material contains inert components (e.g., dietary fiber) that disrupt the stretching and setting of bubble films, tears and holes will appear in the wall of the air cells as well as on the external surfaces of extrudates (Lue et al. 1990). The increasing dietary fiber content caused a decrease in average cell size and an increase in frequency of incomplete flakes and holes on the cell wall. Jin, Hsieh, and Huff (1995) found the size of air cells correlated with the radial expansion of corn meal extrudates including soy fiber, in agreement with the study of Lue, Hsieh, and Huff (1991) on extrudates containing sugar beet fiber. Substituting pectin in the amount of 10% into corn starch reduced the radial expansion of extrudates (Yanniotis, Petraki, and Soumpasi 2007). Yanniotis, Petraki, and Soumpasi (2007) suggested that when non-starch polysaccharides like pectin are present, they have the capacity to hydrate and consequently to compete for and restrict the plasticizer and hence the gelatinization process. Therefore, it was concluded that pectin reduced radial expansion by increasing the melt viscosity and reducing the availability of water for gelatinization. Camire and King (1991) found that cotton linter cellulose reduced expansion more than soy cotyledon fiber did. The non-starch polysaccharides in both types of fiber might bind water more tightly during extrusion than do protein and starch. This binding might inhibit water loss at the die and thus reduce expansion. The other possible reason given by the authors is that as the starch in the corn meal might not have been fully gelatinized in the presence of non-starch polysaccharides, it was less able to support expansion (Camire and King 1991).

It is known that extrusion temperature plays an important role in changing the rheological properties of the extruded melts, which in turn affect the expansion volume (Moraru and Kokini 2003). A negative effect of high temperature on sectional expansion was observed for the extrusion of corn meal with corn bran, barley flour-pomace blends, and rice grits with durum clear flour, partially defatted hazelnut flour, and fruit waste (Mendonça, Grossmann, and Verhé 2000; Altan, McCarthy, and Maskan 2008a, 2008b; Yağcı and Göğüş 2009). A high temperature could lead to lower sectional expansion but not necessarily lower overall expansion because longitudinal expansion might increase. The latter can be especially significant when a slit-die is used, as a gradual pressure drop is encountered that could lead to nucleation of bubbles in the die and their growth along the longitudinal direction (Altan, McCarthy, and Maskan 2008c). An inverse relationship between sectional and longitudinal expansion was reported for the study of extrusion of yellow grit and starch from different sources (Alvarez-Martinez, Kondury, and Harper 1988; Della Valle et al. 1997; Desrumaux, Bouvier, and Burri 1998). Similar observations of the effect of temperature on product expansion have been reported for corn starch, corn grits, and rice flour by Chinnaswamy and Hanna (1988); Ali, Hanna, and Chinnaswamy (1996); and Hagenimana, Ding, and Fang (2006). Chinnaswamy and Hanna (1988) found that the expansion of corn starch increased as the barrel temperature increased from 110°C to 140°C and declined with further increases in temperature. The increase in the expansion of starch with temperature was attributed to its higher degree of gelatinization at such temperatures while the reduction in expansion was attributed to molecular degradation as reported by the authors. At high temperatures, the pressure of saturated vapor was higher than that of melt towards the die exit, and this would favor the start of bubble growth inside the die in the direction of flow and thus higher longitudinal expansion (Della Valle et al. 1997). Hashimoto and Grossmann (2003) also reported a decrease in expansion with respect to increasing extrusion temperature for the extrusion of cassava bran and cassava starch blends. Kokini, Chang, and Lai (1992) found that after a critical temperature, which depends on both the type of starch and the moisture content, expansion decreases with temperature, most likely due to excessive softening and potential structural degradation of the starch melt, which becomes unable to withstand the high vapor pressure and therefore collapses. In addition, expansion phenomena are basically dependent on the viscous and elastic properties of melted dough (Launay and Lisch 1983). Therefore, the elasticity loss with increasing temperature would then be one of the possible reasons for the decrease of sectional expansion. Doğan and Karwe (2003) explained the decrease in sectional expansion index SEI with increasing temperature by the negative effect of temperature on the elasticity of cooked melts. The authors also found that longitudinal expansion appeared to be extensively favored by lower melt viscosity at a higher temperature and a higher moisture level. Yuliani et al. (2006) observed a decrease in expansion at high temperatures and proposed two reasons for this. One reason is due to faster bubble collapse after the initial expansion at the die. At a high temperature melt viscosity decreases, which facilitates bubble growth, but becuase the bubble walls are very thin due to greater expansion at the lower melt viscosity, they cannot withstand the vapor pressure inside, resulting in wall fracture and rapid pressure loss that allows the extrudate to collapse. The other reason attributed by the authors was that at a high temperature, rapid cooling of the extrudate surface could stop the growth of the bubbles, resulting in a decrease in expansion. On the other hand, some authors have approached the case of decreasing expansion with increasing temperature in a different way. Colonna, Tayeb, and Mercier (1989) and Mendonça, Grossmann, and Vehré (2000) reported that the decrease in expansion at higher temperatures could be attributed to the increased dextrinization and weakening of the structure.

Different results have been observed for the effect of screw speed on the expansion of extrudates. In general, screw speed has a positive effect on expansion. An increase in screw speed causes an increase in shear, and thus, a decrease in melt viscosity (Kokini, Chang, and Lai 1992). Obatolu et al. (2005) showed that the expansion ratio was improved by increasing the screw speed from 200 to 300 rpm at a constant moisture content for extruded crab based snacks. However, the screw speed had no significant effect on the expansion of extrudates from corn meal and crab processing by-products in the study of the same group (Murphy et al. 2003). Ryu and Walker (1995) found increases in the expansion ratio of wheat flour extrudates as the screw speed increased from 100 to 160 rpm, but then dramatically decreased at 180 rpm. Ali, Hanna, and Chinnaswamy (1996) reported that overall and radial expansions increased with screw speed, whereas axial expansion showed a reverse trend. On the other hand, other researchers have reported no significant effect of screw speed on the expansion of extrudates (Altan, McCarthy, and Maskan 2008a, 2008b; Stojceska, Ainsworth, Plunkett, and Ibanoglu 2008). The same behavior has been observed by Jin, Hsieh, and Huff (1995); Ding et al. (2005); and Liu et al. (2000) in the extrusion of corn meal with soy fiber, salt and sugar, rice, and oat flour with yellow corn flour. Moraru and Kokini (2003) explained that such differences might be due to significant differences in extrusion conditions such as the type of extruders and screw configuration, temperature, and composition of the feed.

A change in moisture content causes a change in the specific mechanical energy (SME), pressure, and strain applied to the extrudates, resulting in differences in a product's characteristics. Moisture has been found to strongly affect the degree of gelatinization as a reactant in complex interactions with other components (Holay and Harper 1982). Water obviously plays a critical role in the expansion process. Expansion is a function of the amount of shear force during extrusion. Low moisture content causes high shear, which results in higher expansion (Davidson et al. 1984). Chinnaswamy and Hanna (1988) reported that the low moisture content of starch might restrict the material flow inside the extruder barrel, increasing the shear rate and residence time, which would perhaps increase the degree of starch gelatinization and, thus, the expansion. However, when the moisture content of starch is too low (below 14% db), it may create very high shear rates and longer residence time and, thus, might increase the product's temperature. Such conditions are known to cause starch degradation and dextrinization, which would perhaps reduce the expansion (Chinnaswamy and Hanna 1988).

Increased feed moisture content during extrusion may reduce the elasticity of the dough through plasticization of the melt (Ding et al. 2006) and decrease the dough temperature because moisture would reduce the friction between the dough and the screw/barrel, and have a negative impact on the starch gelatinization, thereby reducing the product's expansion (Asare et al. 2004). Ilo et al. (1996) found that the apparent viscosity and the mechanical energy consumption in the extruder were most dependent on the feed moisture content. Higher feed moisture and higher product temperature caused a reduction in the melt viscosity and SME, resulting in an extrudate with a lower starch gelatinization degree and thus decreased sectional expansion. Mendonça, Grossmann, and Verhé (2000) and Obatolu et al. (2005) obtained higher expansion when the moisture content decreased in the extrusion of corn meal with corn bran and corn meal fortified with a crab processing by-product. However, Stojceska et al. (2009) observed that the effect of moisture content on expansion varied in each formulation. In general, increasing feed moisture content decreased the expansion ratio in wheat flour and corn starch extrudates with 10% of brewer's spent grain and corn starch extrudates, including 10% of red cabbage trimming, except for wheat flour and red cabbage extrudates. According to their results, corn starch samples showed a significant change in expansion at feed moistures between 12% and 14% but not between 15% and 17% in all formulations except for the wheat flour and red cabbage samples. The authors explained this difference in expansion between samples as being due to the differences in formulations and characteristics such as the content of amylose and protein. Yağcı and Göğüş (2009) found that increasing the moisture content from 12% to 18% caused an increase in the expansion ratio for most blend formulations of rice grit (67%), durum clear flour (8%–20%), partially defatted hazelnut flour (5%–15%), and fruit waste (3%–7%). Researchers obtained higher expansions with high partially defatted hazelnut flour content (12.5%), whereas an increase in moisture content had no effect on expansion at low concentrations of partially defatted hazelnut flour (5%). The reason for the elevated moisture was attributed to an increase in protein content, which caused more moisture to hydrate. Proteins can affect water distribution in the matrix and change in their macromolecular structure and conformation during extrusion, for example, denaturation, dissociation and alignment of the denatured protein molecules in the direction of flow affect the extensional properties of the extruded melts (Moraru and Kokini 2003).

Bulk Density
Product bulk density is directly related to the extent of extrudate expansion and is a very important parameter in the production of expanded and formed food products (Köksel et al. 2003). Altan, McCarthy, and Maskan (2008a) found that the effect of incorporation of tomato pomace into barley flour on bulk density was dependent on temperature. They obtained the lowest bulk density at higher temperatures with a low level of tomato pomace, whereas the highest value was obtained at lower temperatures with a high level of tomato pomace (Figure 6.3). The same group, however, found that the effect of grape pomace on bulk density of barley-grape pomace extrudates was not dependent on temperature (Altan, McCarthy, and Maskan 2008b).

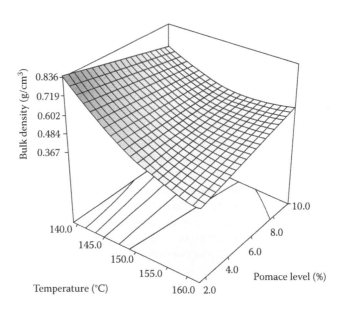

FIGURE 6.3 The effect of pomace level and temperature on the bulk density of barley-tomato pomace extrudates at a screw speed of 175 rpm. (Reprinted from Altan, A. McCarthy, K.L. and Maskan, M. 2008a. Evaluation of snack foods from barley-tomato pomace blends by extrusion processing. *J Food Eng* 84: 231–242. With permission.)

The bulk density of extrudates increased with an increase in percentage of grape pomace (Figure 6.4). An increase in bulk density with an increase in the level of pomace was attributed to the increasing fiber content of the feed material. This is because the presence of fiber particles tended to rupture the cell walls before the gas bubbles had expanded to their full potential (Lue, Hsieh, and Huff 1991). Colonna and Mercier (1983) also reported that partially molten starch granules adhered to the cellulosic walls, leading to a composite wall of cellulose, gelatinized starch, and cellular protein. The formation of this complex wall should restrict the product's expansion ability (Chang et al. 1998). Sugar, from pomace, could be another reason for an increase in the bulk density of barley-grape pomace extrudates. Sugar could limit the availability of water. Limited water might also hinder the gelatinization of starch, which could be another factor in increasing bulk density. However, the actual mechanism has more to do with the plasticization effect of sugar leading to lower melt temperatures and thus reducing the vapor pressure of water (Altan, McCarthy, and Maskan 2008b). Jin, Hsieh, and Huff (1994) reported that the addition of sugar reduced the product temperature, which might have also decreased or delayed the starch gelatinization. A similar effect of fiber has been observed for the extrusion of yellow corn with wheat and oat fiber, corn meal and sugar beet fiber, corn meal with soy fiber, salt and sugar, jatobá flour and cassava starch blends, and corn starch with pectin and wheat fiber (Hsieh et al. 1989; Lue, Hsieh, and Huff 1991, 1994; Chang et al. 1998; Yanniotis, Petraki, and Soumpasi 2007).

The bulk density of rice based extrudates was most dependent on partially defat-ted hazelnut flour content rather than fruit waste content in the study of Yağcı and

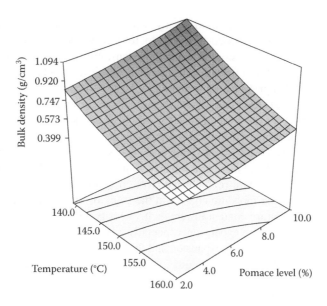

FIGURE 6.4 The effect of pomace level and temperature on the bulk density of barley-grape pomace extrudates at a screw speed of 175 rpm. (Reprinted from Altan, A., McCarthy, K.L. and Maskan, M. 2008b. Twin-screw extrusion of barley-grape pomace blends: extrudate characteristics and determination of optimum processing conditions. *J Food Eng* 89: 24–32. With permission.)

Göğüş (2008). Murphy et al. (2003) observed an increase in bulk density as the crab processing by-product content increased, either wet or dry type. The decrease in the expansion and increase in the bulk density were obtained at increased brewer's spent grain levels in the formulation (Ainsworth et al. 2007). Sun and Muthukumarappan (2002) studied extruded soy based extrudates. They reported that increasing the content of defatted soy flour (10%–30%) into corn flour decreased the expansion ratio and increased the bulk density of the product. The addition of 10% rice bran caused a small increase in the product diameter, but further increases in rice bran content reduced the extrudate radial expansion and increased the bulk density (Grenus, Hsieh, and Huff 1993). Liu et al. (2000) reported that increases in oat flour percentage resulted in an increase in oat-corn puff bulk density. Hsieh et al. (1989) observed increases in the bulk density of both wheat fiber and oat fiber–containing corn meal extrudates when the fiber content was increased in corn meals. Lue et al. (1990) studied the microstructure of extruded corn meal with oat fiber and wheat fiber. They indicated that the degree of extrudate expansion was dependent on the source and amount of dietary fiber and associated with the sizes of the air cells inside the extrudates and with the external structure of the products. Increasing the fiber content produced extrudates with a denser structure of reduced average cell size, an increased number of holes on the cell wall, as well as an increased number of apertures on the surface of extrudates. They concluded that the overall effect was a decrease in radial expansion and an increase in bulk density. Moore et al. (1990) found that the density of extrudates increased as the concentration of the bran increased. The authors observed that the cell number per pixel area increased greatly while the average cell size decreased as the bran concentration increased from 0% to 16%. It was also explained that bran interfered with bubble expansion, reducing the extensibility of the cell walls and causing the premature rupture of steam cells at a critical thickness related to the particle size of bran, and this prevents the formation of large cells (Moore et al. 1990). Lue, Hsieh, and Huff (1991) found the size of the air cells in the extrudate correlated with the radial expansion of the corn meal–sugar beet fiber extrudate; as the radial expansion decreased, the sizes of the air cells decreased.

The bulk density of the wheat flour and corn starch extrudates fortified with brewer's spent grain (10%) increased linearly with increasing feed moisture content, but extrudates fortified with red cabbage (10%) had no significant change in bulk density with increasing moisture content in the study of Stojceska et al. (2009). An increase in feed moisture content lowers melt viscosity, and this causes the bubbles to collapse after maximum expansion, especially when the wall of bubbles becomes very thin. This results in lower expansion and thus increases in bulk density (Kokini, Chang, and Lai 1992). Yağcı and Göğüş (2008) observed that the effect of moisture content on the bulk density was dependent on the composition of the extrudates. They obtained denser extrudates with increasing feed moisture at high levels of fruit waste and low levels of partially defatted hazelnut flour additions. The decrease in bulk density with increasing moisture content has also been observed by Asare et al. (2004); Baik, Powers, and Nguyen (2004); and Ding et al. (2006).

An increase in the barrel temperature increases the degree of super-heating of water in the extruder, encouraging bubble formation and also a decrease in melt viscosity leading to reduced density (Ding et al. 2006). Bulk density and expansion are inversely related and a high bulk density is associated with a low expansion (Rayas-Duarte, Majewska, and Doetkott 1998; Suknark, Philips, and Chinnan 1998). The bulk density of barley based extrudates with tomato and grape pomace decreased when the temperature increased (Altan, McCarthy, and Maskan 2008a, 2008b). An increase in the barrel temperature decreases melt viscosity and the reduced viscosity effect favors bubble growth during extrusion. The higher temperature provides a higher potential energy for flash-off of super-heated water from extrudates as they leave the die, also leading to greater expansion and hence giving a low bulk density (Koksel et al. 2004). Bhattacharya and Choudhury (1994) also reported an increase in temperature markedly reduced the bulk density of rice extrudates and showed a curvilinear relationship. A decrease in bulk density with an increasing temperature was observed for the extrusion of rice by Hagenimana, Ding, and Fang (2006) and Guha and Ali (2006), in agreement with Ilo, Liu, and Berghofer (1999) and Singh, Sekhon, and Singh (2007) for rice-amaranth extrudates and rice-pea grit extrudates. Guha and Ali (2006) explained that higher barrel temperatures increase the extent of gelatinization and also the content of super-heated steam, which causes the rice extrudate to expand more, leading to the production of a low density product.

Baik, Powers, and Nguyen (2004) found that the bulk density of barley extrudates was not significantly affected by screw speed. It was also reported that screw speed had no significant effect on the bulk density of rice extrudates including rice bran, rice flour, amaranth extrudate, and oat-corn puff (Grenus, Hsieh, and Huff 1993; Ilo, Liu, and Berghofer 1999; Liu et al. 2000; Altan, McCarthy, and Maskan 2008a,

2008b). However, other authors have reported a significant effect of screw speed on bulk density. Seker (2005) found a reduction in bulk density of starch-soy protein extrudates with increasing screw speed. Similar findings were also observed for corn meal extrudates including soy fiber and rice extrudates by Jin, Hsieh, and Huff (1994) and Ding et al. (2006). On the other hand, Sun and Muthukumarappan (2002) found an increase in the bulk density of soy based extrudates with increasing screw speed. Yağcı and Göğüş (2008) also observed the same result for the bulk densities of most of the extrudate compositions. As has been mentioned before, these differences can be due to the significant differences in the extrusion conditions, such as the type of extruders, screw configuration, temperature, and composition of the feed.

Color

It is obvious that color will change by the incorporation of food industry by-products into formulations and, also, extrusion conditions affect the color of the finished product. Therefore, color is an important quality factor directly related to the acceptability of food products, and is an important physical property to report for extrudate products. The color of a product should be at an acceptable level where the addition of by-products into formulations is concerned.

The incorporation of tomato pomace into barley based extrudates had a significant effect on product color parameters (Altan, McCarthy, and Maskan 2008a). An increase in tomato pomace level decreased the lightness (*L* value) of the samples and increased the redness (*a* value) due to the lycopene pigment in the tomato pomace. Liu et al. (2000) reported that the redness of the oat-corn extrudate was enhanced with the oat flour and attributed to the higher redness of oat flour than corn flour. The increase in darkness could be attributed to the darkness of the bran compared to the nearly white rice flour. In the study of Wu, Huff, and Hsieh (2007), the color of extruded flaxseed-corn puff with respect to lightness, redness, and yellowness was affected significantly by the flaxseed meal content. They obtained a darker (lower in lightness), redder, but less yellow extrudate with the higher a flaxseed meal content. Sun and Muthukumarappan (2002) found that the L^* value decreased with increasing defatted soy flour content, whereas the a^* and b^* values increased significantly with increasing soy flour content. The lower value of brightness of ready-to-eat products was observed with an increasing level of brewer's spent grain in agreement with other studies (Stojceska, Ainsworth, Plunkett, and Ibanoglu 2008). Yağcı and Göğüş (2009) obtained a slightly darker product when the partially defatted hazelnut flour ratio was increased in the formulation and there was a decrease in the lightness of the extrudates with a further increase in waste ratio, more than 5% waste content. However, the redness and yellowness of the extrudates decreased with the addition of partially defatted hazelnut flour, whereas they increased with an increasing addition of fruit waste. The authors explained the increase in darkness to be a result of the darkness of the partially defatted hazelnut flour compared to the nearly white rice and durum clear flour. Altan, McCarthy, and Maskan (2008b) found a reduction in lightness with increasing grape pomace level in blends of barley flour and grape pomace. This was attributed to the occurrence of a browning reaction such as the Maillard reaction and caramelization, because of the contribution of more sugar (from the grape pomace), which favored the browning reaction. It was proposed that the increases in the redness and yellowness values of the extrudates might be associated with the Maillard reaction and yellowish pigments in the pomace with increasing grape pomace levels. Sacchetti et al. (2004) observed the snack's lightness ($L*$) tended to decrease with an increase in chestnut flour (which is dark) content in the initial blend. They found that the extrusion temperature appeared to have little effect on the product's $L*$ and $a*$ values with low chestnut flour, whereas the extrusion temperature caused product browning with an increased level of chestnut flour. This was attributed to the high reducing sugars content of chestnut flour, which could promote color changes due to the Maillard reaction development according to extrusion temperature. Ilo and Berghofer (1999) also reported that the changes in lightness and redness during the extrusion cooking of yellow maize grits were due to the effects of browning reactions. Murphy et al. (2003) found a strong trend for both lightness and yellowness values to decrease as the level of crab processing by-product increased. However, the color of the extrudates was not affected by the extrusion conditions. The effect of temperature on the lightness of the extrudates was not significant in the study of Altan, McCarthy, and Maskan (2008a, 2008b) and Yağcı and Göğüş (2009). However, other researches have reported significant effects of temperature on lightness of extrudates.

Apruzzese, Balke, and Diosady (2000) observed that increased temperature at constant screw speed provided a darker product. A high temperature range, where Maillard reactions become important, results in increased browning of the final product (Apruzzese, Balke, and Diosady 2000) and hence a reduction in lightness. The higher extrusion temperature resulted in darker products, probably due to the Maillard reaction favored by the high temperature and relatively low water in the extruder (Pelembe, Erasmus, and Taylor 2002). Bhattacharya, Sivakumar, and Chakraborty (1997) reported that the brightness of rice based extrudates decreases as a result of extrusion processing, particularly with temperature. They observed that an increase in temperature from 100°C to approximately 140°C markedly decreased the brightness of the samples, but beyond 140°C the decrease in brightness values was rather low. Gutkoski and El-Dash (1999) found that as the extrusion temperature increased, the luminosity ($L*$) was lowered to a minimum at 120°C and then increased again.

The effect of screw speed on color parameters was not significant for barley based extrudates (Altan, McCarthy, and Maskan 2008a, 2008b). The study of Liu et al. (2000) revealed that screw speed had no significant effect on the lightness and yellowness of the extrudate but significantly enhanced the redness of the oat-corn puff extrudate. Grenus, Hsieh, and Huff (1993) also observed that an increase in screw speed had no significant effect on the darkness of the extrudate. Ilo and Berghofer (1999) found that lightness and redness were markedly dependent on the barrel temperature and feed moisture content, whereas the screw speed was not significant. In contrast to these results, increasing the screw speed caused an increase in the lightness of rice based extrudates, including partially defatted hazelnut flour and fruit waste, with increasing screw speed at a low waste ratio. However, the lightness of the extrudates increased up to 240 rpm screw speed and then decreased when the waste ratio was increased in the formulation (Yağcı and Göğüş 2009). An increase of screw speed reduces residence time and hence decreases the changes in the color of the product. On the other hand, increasing the screw speed increases the shear and temperature and can lead to more browning (Berset 1989).

The color of extrudates developed from food by-products showed a different trend with increasing moisture content. Yağcı and Göğüş (2009) found improvement in the lightness of extrudates from rice grits and food waste with increasing moisture content from 12% to 18%. Stojceska et al. (2009) reported a decrease in the lightness of wheat flour and corn starch based extrudates with brewer's spent grain and red cabbage trimming, whereas the lightness increased for extrudates from corn starch with brewer's spent grain and wheat flour with red cabbage when the moisture content was increased from 12% to 17%. Liu et al. (2000) observed a decrease in the lightness of oat-corn puff with an increase in moisture content from 18% to 21%. This was attributed to a reduction in expansion. Later, Altan, McCarthy, and Maskan (2008a) found a correlation between expansion and the lightness of extrudates. It was explained that the expanded product gives more bright color in extrudates due to air cells. The reduction in redness with increasing moisture content has been reported for different formulations with food by-products (Obatolu et al. 2005; Stojceska et al. 2009; Yağcı and Göğüş 2009).

FUNCTIONAL PROPERTIES

Water Absorption Index

Water absorption depends on the availability of hydrophilic groups that bind water molecules and on the gel-forming capacity of macromolecules (Gomez and Aguilera 1983). Water absorption (WAI) and solubility (WSI) indices have been used to estimate the functional characteristics of extrudates developed from by-products. The incorporation of tomato pomace or grape pomace into barley flour resulted in a decrease of the WAI of extrudates (Altan, McCarthy, and Maskan 2008a, 2009b). Similar results have been reported by Yağcı and Göğüş (2008) from a study of the production of partially defatted hazelnut flour-rice based extrudates. Stojceska, Ainsworth, Plunkett, Ibanoglu, and Ibanoglu (2008) found a negative correlation between the level of cauliflower and the WAI of extrudates. In another study by the same research group, control extrudates had a higher WAI than those with brewer's spent grain (Stojceska, Ainsworth, Plunkett, and Ibanoglu 2008). The reduction in the WAI of extrudates by the addition of by-products has been attributed to the relative decrease in starch content and competition of absorption of water between pomace and available starch (Altan, McCarthy, and Maskan 2008a). Artz, Warren, and Villota (1990) found a decrease in the water holding capacity as a result of an increase in the ratio of fiber/corn starch in the extrusion of corn fiber and corn starch blends. They concluded that gelatinized corn starch has a much greater water holding capacity than either hemicellulose or cellulose, the major components of corn fiber; thus, any reduction in the gelatinized starch should reduce the water holding capacity of the extrudate. A decrease in the water absorption capacity of potato fiber with extrusion was observed by Camire and Flint (1991). In addition, Singh, Sekhon, and Singh (2007) observed a decrease in the WAI with the addition of pea grits in the extrusion of rice. They explained that a decrease in the WAI was due to the dilution of starch in the rice pea blends. The addition of hull to the high starch protein fraction of pinto bean flour resulted in a significant decrease of the WAI, probably as a result of the reduction of starch in the starting material (Gujska and Khan 1991).

Jin, Hsieh, and Huff (1995) also reported that increasing the fiber content from 0% to 20% resulted in products with a low WAI. However, Chang et al. (1998) indicated that as the concentration of jatobá flour increased, the WAI continuously increased at higher moisture contents. They explained that a moderate extrusion treatment disrupts structures and therefore creates pores that water can penetrate.

Increasing the temperature from 140°C to 160°C decreased the WAI of barley-tomato pomace extrudates (Altan, McCarthy, and Maskan 2008a). The decrease in the WAI with increasing temperature was probably due to the decomposition or degradation of starch molecules. Similar observations were reported by Pelembe, Erasmus, and Taylor (2002) at a high extrusion temperature (165°C). Mercier and Feillet (1975) reported that the WAI for waxy corn and other starches decreased with an increasing extrusion temperature from 70°C to 225°C. They indicated that WAI decreases with the onset of dextrinization. Badrie and Mellowes (1991) found that increasing the temperature from 100°C–105°C to 120°C–125°C lowered the WAI of cassava extrudates. Lee, Ryu, and Lim (1999) found that in the range of 80°C–100°C, WAI increased rapidly with increasing temperatures up to 90°C, it decreased after 90°C. Guha, Ali, and Bhattacharya (1997) observed a decrease in the WAI when the temperature was increased in the range of 80°C–120°C. It has been attributed to the degradation of starch that causes a reduction in the water holding capacity of the molecules as a result of a decrease in molecular size. The lower temperature causes more undamaged polymer chains and a greater availability of hydrophilic groups for binding more water and results in higher values of the WAI (Gomez and Aguilera 1983; Guha, Ali, and Bhattacharya 1997). Similar results were reported by Ding et al. (2005) and Ding et al. (2006) for rice and wheat extrudates. Ding et al. (2006) also stated that the WAI decreases with increasing temperature if dextrinization or starch melting prevails over the gelatinization phenomenon. Hashimoto and Grossmann (2003) showed that increasing the extrusion temperature from 150°C to 180°C decreased the WAI of cassava bran/cassava starch extrudates, which was probably due to an increase in starch degradation. A decrease in the WAI with an increase in extrusion temperature at low moisture content was also reported by Kirby et al. (1988). Opposite results have been observed for the WAI of extruded whole pinto bean meal, quinoa extrudates, and rice based exrudates including pea grits (Baladrán-Quintana et al. 1998; Doğan and Karwe 2003; Singh, Sekhon, and Singh 2007; Yağcı and Göğüş 2008). They found that the WAI increased with increasing temperature.

A number of studies have shown that the WAI increased by increasing the moisture content of the feed material. Yağcı and Göğüş (2008) observed an interaction effect of feed moisture and partially defatted hazelnut flour content. The WAI of extrudates did not affect the feed moisture content at a high level of partially defatted hazelnut flour, whereas increasing the moisture content from 12% to 18% caused an increase in the WAI of rice based extrudates at lower partially defatted hazelnut flour content. Stojceska et al. (2009) found an increased WAI in extrudates of wheat flour and corn starch with red cabbage and brewer's spent grain as the feed moisture level increased from 12% to 17%. According to Chang et al. (1998), protein denaturation, starch gelatinization, and swelling of fiber can be responsible for the increased WAI of extruded products, and a moderate extrusion condition disrupts structures, which create pores that water can penetrate. Hagenimana, Ding, and Fang (2006) obtained the highest values of the WAI between 19% and 22% moisture content, which results in lower degradation of the starch granules because water acts as a plasticizer during extrusion cooking.

A negative effect on the WAI has been observed at high screw speeds. The WAI of barley-pomace extrudates was affected by the interaction of grape pomace ratio and screw speed (Altan, McCarthy, and Maskan 2009b). The study showed that the WAI decreased with increasing screw speed at a low grape pomace level but at a high pomace level, increasing the screw speed led to a rise in the WAI. The reduction in the WAI at low pomace levels has been attributed to starch degradation or dextrinization under high shear conditions, whereas the increase of the WAI at high pomace levels was due to the structural modification of the grape pomace components with increasing screw speed. Yağcı and Göğüş (2008) observed a decrease in the WAI with an increasing screw speed for rice based extrudates enriched with partially defatted hazelnut flour and food waste. Hashimoto and Grossmann (2003) found that the WAI of cassava bran/cassava starch extrudates decreased and then increased when the screw speed increased from 120 to 150 and to 180 rpm. Guha, Ali, and Bhattacharya (1997) reported that samples extruded at lower screw speeds showed relatively high WAI values than at higher screw speeds. This was attributed to the high residence time at low screw speeds permitting enhanced extent of cooking. Jin, Hsieh, and Huff (1995) stated that at low shear rates (low screw speeds), there were more undamaged polymer chains and a greater availability of hydrophilic groups that could bind more water molecules and result in higher values of the WAI. Mezreb et al. (2003) observed an increase in the WAI of wheat extrudates when the screw speed was changed from 200 to 300 rpm but decreased at 500 rpm. The reduction in the WAI was explained by low residence time at high screw speed to allow for sufficient gelatinization. Badrie and Mellowes (1991) also observed the lower WAI of cassava extrudate when increasing the screw speed from 425 to 520 rpm and the higher WAI of cassava extrudate due to further increasing the screw speed from 520 to 560 rpm. This could be attributed to an increase in the shear rate that results in structural modifications for the lower value of WAI and to the shorter residence time for the higher value of WAI with increasing screw speeds, as opposed to Mezreb et al. (2003).

Water Solubility Index

Water solubility gives information about degradation while water absorption is more related to the swelling capability of granules. Different results have been observed for the effect of the incorporation of food by-products on the functionality of extrudates. In the studies of Stojceska, Ainsworth, Plunkett, and Ibanoglu (2008) and Stojceska, Ainsworth, Plunkett, Ibanoglu, and Ibanoglu (2008), the WSI of extrudates was not affected by the different levels of brewer's spent grain and cauliflower additions. Yağcı and Göğüş (2008) found that the WSI changed as a function of the partially defatted hazelnut content. The increasing level of pomace raised the WSI of barley-tomato pomace and barley-grape pomace extrudates (Altan, McCarthy, and Maskan 2008a, 2009b). All these changes could be related to the modification of fiber, coming from pomace, due to extrusion, and the release of low molecular weight compounds that cause an increase in the WSI. Larrea, Chang, and Martinez-Bustos (2005b) extruded orange pulp for use as a source of fiber in the preparation of biscuit-type cookies. They observed that the WSI of the extruded pulp increased by extrusion. Sacchetti et al. (2004) found the WSI increased as the percentage of chestnut flour was increased. Jin, Hsieh, and Huff (1995) observed an increase in the WSI of corn meal extrudates as the fiber content increased from 0% to 20%. Increased feed moisture resulted in the decrease in the WSI of extrudates (Yağcı and Göğüş 2008).

An increase in the WSI with increasing temperature has been reported (Gujral, Singh, and Singh 2001; Yuliani et al. 2006; Hagenimana, Ding, and Fang 2006; Singh, Sekhon, and Singh 2007). However, the WSI of barley-tomato pomace and barley-grape pomace extrudates decreased with increasing temperature, which is in agreement with the work of Gutkoski and El-Dash (1999) and Yağcı and Göğüş (2008) in extruded oat products and rice based extrudates with hazelnut flour and food waste. The authors suggested that increasing the temperature increases the degree of gelatinization that can increase the amount of soluble starch and cause an increase in the WSI (Ding et al. 2005). An increase in the WSI has been attributed to an increase in the amount of dextrinized starch during extrusion cooking. On the other hand, it was proposed that the molecular interactions between degraded starch, protein, and lipid components causing an increase in molecular weight resulted in a decrease of the solubility (Doğan and Karwe 2003; Altan, McCarthy, and Maskan 2009b). Hagenimana, Ding, and Fang (2006) reported that the combination of harsh conditions and low moisture content caused an increase in the amount of degraded starch granules, resulting in an increased formation of water-soluble products.

The WSI increased with increasing screw speed for barley-tomato pomace extrudates (Altan, McCarthy, and Maskan 2008a). Jin, Hsieh, and Huff (1995) obtained a higher WSI when the screw speed was increased from 150 to 350 rpm for corn meal extrudates. Gujral, Singh, and Singh (2001) found similar effects of screw speed for the WSI of sweet corn and flint corn grits extrudates in the range of 100 to 150 rpm. The solubility of yam flour increased with screw speed, from 40 to 280 rpm (Sebio and Chang 2000). Jin, Hsieh, and Huff (1995) reported that the WSI depends on the quantity of soluble molecules, which is related to the degradation. Lee, Ryu, and Lim (1999) also stated that water solubility usually increases when starch chains degrade into smaller fragments. Guha, Ali, and Bhattacharya (1997) reported that a higher screw speed resulted in more fragmentation than a lower screw speed and thus increased the WSI of rice extrudates. The higher WSI of extrudates with increasing screw speeds was related to increasing SME input with screw speed. A higher SME input causes greater restriction to material flow, resulting in the breakdown of polymers to small molecules with higher solubility (Choudhury and Gautam 1998). Smith (1992) observed the increase in WSI with decreasing molecular weight or that the molecular weight falls with increasing SME. Mezreb et al. (2003) found that the WSI increased as the screw speed increased from 200 to 300 rpm for wheat extrudates and from 300 to 500 rpm for corn extrudates. The authors reported that the increase of screw speed induced a sharp increase of SME and the high mechanical shear degraded macromolecules, so the molecular weight of starch granules decreased and hence increased the WSI. As the water solubility increases there is a linear decrease in intrinsic viscosity, which reflects a decrease in the average molecular weight of the amylose and amylopectin chains (Kirby et al. 1988). With the agreement of other authors, Kirby et al. (1988) also found greater water solubility with large mechanical energy inputs. Altan, McCarthy, and Maskan (2008a) found a positive correlation between the WSI and the SME for barley-tomato pomace extrudates. A similar correlation also has been reported by Kirby et al. (1988) and Choudhury and Gautam (1998). A negative correlation between the WSI and the WAI was observed for rice based extrudates and barley-grape pomace extrudates (Yağcı and Göğüş 2008; Altan, McCarthy, and Maskan 2009b). This agrees well with the results of Badrie and Mellowes (1991), who found a negative correlation between the WSI and the WAI for cassava extrudates. Iwe (1998) studied the effects of the extrusion process on the functional properties of mixtures of full-fat soybean and sweet potato. He found that the WSI of samples decreased as the WAI increased. Van der Burgt, Van der Woude, and Janssen (1996) reported the same relationship between the WAI and the WSI. They plotted the WSI against the WAI and saw a continuous decrease of the WSI with an increase in the WAI. Kirby et al. (1988) also observed a strong correlation between the two indices for maize grits extrudates.

TEXTURAL PROPERTIES

Hardness

Texture is one of the most important sensory attributes of extruded products being developed as snack foods. The successful development of food products requires both a comprehensive understanding of texture as perceived by the consumer and appropriate measurement methods. Hardness, crispness, and brittleness have been used to evaluate the textural properties of extrudates developed from food by-products (Altan, McCarthy, and Maskan 2008a, 2008b, 2008c). The textural property of an extrudate is determined by measuring the force required to break the extrudate (Singh, Hoseney, and Faubion 1994; Altan 2008). The maximum peak force obtained from a texture analyzer gives the hardness of the product (Stojceska, Ainsworth, Plunkett, and Ibanoglu 2008; Stojceska, Ainsworth, Plunkett, Ibanoglu, and Ibanoglu 2008).

The incorporation of any type of food by-product into extrudates changes the texture of the extrudates. Whey products incorporated in amounts of 25% and 50% into extrudates of corn, rice, or potato responded differently on breaking strength, used for a textural indicator, depending on the type of flour (Onwulata, Konstance et al. 2001). Whey protein concentrate in the amount of 25% reduced the breaking strength, but increasing the level of whey protein concentrate to 50% increased the breaking strength of the corn based extrudates. On the other hand, adding sweet whey solids at both concentrations increased the breaking strength of the potato extrudates. In rice based extrudates, even a high concentration of whey protein concentrate (50%) produced softer and crispier extrudates than extrudates not containing whey proteins as opposed to sweet whey solids. Even though Ainsworth and co-workers reported scattered data for the hardness of extrudates with the addition of brewer's spent grain at a level of 0%–30% and screw speeds from 100 to 300 rpm, the incorporation of brewer's spent grain up to 20% did not affect the hardness of the extrudates at a screw speed of 300 rpm (Ainsworth et al. 2007). Stojceska Ainsworth, Plunkett, and Ibanoglu (2008) obtained high values of hardness at 30% brewer's spent grain with no addition of corn starch besides wheat flour. It has been observed that a mean cell area decreases with the addition of brewer's spent grain. However, the hardness of extrudates could be reduced with an increase in softness occurring as corn starch levels are increased. In the incorporation of cauliflower trimmings into wheat based extrudates (Figure 6.5), the total cell area in extrudates was negatively related but the wall thickness of the cells was positively affected by the increasing level of cauliflower by-products (Stojceska, Ainsworth, Plunkett, Ibanoglu, and Ibanoglu 2008).

Altan, McCarthy, and Maskan (2008a) found that changes in the hardness of extrudates as a function of tomato pomace level were dependent on temperature. A decrease in die temperature with an increasing level of tomato pomace increased the product hardness, whereas lower product hardness was obtained at higher temperatures. In the study carried out by the same research group, increasing the grape pomace ratio resulted in a peak force of extrudates with respect to quadratic effects (Altan, McCarthy, and Maskan 2008b). Yağcı and Göğüş (2009) found an improvement in the texture of extrudates with the addition of partially defatted hazelnut flour up to ~12%. However, the hardness of the extrudates increased when the level of partially defatted hazelnut flour was increased beyond this value. The former has been attributed to the addition of oil, the lubricant effect in extrusion cooking, from partially defatted hazelnut flour to the blends, which improved the extrusion process and texture, and the latter could be excess fat, which reduces expansion, or increasing protein content of blends and increased hardness of extrudates. Onwulata, Smith et al. (2001) found reduced expansion and breaking strength with the inclusion of fiber at 125 g/kg. They reported that reduced expansion and increased breaking strength are characteristics of fiber substituted products resulting from reduced elasticity due to the presence of the fiber. Liu et al. (2000) observed that extrudates with higher oat flour contents had a higher hardness than extrudates with lower oat flour contents possibly due to increasing the fiber level with oat. Hsieh et al. (1989) reported that increasing the fiber content in corn meal decreased the radial expansion and increased the bulk density and breaking force for the extrudates. Yanniotis, Petraki, and Soumpasi (2007) found increased hardness and decreased porosity while adding more fiber. The change of hardness with fiber is related to the effect of fiber on the cell wall thickness. Fiber causes a less porous matrix, thicker cell wall, and harder extrudate (Yanniotis, Petraki, and Soumpasi 2007). Ahmed (1999) obtained increased bulk densities

| 0% | 5% | 10% | 15% | 20% |

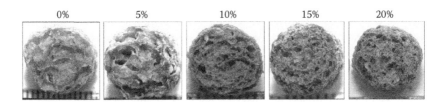

FIGURE 6.5 Images of extrudates containing different levels of cauliflower. (Reprinted from Stojceska, V., Ainsworth, P., Plunkett, A., Ibanoglu, E. and Ibanoglu, S. 2008. Cauliflower by-products as a new source of dietary fiber, antioxidants and proteins in cereal based ready-to-eat expanded snacks. *J Food Eng* 87: 554–563. With permission.)

and breaking strengths of extrudates over the flax addition range 5% to 20%. Jin, Hsieh, and Huff (1995) also reported that the breaking strength increased with increasing sugar and soy fiber contents. They attributed this to a more compact extrudate with thick cell walls and small air cell size by the incorporation of sugar and fiber into corn meal. The authors found a negative correlation between breaking strength and the radial expansion of extrudates. The incorporation of fiber in extruded product results in increased density and hardness of the extrudates (Onwulata et al. 2000).

A decrease in the hardness of products has been observed by increasing the temperature for extrudates developed from food by-products (Altan, McCarthy, and Maskan 2008a, 2008b). For instance, Altan, McCarthy, and Maskan (2008a) found that a decrease in die temperature increased the product hardness, which is in line with bulk density where an increase in density, giving a maximum at about 133°C, 175 rpm screw speed, and 6% tomato pomace level. Sebio and Chang (2000) obtained low hardness of yam flour extrudates with the samples extruded at higher barrel temperatures. Lee, Ryu, and Lim (1999) reported that the breaking stress increased as the barrel temperature decreased for corn starch extrudates. Doğan and Karwe (2003) found increased hardness of quinoa extrudates as die temperature decreased. Yuliani et al. (2006) reported that increasing the temperature resulted in a decrease in hardness of extrudates from mixtures of starch and D-limonene. Yao et al. (2006) found that extrusion at higher temperatures resulted in cereal with less hardness for extrudates from two oat lines. Mendonça, Grossmann, and Verhé (2000) obtained lower hardness at high temperatures for corn meal extrudates including corn bran. Similarly, Ryu and Walker (1995) observed that the breaking strength and bulk density decreased over the temperature range of 140°C to 160°C in the extrusion cooking of wheat flour. Lee, Ryu, and Lim (1999) found that the breaking stress increased as the barrel temperature decreased and a negative correlation with expansion ratio, indicating that extrudates became brittle when extrusion was performed at high temperatures. It is suggested that increasing temperature decreases the melt viscosity and increases the vapor pressure of water favoring bubble growth, which is the driving force for expansion and decreases the bubble wall thickness. This results in a lower bulk density and hence a lower hardness of the extrudates (Ding et al. 2005, 2006; Yuliani et al. 2006; Altan, McCarthy, and Maskan 2008b). The extrudate with a higher density would have relatively thicker cell walls and an overall lower porosity (Barrett et al. 1994), which is directly related to the hardness of the samples. Agbisit et al. (2007) observed an increase in cell diameter and a decrease in cell number density with higher overall expansion. The authors found high negative correlations between mechanical properties and the average cell diameter. Positive correlations between hardness and the bulk density of extrudates have been observed in different studies (Sacchetti, Pittia, and Pinnavaia 2005; Altan, McCarthy, and Maskan 2008a, 2008b). A high density product naturally offers high hardness, evident by the high correlation between the product density and the hardness. Rayas-Duarte, Majewska, and Doetkott (1998) found a negative correlation between the expansion index and the breaking strength, which means that the decreased breaking strength of extrudates was associated with a high expansion index and a low bulk density. Veronica, Olusola, and Adebowale (2006) observed a negative relationship among expansion ratio, hardness, and breaking strength for puffed snacks from a maize-soybean mixture. They found increased breaking strength and hardness values when expansion decreased.

The hardness of extrudates slightly decreased with increasing screw speed during the extrusion of barley flour and tomato pomace blends (Altan, McCarthy, and Maskan 2008a). Liu et al. (2000) found that the hardness of the extruded oat-corn flour increased as the screw speed decreased. The high screw speed caused an increase in product temperature, which usually leads to a higher expansion and a decrease in hardness. Wu, Huff, and Hsieh (2007) observed a decrease in hardness of extrudate from a flaxseed-corn meal blend with increasing screw speed from 200 to 400 rpm. Increasing the screw speed would increase the SME, which has a positive influence on the expansion index and would cause a reduction in hardness. Ryu and Ng (2001) found that with the increase in the SME input, the apparent elastic moduli and breaking strength in bending and in compression were decreased for wheat flour and whole corn meal extrudates. They concluded that extrudates that puffed at a higher SME input could have a softer, more brittle and crispier texture since the apparent elastic modulus and breaking strength were relatively lower than those of a lower SME input. Ding et al. (2006) found a decreased hardness of wheat extrudate with increasing screw speed, especially at higher barrel temperatures. They attributed this to the lower melt viscosity of the mix obtained due to increasing the screw speed and resulting in a less dense, softer extrudate. Yağcı and Göğüş (2009) reported an interaction effect of screw speed and food waste content on the hardness of extrudates. According to the results, the hardness of extrudates increased with increasing screw speed at low and high fruit waste contents, but no significant effect of screw speed was observed at an intermediate waste content (5%).

In the study of Stojceska et al. (2009), the extrusion of wheat flour with brewer's spent grain and red cabbage gave a harder product than corn starch extrudates. An increase in feed moisture caused a significant increase in the hardness of wheat based extrudates with brewer's spent grain. However, there was no significant difference in the hardness of wheat based extrudates with red cabbage trimming. The reason for the increasing hardness was attributed to reduced expansion and a higher total dietary fiber level. The effect of feed moisture on hardness was more evident at feed moisture levels between 14% and 17% for wheat and corn based extrudates with brewer's spent grain. Obatolu et al. (2005) found an increase in fracture stress as the moisture content increased for crab based snack products.

Crispness

Crispness is associated with a low-density cellular structure that is brittle and generates a high-pitched noise when fractured (Le Meste et al. 2002). Vincent (1998) reported that crispness may be associated with a rapid drop in force that is associated with a rapid propagation of fracture, which, in turn, necessitates that the material be brittle. The slope of the force–distance curve before the first major fracturability peak has been measured as the crispness of the extrudates (Jackson, Bourne, and Barnard 1996; Altan, McCarthy, and Maskan 2008b). The lower the slope, the crisper the product has been considered for extrudates developed from barley flour and by-products (Altan 2008). The slope decrease with increasing temperature (Figure 6.6) means an increase in the crispness of barley based extrudates (Altan 2008; Altan, McCarthy,

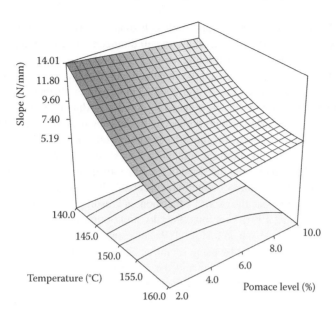

FIGURE 6.6 The effect of pomace level and temperature on the slope of barley-tomato pomace extrudates at a screw speed of 175 rpm. (Data taken from Altan, A. 2008. Production and properties of snack foods developed by extrusion from composite of barley, and tomato and grape pomaces. PhD dissertation, University of Gaziantep, Turkey, pp. 211.)

and Maskan 2008b). The crispness of the extrudate is related to the expansion and cell structure of the product. It has been reported that the parameters controlling the mechanical properties of cellular material such as density, cell wall thickness, cell size, and cell number are expected to predict the product crispness (Hutchinson, Mantle, and Smith 1989; Guraya and Toledo 1996; Roudaut et al. 2002).

In the study of Altan, McCarthy, and Maskan (2008a, 2008b), increasing the temperature decreased the peak force and bulk density of extrudates. It was noted that a progressive increase in temperature resulted in pores in the structure due to the formation of air cells, and the surface appeared flaky and porous and hence had decreased hardness (Bhattacharya and Choudhury 1994). Duizer and Winger (2006) reported that less force is required in breaking a product that is very crisp. Therefore, it can be expected that the crispness of extrudates will increase with increasing temperature because it is related to the cellular structure of the extrudates and thus bulk density and hardness. Strong positive correlations have been found between the slope and bulk density as well as the peak force for extrudates from barley flour-tomato pomace and barley flour-grape pomace blends (Altan 2008). It was already suggested that increasing the temperature decreases the melt viscosity, which favors bubble growth and produces low density products with small and thin cells, thus increasing the crispness of the extrudate (Ding et al. 2005). Rayas-Duarte, Majewska, and Doetkott (1998) also stated that low breaking strength values are usually related to a large number of small cells per unit area with thinner cell walls, resulting in a crispy texture. Agbisit et al. (2007) found that both average crushing force and crispness work had high negative correlations with cell diameter. Both crushing force and crispness work increased with a decrease in cell diameter, indicating that more force and work are needed to deform/fracture smaller-size cells.

Increasing screw speed decreased the slope and therefore increased the crispness of barley-tomato pomace extrudates (Altan 2008). This has been attributed to a decrease in extrudate hardness when the screw speed is increased. Altan (2008) obtained a low value of crispness for barley-tomato pomace extrudates by increasing the temperature with a decreasing pomace level (Figure 6.6). Altan, McCarthy, and Maskan (2008b) found that increasing grape pomace content in the extrusion cooking of barley flour–grape pomace blends increased the slope of the force–time curve and resulted in a less crispy extrudate. Fiber reduces the cell size, probably by causing premature rupture of gas cells, which reduces the overall expansion and results in a less porous structure (Lue, Hsieh, and Huff 1991; Yanniotis, Petraki, and Soumpasi 2007) and therefore a less crispy texture. Jin, Hsieh, and Huff (1995) investigated the structure of corn meal extrudates including soy fiber by using scanning electron microscopy. It was revealed that the air bubbles became smaller and the cell became thicker as the fiber levels were increased. They also reported that increasing the fiber content resulted in a less expanded, more compact textured extrudate. Lue et al. (1990) observed that increasing the dietary fiber content decreased the average cell size and increased the frequency of incomplete flakes and holes on the cell wall. They also concluded that these holes and incomplete flakes were indicative of poor bubble formation during puffing. Moore et al. (1990) reported that the effect of bran on cell expansion and cell structure results from the fact that bran particles reduce the extensibility of cell walls, creating more broken and small cells that cause the premature rupture of the cell walls.

Brittleness

The distance that is required to break extrudates has been measured as brittleness and evaluated as the shortest distance being the most brittle product (Altan, McCarthy, and Maskan 2008b, 2008c). The brittleness of extrudates was mostly affected by temperature and pomace level in the studies of Altan (2008) and Altan, McCarthy, and Maskan (2008b). Mendonça, Grossmann, and Verhé (2000) evaluated the fracturability of extrudates by measuring the force at the first yield break of snacks using a texture analyzer. The fracturability decreased with increasing glycerol monostearate and decreasing moisture content and, also, lower product hardness was obtained at higher temperatures and lower moisture content. The best conditions for desirable snacks, which were low values of fracturability and hardness, were determined as low moisture content, high temperature, and high glycerol monostearate content in the study. The breaking distance of barley-tomato pomace extrudates was lower at a low pomace level with a high temperature. The decrease in brittleness (high in distance) for barley-tomato pomace extrudates could be explained by the changes in the cell size distribution and orientation which are associated with expansion, bulk density and cell wall properties of extrudates (Altan 2008). Studies showed that increased fiber produced a thicker cell wall, reduced the average cell size, decreased expansion, and increased the density and hardness or breaking strength of the extrudates, as reported earlier (Hsieh et al. 1989; Jin, Hsieh, and Huff 1995; Lue et al. 1990; Hsieh et al. 1991; Lue, Hsieh, and Huff 1991; Yanniotis, Petraki, and Soumpasi

2007). Thick cell walls are inherently less fragile and less likely to rupture than thin cell walls (Barrett and Ross 1990). However, Altan, McCarthy, and Maskan (2008b) observed a decrease in the distance of barley-grape pomace extrudates with an increasing grape pomace level and hence an increased brittleness of the samples, which is not in line with bulk density, hardness, and slope (crispness). Similarly, Chanvrier et al. (2007) found small distance values for wholemeal breakfast cereals; that is, they were more brittle. They suggested that this might be due to the high amount of particles (bran, insoluble fibers) in the products, which favors the initiation of product fracture. They also observed by light microscopy that the cohesion between the starch matrix and the particles appeared to be weak.

SENSORIAL PROPERTIES

Raw materials undergo physical and chemical modifications such as gelatinization, breakdown of starch, denaturation of proteins, and interactions between them resulting from high temperatures and pressures with combined of shearing effect during extrusion. These changes affect the sensory properties such as appearance, aroma, flavor, and texture of the extruded products. Therefore, sensory properties are important for extruded food products being developed as new brands.

Both descriptive sensory analysis and the hedonic scale have been used to evaluate the sensory properties of extruded food produced with different raw materials. A detailed study has been done for the effects of extrusion parameters on the sensory properties of corn meal extrudates (Chen et al. 1991). A descriptive sensory analysis has been used to characterize the appearance, aroma, taste, and texture of extrudates. Temperature was the most significant factor affecting the appearance, aroma and flavor, denseness, crispness, chewiness, and hardness of the extrudates of toasted corn. For example, a maximum Munsell color value was obtained at low temperatures and high moisture contents with a minimum color value occurring at high temperatures and low moisture. Higher airiness scores were observed at medium temperatures than at low and high temperatures. As a result of sensory analyses, raw flour taste and boiled corn taste had lower scores at the range of extrusion variables applied in the study. On the other hand, toasted corn taste increased with increasing temperature and decreasing moisture content. The denseness of extrudates increased at low temperatures with decreasing screw speed. Liu et al. (2000) studied the effect of screw speed, moisture content, and different percentages of oat flour on the sensory properties of oat-corn puffs using descriptive sensory analysis to evaluate the flavor, texture, aftertaste, and appearance of products. Decreasing the moisture content and increasing the screw speed resulted in increased product temperature, which was correlated with the attributes of a more expanded product such as lightness, crispness, shininess, and an open cell structure. The results also showed that oat associated flavor attributes such as those for raw oat flour or toasted oat flour were related to the extrudates with a high percentage of oat flour while corn flavor was found to be more process dependent. For example, corn-related flavors developed at high screw speeds and high product temperatures. The percent of oat flour was positively correlated with some visual attributes such as roughness, compact, dry surface, curving, and an irregular shape of the extrudate. The effects of adding pigeon pea flour to cassava flour on the physico-chemical and sensory qualities of enrobed flavored extrudates have been investigated by Rampersad, Badrie, and Comissiong (2003). In this research, trained panelists rated the texture of unflavored cassava/pigeon pea extrudates to determine the most suitable textured product in the first stage. In the second stage, attributes of color, odor, flavor, texture, and overall acceptability of enrobed flavored products were evaluated by hedonic testing using consumer panelists. The authors found that the effect of pigeon pea addition to cassava flour had a significant influence on the sensory texture of the extrudates. All enrobed flavored extrudates were liked moderately to very much in overall acceptability, with the chocolate extrudates having the highest scores for color and flavor.

The sensory properties of corn based extrudates enriched with corn bran were evaluated using a seven-point hedonic scale for appearance, palatability, and general acceptability (Mendonça, Grossmann, and Verhé 2000). The appearance score was affected mostly by temperature and feed moisture and slightly influenced by corn bran content. Extrudates produced at high temperature and moisture content gave lower values on the appearance attribute. The scores for appearance decreased with the increase of corn bran content in the extrudates. The palatability score decreased when the moisture content was increased and increased with increasing temperature. A decrease in the general acceptability score was obtained with the increase of moisture or corn bran. Sensory analysis has been done on five samples with different levels of cauliflower trimming (0%–20%) to assess the flavor acceptability of extrudates (Stojceska, Ainsworth, Plunkett, Ibanoglu, and Ibanoglu 2008). The results indicated that extrudates with 0%–10% of cauliflower were more acceptable than products containing 15%–20% of cauliflower. In the study of Yağcı and Göğüş (2008), the number of samples to be assessed sensorially was selected according to the expansion of the product at first. Later samples were chosen at minimum and maximum partially defatted hazelnut flour contents with each fruit waste ratio to evaluate bitterness, off-flavor, air cell homogeneity, and overall acceptability by semi-trained panelists. They observed slight bitterness in the extrudates containing higher partially defatted hazelnut flour levels (12.2%–15%). The air cell homogeneity scores of samples were in the range of 5.2 to 6 meant to good homogeneity. The scores for overall acceptability decreased at high levels of partially defatted hazelnut flour (12.5%–15%). In a later study of the same group, extrudates developed from blends of durum clear flour, partially defatted hazelnut flour, and fruit waste were also assessed for flavor (orange and hazelnut flavors), color, and texture (hardness, crispness, breakability) (Yağcı and Göğüş 2009). According to the results, the panelists did not find a characteristic orange and hazelnut flavor in most of the products. Extruded foods containing lower levels of partially defatted hazelnut flour (5%–7%) had higher color scores than the high hazelnut flour content (12.2%–15%). A lower sensory score for crispness and breakability was observed for extrudates having 5% fruit waste and 12.2% partially defatted hazelnut flour.

Sensory analyses of selected extrudates developed from barley flour-pomace blends were carried out by semi-trained panelists for appearance (color and porosity), texture (hardness, crispness and brittleness), taste (bran flavor, tomato flavor and brittleness), off-odor, and overall acceptability (Altan 2008; Altan, McCarthy, and Maskan 2008a, 2008b). Extrudates containing 10% tomato pomace had the highest score for color but received a low score for porosity. A good negative correlation was observed between porosity and distance (brittleness) from the force–distance curve (Altan

2008). It is expected that the higher the porosity of samples, the lower the breaking distance (higher brittleness) of the extrudates. Vincent (1998) also suggested that if the material is brittle, the fracture will travel quickly, resulting in a sudden unloading of the muscles; this is seen as a sudden drop in the load on a force–deflection curve. Panelists perceived weak tomato flavor for extrudates with the highest level of pomace (12.73%). No bitter taste and off-odor were detected for barley-tomato pomace extrudates (Altan, McCarthy, and Maskan 2008a). A higher score of overall acceptability was obtained for extrudates at 10% tomato pomace, temperature of 160°C, and 200 rpm and the lowest in extrudate with no pomace addition, 150°C and 175 rpm. Altan, McCarthy, and Maskan (2008b) obtained a decrease in the sensory score for color when the grape pomace level was increased. A positive correlation was found between porosity and color. It has been explained that porosity is directly related to the expansion of a product and therefore less expansion produces a compact structure which appears to be dull and results in a low score in sensory color. The panelists detected a more sweet taste in extrudates at 12.73% grape pomace level. High correlations were found between sensory attributes and texture data. An increase in hardness and crispness measured instrumentally resulted in a decrease in the preference of the panelists in sensory hardness. Extrudates with 2% and 10% grape pomace levels gave a higher score for crispness (5.56–5.59) while 12.73% of grape pomace had a lower preference in crispness (3.31). The lowest score (2.93) for brittleness was observed for extrudates at high pomace (12.73%), but the highest was obtained at a 2% pomace level. The overall acceptability of extrudates at 2% pomace level had a higher preference (5.53) but was not significantly different from extrudates with 6% and 10% pomace levels.

Camire et al. (2002) studied the stability and acceptability of blueberry and grape anthocyanins in extruded cereals. They used a nine-point hedonic scale for overall acceptability and acceptability of color, sweetness, hardness, and flavor. In their results the addition of fruit juice concentrate significantly affected overall acceptability and acceptability of color, sweetness, and flavor but not hardness acceptability. Ahmed (1999) studied the effect of flaxseed flour on the chemical, physical, microstructural, and sensory qualities of a corn based snack. A hedonic scale was used for the product's sensory attributes of appearance, aroma, taste, texture, and color. Sensory evaluation showed that the total score gradually decreased by increasing the proportion of added flaxseed but was still acceptable for the panelists. Veronica, Olusola, and Adebowale (2006) fortified a maize based snack with partially defatted soybean and analyzed it for physical, chemical, and sensory characteristics. The extruded samples were subjected to two different sensory evaluations. A nine-point hedonic scale was used to determine the preference in color, flavor, taste, and overall acceptability, while a five-point "just-about-right" scale was used to assess sensory crispness and puffiness of the product. The results showed that incorporating partially defatted soybean in a maize based snack had a positive effect on the chemical properties but a negative effect on the physical and sensory characteristics.

NUTRITIONAL PROPERTIES

The effect of the extrusion process on nutritional changes of food has been reviewed recently (Singh, Gamlath, and Wakeling 2007). The current interest is how the extrusion process and addition of food by-products from different sources impact on nutritional or non-nutrient healthful components such as phytochemicals of newly developed extruded products.

In a study by Tahvonen et al. (1998), the incorporation of black currant press residue from 20% to 40% enhanced the proportions of α- and γ-linolenic acids and stearidonic acid by 16%–20% in the product. No extensive losses of nutritionally valuable fatty acids from the black currant press residue were observed during the extrusion process. The mineral contents of the product (Ca, Mg, and Fe) increased with the increasing amount of press residue. In the model product, incorporation of black currant residue (30%) with combinations of 10% oat flour, 20% oat bran and 30% potato starch, resulted in a total fat content of 3%, where the fatty acid composition was linoleic acid 43 mol-%, oleic acid 20 mol-%, α-linolenic acid 12 mol-%, γ-linolenic acid 9 mol-%, and steridonic acid 2 mol-%. The total dietary fiber and mineral contents of the product were approximately 20% and 1.9 g/kg Ca, 1 g/kg Mg, and 59 mg/kg Fe, respectively. The addition of crab processing by-products (40%) into corn meal increased the protein content of the extrudates by 70% when compared to control extrudates (Murphy et al. 2003). The ash content of extrudates increased with the increasing ratio of crab processing by-product due to the calcium content of the crab shell ash. An increase in the calcium content of the extrudates was observed. It has been concluded that all the extrudates formulated with dry or wet crab processing by-products contained more than 240 mg Ca per 28 g serving. An increase in crude protein, crude fat, and fiber contents of extrudates has been observed by replacing maize flour with brewer's spent grain (Ainsworth et al. 2007). The authors did not find any difference in the total antioxidant capacity and total phenolic compounds between samples, either by changing the screw speed or adding brewer's spent grain. The phytic acid content of the products increased with the increasing level of brewer's spent grain when compared with the control sample but was not changed by screw speed. However, a reduction in phytic acid of peas with cooking and extrusion methods has been observed in different studies (Habiba 2002; Abd El-Hady and Habiba 2003). Phytic acid has been known as an antinutrient; it reduces the absorption of minerals by forming a complex with them. Protein digestibility is considered to be the most important factor besides the quantity and availability of essential amino acids when evaluating protein nutritional value (Singh, Gamlath, and Wakeling 2007). Ainsworth et al. (2007) found increased protein *in vitro* digestibility in control samples without brewer's spent grain with increasing screw speed. However, the screw speed did not influence protein digestibility for samples containing brewer's spent grain. Increasing the level of brewer's spent grain raised the resistant starch of the extrudates. The extrusion of cauliflower trimmings with a combination of wheat flour resulted in increased total phenolics at a 0%–20% cauliflower level and total antioxidant capacity at a level of 0%–10% cauliflower in the final product (Stojceska, Ainsworth, Plunkett, Ibanoglu, and Ibanoglu 2008). However, the total dietary fiber content and protein digestibility of the products decreased after the extrusion process. The reduction in fiber was explained by the degradation of pectic substances due to solubilization under heat and moisture. The fiber and protein contents of the finished product changed from 5.6% to 11.6% and 16.3% to 20.4%, respectively, at a level of 0%–20% cauliflower by-product.

In a study of incorporation of brewer's spent grain and red cabbage into wheat flour and corn flour, changing the moisture content from 12% to 17% during the extrusion process had no effect on the total antioxidant capacity and total phenolic content of the extrudates (Stojceska et al. 2009). No significant effect of the extrusion process on antioxidant capacity and phenolic contents has been observed for extrudates containing brewer's spent grain, but extrudates containing red cabbage had a slight increase in antioxidant activity and phenolic content after extrusion processing. Similarly, the phtytic acid content of extrudates did not change with extrusion cooking. In general, an increase in the total dietary fiber content of the extrudates was obtained especially in corn starch with brewer's spent grain containing extrudates. According to the authors, the high amylose content of the sample having corn starch resulted in the formation of more retrograded amylose that is a mechanism for resistant starch formation.

The preliminary studies on the extrusion processing of fruits and vegetable by-products with barley flour showed that there was a significant difference in fiber level as a function of the type of pomace (tomato, grape, and pomegranate pomaces) (Vandeven et al. 2007). Extrudates containing grape pomace had a lower fiber content than the extrudates with tomato and pomegranate pomaces. In the detailed research of Altan, McCarthy, and Maskan (2009a), the antioxidant activity of extrudates from tomato pomace and barley flour blends decreased with increasing screw speed at low pomace levels. However, an increase in antioxidant activity was obtained when the tomato pomace ratio increased at high screw speeds. This was attributed to the contribution of antioxidants present in the tomato pomace. On the other hand, the total phenolics of samples changed between 2.09 and 2.81 mg ferulic acid per gram dry sample. Extrusion cooking decreased the total phenolic content of the extrudates. Antioxidant activity of grape pomace extrudates was affected by screw speed and interaction of temperature and grape pomace level. An increase in the antioxidant activity of samples was obtained under conditions of high temperature and grape pomace level. It has been attributed to the high molecular weight products of the Maillard reaction that are formed at higher temperatures and act as antioxidants. The reduction in antioxidant activity of grape pomace extrudates was explained by the destruction of the present antioxidants and phenolic compounds due to the high shear conditions.

The effect of extrusion processing on the procyanidin content of grape seed and pomace, grape juice processing by-products, as well as the total anthocyanin content of grape pomace has been investigated recently (Khanal, Howard, and Prior 2009). As a result of this research, an enhancement has been observed by a substantial increase in monomer and dimer contents of grape seeds after extrusion processing due to the conversion of some higher level oligomers and polymers into their lower oligomer counterparts. However, trimer contents were either reduced or not affected by extrusion. It has been observed that the monomer contents of grape seeds increased with increasing temperature from 160°C to 170°C, but a further increase in temperature did not increase the monomer content. In the case of grape pomace, the contents of monomers, dimers, trimers and tetramers increased at 170°C and 200 rpm. An increase in temperature from 160°C to 170°C with an increasing residence time increased monomers through tetramers in grape pomace. The extrusion process increased the total procyanidin contents in grape pomace by 49% while procyanidin contents in grape seed were either reduced or remained similar in grape seed. The reduction has been observed in the total anthocyanin content of grape pomace from 18% to 53% after extrusion. An increase in the extrusion temperature decreased the anthocyanin content but decreasing the screw speed increased the anthocyanin loss due to the high residence time of the material inside the extruder.

The effect of extrusion conditions and pomace levels added on β-glucan contents of extrudates has been studied (Altan, McCarthy, and Maskan 2009a). The results showed that extrusion cooking and pomace level for either grape or tomato pomace decreased the β-glucan content of extrudates. The reduction in β-glucan content by extrusion has been attributed to the destruction of β-glucan under high shear conditions. The effect of temperature (Figure 6.7) and screw speed on the β-glucan content of grape pomace extrudates has been found to be significant, but only the tomato pomace level affected the β-glucan content of the extrudates developed from barley flour and tomato pomace. It has been reported that extrudates would provide 2.8 g and 3.2 g β-glucan per 100 g serving at 10% tomato pomace and grape pomace levels.

Gelatinization destroys the compact granular structure by breaking inter- and intramolecular hydrogen bonds and allowing different degrees of swelling and absorption of water; as a result, fully hydrated starch molecules leach from the granule. Therefore, the availability of starch granules to digestive enzymes increases to different levels with increasing degrees of gelatinization (Holm et al. 1988). Many researchers have reported a positive effect of the extrusion process on starch digestibility (Guha, Ali,

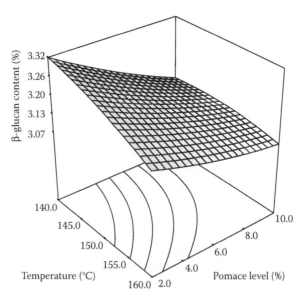

FIGURE 6.7 The effect of pomace level and temperature on the β-glucan content of barley-grape pomace extrudates at a screw speed of 175 rpm. (Data taken Altan, A. 2008. Production and properties of snack foods developed by extrusion from composite of barley, and tomato and grape pomaces. PhD dissertation, University of Gaziantep, Turkey, pp. 211.)

and Bhattacharya 1997; Alonso et al. 2000; Bryant et al. 2001; Hagenimana, Ding, and Fang 2006; Altan, McCarthy, and Maskan 2009b). The improvement in *in vitro* starch digestibility after extrusion has been attributed to the increased susceptibility of starch to amylase resulting from the loss of structural integrity due to the shearing action developing heat through the dissipation of mechanical energy and the hydration of starch granules and the degree of gelatinization (Hagenimana, Ding, and Fang 2006; Altan, McCarthy, and Maskan 2009b). The starch digestibility of extrudates from tomato pomace decreased when the extrusion temperature increased (Altan, McCarthy, and Maskan 2009b). The authors obtained a low value of digestibility of extrudates at a temperature of 167°C, 175 rpm screw speed, and 6% tomato pomace level. The retrogradation or reassociation of gelatinized starch or the formation of an amylose–lipid complex, a starch–protein complex, or starch has been shown to be a possible reason for the reduction in digestibility (Guha, Ali, and Bhattacharya 1997; Altan, McCarthy, and Maskan 2009b). Increasing both the tomato and grape pomace ratios in the blends decreased the digestibility of the extrudates (Altan, McCarthy, and Maskan 2009b). The reduction in digestibility could be due to fiber in the pomace, which tends to reduce starch digestibility by trapping starch granules within a viscous protein–fiber–starch network. The reduction in digestibility may result from proteins, and lipids in extrudates coming from pomaces. Because the presence of protein bodies around starch granules can restrict granule swelling and starch gelatinization, it also reduces the susceptibility to enzymatic attack.

CONCLUSION AND FUTURE TRENDS

So far, food processing by-products, especially fruit and vegetable by-products, have been studied for non-nutrient compounds such as antioxidants, phenolics, and dietary fiber, all of which have health benefits. However, there have been few studies using by-products directly for the development of different products, e.g., pie fillings, crackers, cookies, and bread. Recently, there has been great interest in the transformation and upgrading of food processing-derived plant based by-products with a combination of different cereals into value-added food products to minimize the environmental impact of these by-products and to utilize their valuable constituents. The results of different research suggest that it is possible to produce extruded foods from blends of by-products with different cereals. The incorporation of by-products is a suitable means to enrich extruded products with dietary fiber or bioactive compounds. The identification of the nutraceutical potential of natural compounds is a growing interest, and the evaluation of food processing by-products with the extrusion process has the potential to be able to develop more commercial applications. However, further research is needed on *in vitro* and *in vivo* studies for the evaluation of bioavailability and toxicology of non-nutrient compounds incorporated from by-products. Extrusion is a promising process that allows the utilization and co-processing of various by-products. It provides simultaneous mixing, cooking, and texturization of different recipes under the shearing effects of screws with thermal energy generated by viscous dissipation that modifies the rheological properties of raw materials due to the physical and chemical changes of biopolymers. Therefore, it is essential to investigate the rheological properties of dough and relationships between compositional changes on extrudate quality. Furthermore, more research is necessary to focus on the degradation/retention and stability of nutrients and phytochemicals, and also interactions of phytochemicals with other food ingredients during extrusion processing and storage. These research efforts and the optimization of extrusion processing will bring progress in the utilization of by-products for the development of new products in the markets.

REFERENCES

Abd El-Hady, E.A. and Habiba, R.A. 2003. Effect of soakingand extrusion conditions on antinutrients and protein digestibility of legume seeds. *Lebens-Wiss Technol* 36: 285–293.

Agbisit, R., Alavi, S., Cheng, E., Herald, T. and Trateri, A. 2007. Relationships between microstructure and mechanical properties of cellular cornstarch extrudates. *J Texture Studies* 38: 199–219.

Ahmed, Z.S. 1999. Physico-chemical, structural and sensory quality of corn-based flax-snack. *Nahrung* 43: 253–258.

Ainsworth, P., Ibanoğlu, S., Plunkett, A., Ibanoğlu, E. and Stojceska, V. 2007. Effect of brewers spent grain addition and screw speed on the selected physical and nutritional properties of an extruded snack. *J Food Eng* 81: 702–709.

Ali, Y., Hanna, M.A. and Chinnaswamy, R. 1996. Expansion characteristics of extruded corn grits. *Lebens-Wiss Technol* 29: 702–707.

Alonso, R., Grant, G., Dewey, P. and Marzo, F. 2000. Nutritional assessment in vitro and in vivo of raw and extruded peas (Pisum sativum L.). *J Agric Food Chem* 48: 2286–2290.

Altan, A. 2008. Production and properties of snack foods developed by extrusion from composite of barley, and tomato and grape pomaces. PhD dissertation, University of Gaziantep, Turkey, pp. 211.

Altan, A., McCarthy, K.L. and Maskan, M. 2008a. Evaluation of snack foods from barley-tomato pomace blends by extrusion processing. *J Food Eng* 84: 231–242.

Altan, A., McCarthy, K.L. and Maskan, M. 2008b. Twin-screw extrusion of barley-grape pomace blends: extrudate characteristics and determination of optimum processing conditions. *J Food Eng* 89: 24–32.

Altan, A., McCarthy, K.L. and Maskan, M. 2008c. Extrusion cooking of barley flour and process parameter optimization by using response surface methodology. *J Sci Food Agric* 88: 1648–1659.

Altan, A., McCarthy, K.L. and Maskan, M. 2009a. Effect of extrusion processing on antioxidant activity, total phenolics and β-glucan content of extrudates developed from barley-fruit and vegetable by-products. *Int J Food Sci Technol* 44: 1263–1271.

Altan, A., McCarthy, K.L. and Maskan, M. 2009b. Effect of extrusion cooking on functional properties and in vitro starch digestibility of barley based extrudates from fruit and vegetable by-products. *J Food Sci* 79: E77–E86.

Alvarez-Martinez, L., Kondury, K.P. and Harper, J.M. 1988. A general model for expansion of extruded products. *J Food Sci* 53(2): 609–615.

Apruzzese, F., Balke, S.T. and Diosady, L.L. 2000. In-line colour and composition monitoring in the extrusion cooking process. *Food Res Int* 33: 621–628.

Arhaliass, A., Bouvier, J.M. and Legrand, J. 2003. Melt growth and shrinkage at the exit of the die in the extrusion-cooking process. *J Food Eng* 60: 185–192.

Arora, A., Zhao, J. and Camire, M.E. 1993. Extruded potato peel functional properties affected by extrusion conditions. *J Food Sci* 58(2): 335–337.

Artz, W.E., Warren, C.C. and Villota, R. 1990. Twin screw extrusion modification of corn fiber and corn starch extruded blend. *J Food Sci* 55: 746–750, 754.

Asare, E.K., Sefa-Dedeh, S., Sakyi-Dawson, E. and Afoakwa, E.O. 2004. Application of response surface methodology for studying the product characteristics of extruded rice-cowpea-groundnut blends. *Int J Food Sci Nutr* 55: 431–439.

AWARENET, 2004. Handbook for the prevention and minimization of waste and valorization of by-products in European agro-food industries. Deposito Legal: BI-223-04.

Badrie, N. and Mellowes, W.A. 1991. Effect of extrusion variables on cassava extrudates. *J Food Sci* 56: 1334–1337.

Baik, B.K., Powers, J. and T. Nguyen, L. 2004. Extrusion of regular and waxy barley flours for production of expanded cereals. *Cereal Chem* 81: 94–99.

Baladrán-Quintana, R.R., Barbosa-Canovas, G.V., Zazueta-Morales, J.J., Anzaldua-Morales, A. and Quintero-Ramos, A. 1998. Functional and nutritional properties of extruded whole pinto bean meal (Phaseolus vulgaris L.). *J Food Sci* 63: 113–116.

Barrett, A.H., Cardello, A.V., Lesher, L.L. and Taub, I.A. 1994. Cellularity, mechanical failure and texture perception of corn meal extrudates. *J Texture Studies* 25: 77–95.

Barrett, A.H. and Ross, E.W. 1990. Correlation of extrudate infusibility with bulk properties using image analysis. *J Food Sci* 55: 1378–1382.

Baysal, T., Ersus, S. and Starmans, D.A.J. 2000. Supercritical CO_2 extraction of β-carotene and lycopene from tomato paste waste. *J Agric Food Chem* 48: 5507–5511.

Berset, C. (1989). Color. In *Extrusion cooking*, eds. C. Mercier, P. Linko, and J.M. Harper, 371–385. St. Paul, MN: American Association of Cereal Chemists Inc.

Bhattacharya, S. and Choudhury, G.S. 1994. Twin-screw extrusion of rice flour: Effect of extruder length-to-diameter ratio and barrel temperature on extrusion parameters and product characteristics. *J Food Process Pres* 18: 389–406.

Bhattacharya, S. and Prakash, M. 1994. Extrusion of blends of rice and chick pea flours: A response surface analysis. *J Food Eng* 21: 315–330.

Bhattacharya, S., Sivakumar, V. and Chakraborty, D. 1997. Changes in CIELab colour parameters due to extrusion of rice-greengram blend: a response surface approach. *J Food Eng* 32: 125–131.

Bobek, P., Ozdin, L. and Hromadova, M. 1998. The effect of dried tomato grape and apple pomace on the cholesterol metabolism and antioxidative enzymatic system in rats with hypercholesterolemia. *Nahrung* 42: 317–320.

Bryant, R.J., Kadan, R.S., Champagne, E.T., Vinyard, B.T. and Boykin, D. 2001. Functional and digestive characteristics of extruded rice flour. *Cereal Chem* 78: 131–137.

Camire, M.E. and Flint, S.I. 1991. Thermal processing effects on dietary fibre composition and hydration capacity in corn meal, oat meal, and potato peels. *Cereal Chem* 68: 645–647.

Camire, M.E. and King, C.C. 1991. Protein and fiber supplementation effects on extruded cornmeal snack quality. *J Food Sci* 56: 760–763, 768.

Camire, M.E., Chaovanalikit, A., Dougherty, M.P. and Briggs, J. (2002). Blueberry and grape anthocyanins as breakfast cereal colorants. *J Food Sci* 67: 438–441.

Camire, M.E., Violette, D., Dougherty, M.P. and McLaughlin, M.A. 1997. Potato peel dietary fiber composition: Effects of peeling and extrusion cooking processes. *J Agric Food Chem* 45: 1404–1408.

Cantarelli, P.R., Regitano-d'arce, M.A.B. and Palma, E.R. 1993. Physicochemical characteristics and fatty acid composition of tomato seed oils from processing wastes. *Scienta Agricola, Piracicaba* 50(1): 117–120.

Carson, K.J., Collins, J.L. and Penfield, M.P. 1994. Unrefined, dried apple pomace as a potential food ingredient. *J Food Sci* 59(6): 1213–1215.

Chang, Y.K., Silva, M.R., Gutkoski, L.C., Sebio, L. and Da Silva, M.A.A.P. 1998. Development of extruded snacks using jatobá (Hymenaea stigonocarpa Mart) flour and cassava starch blends. *J Sci Food Agric* 78: 59–66.

Chanvrier, H., Appelqvist, I.A.M., Bird, A.R., Gilbert, E., Htoon, A., Li, Z., Lillford, P.J., Lopez-Rubio, A., Morell, M.K. and Topping, D.L. 2007. Processing of novel elevated amylose wheats: Functional properties and starch digestibility of extruded products. *J Agric Food Chem* 55: 10248–10257.

Chen, J., Serafin, F.L., Pandya, R.N. and Daun, H. 1991. Effects of extrusion conditions on sensory properties of corn meal extrudates. *J Food Sci* 56: 84–89.

Chidambara Murthy, K.N., Singh, P.R. and Jayaprakasha, K.G. 2002. Antioxidant activities of grape (Vitis vinifera) pomace extracts. *J Sci Food Agric* 50: 5909–5914.

Chinnaswamy, R. and Hanna, M.A. 1988. Optimum extrusion-cooking conditions for maximum expansion of corn starch. *J Food Sci* 53: 834–840.

Choudhury, G.S. and Gautam, A. 1998. Comparative study of mixing elements during twin-screw extrusion of rice flour. *Food Res Int* 31: 7–17.

Colonna, P. and Mercier, C. 1983. Macromolecular modifications of manioc starch components by extrusion cooking with and without lipids. *Carbohydrate Polymers* 3: 87–108.

Colonna, P., Tayeb, J. and Mercier, F. 1989. Extrusion cooking of starch and starchy products. In *Extrusion cooking*, eds. C. Mercier, P. Linko and J.M. Harper, 247–319. St. Paul, MN: American Association of Cereal Chemists Inc.

Davidson, V., Paton, D., Diosady, L.L. and Rubin, L.J. 1984. A model for mechanical degradation of wheat starch in a single screw extruder. *J Food Sci* 49: 1154–1157.

Del Valle, M., Camara, M. and Torija, M.-E. 2006. Chemical characterization of tomato pomace. *J Sci Food Agric* 86: 1232–1236.

Della Valle, G., Vergnes, B., Colonna, P. and Patria, A. 1997. Relations between rheological properties of molten starches and their expansion behaviour in extrusion. *J Food Eng* 31: 277–296.

Desrumaux, A., Bouvier, J.M. and Burri, J. 1998. Corn grits particle size and distribution effects on the characteistics of expanded extrudates. *J Food Sci* 63: 857–863.

Ding, Q.B., Ainsworth, P., Plunkett, A., Tucker, G. and Marson, H. 2006. The effect of extrusion conditions on the functional and physical properties of wheat-based expanded snacks. *J Food Eng* 73: 142–148.

Ding, Q.B., Ainsworth, P., Tucker, G. and Marson, H. 2005. The effect of extrusion conditions on the physicochemical properties and sensory characteristics of rice-expanded snacks. *J Food Eng* 66: 283–289.

Doğan, H. and Karwe, M.V. 2003. Physicochemical properties of quinoa extrudates. *Food Sci Technol Int* 9: 101–114.

Duizer, L.M. and Winger, R.J. 2006. Instrumental measures of bite forces associated with crisp products. *J Texture Studies* 37: 1–15.

Durham, R.J. and Hourigan, J.A. 2007. Handbook of waste management and co-product recovery in food processing. In *Waste management and co-product recovery in dairy processing*, ed. K. Waldron, vol 1, 332–387. Abington, Cambridge, England: Woodhead Publishing Limited; Boca Raton, Florida, USA: CRC Press LLC.

Figuerola, F., Hurtado, M., Estévez, A.M., Chiffelle, I. and Asenjo, F. 2005. Fibre concentrates from apple pomace and citrus peel as potential fibre sources for food enrichment. *Food Chem* 91: 395–401.

Gibert, O. and Rakshit, S.K. 2005. Cassava starch snack formulation using functional shell fish by-products: mechanical, sorption and geometric properties. *J Sci Food Agric* 85: 1938–1946.

Goldhor, S. and Regenstein, J. 2007. Handbook of waste management and co-product recovery in food processing. In *Waste management and co-product recovery in fish processing*, ed. K. Waldron, vol 1, 388–416. Abington, Cambridge, England: Woodhead Publishing Limited; Boca Raton, Florida, USA: CRC Press LLC.

Gomez, M.H. and Aguilera, J.M. 1983. Changes in the starch fraction during extrusion-cooking of corn. *J Food Sci* 48: 378–381.

Grenus, K.M., Hsieh, F. and Huff, H.E. 1993. Extrusion and extrudate properties of rice flour. *J Food Eng* 18: 229–245.

Guha, M., Ali, S.Z. and Bhattacharya, S. 1997. Twin-screw extrusion of rice flour without die: Effect of barrel temperature and screw speed on extrusion and extrudate characteristics. *J Food Eng* 32: 251–267.

Guha, M. and Ali, Z. 2006. Extrusion cooking of rice: Effect of amylose content and barrel temperature on product profile. *J Food Process Pres* 30: 706–716.

Gujral, H.S., Singh, N. and Singh, B. 2001. Extrusion behaviour of grits from flint and sweet corn. *Food Chem* 74: 303–308.

Gujska, E. and Khan, K. 1991. Functional properties of extrudates from high starch fractions of navy pinto beans and corn meal blended with legume high protein fractions. *J Food Sci* 56: 431–435.

Guraya, H.S. and Toledo, R.T. 1996. Microstructural characteristics and compression resistance as indices of sensory texture in a cruncy snack product. *J Texture Studies* 27: 687–701.

Gutkoski, L.C. and El-Dash, A.A. 1999. Effect of extrusion process variables on physical and chemical properties of extruded oat products. *Plant Foods Human Nutr* 54: 315–325.

Habiba, R.A. 2002. Changes in anti-nutrients, protein solubility, digestibility, and HCl-extractability of ash and phosphorus in vegetable peas as affected by cooking methods. *Food Chem* 77: 187–192.

Hagenimana, A., Ding, X. and Fang, T. 2006. Evaluation of rice flour modified by extrusion cooking. *J Cereal Sci* 43: 38–46.

Hashimoto, J.M. and Grossmann, M.V.E. 2003. Effects of extrusion conditions on quality of cassava bran/cassava starch extrudates. *Int J Food Sci Technol* 38: 511–517.

Holay, S.H. and Harper, J.M. 1982. Influence of extrusion shear environment on plant protein texturization. *J Food Sci* 47: 1869–1874.

Holm, J., Lundquist, M.D., Björck, I., Eliasson, A.-C. and Asp, N.-G. 1988. Degree of starch gelatinization, digestion rate of starch in vitro, and metabolic response in rats. *American J Clinical Nutr* 47: 1010–1016.

Hsieh, F., Huff, H.E., Lue, S. and Stringer, L. 1991. Twin screw extrusion of sugar-beet fiber and corn meal. *Lebens-Wiss Technol* 24: 495–500.

Hsieh, F., Mulvaney, S.J., Huff, H.E., Lue, S. and Brent, J. 1989. Effect of dietary fiber and screw speed on some extrusion processing and product variables. *Lebens-Wiss Technol* 22: 204–207.

Hutchinson, R.J., Mantle, S.A. and Smith, A.C. 1989. The effect of moisture content on the mechanical properties of extruded food foams. *J Material Sci* 24: 3249–3253.

Ilo, S. and Berghofer, E. 1999. Kinetics of colour changes during extrusion cooking of maize gritz. *J Food Eng* 39: 73–80.

Ilo, S., Liu, Y. and Berghofer, E. 1999. Extrusion cooking of rice flour and amaranth blends. *Lebens-Wiss Technol* 322: 79–88.

Ilo, S., Tomschik, U., Berghofer, E. and Mundigler, N. 1996. The effect of extrusion operations on the apparent viscosity and the properties of extrudates in twin-screw extrusion cooking of maize grits. *Leben-Wiss Technol* 29: 593–598.

Iwe, M.O. 1998. Effects of extrusion cooking on functional properties of mixtures of full-fat soy and sweet potato. *Plant Foods Human Nutr* 53: 37–46.

Jackson, J.C., Bourne, M.C. and Barnard, J. 1996. Optimization of blanching for crispness of banana chips using response surface methodology. *J Food Sci* 61: 165–166.

Jayaprakasha, G.K., Singh, R.P. and Sakariah, K.K. 2001. Antioxidant activity of grape seed (Vitis vinifera) extracts on peroxidation models in vitro. *Food Chem* 73: 285–290.

Jin, Z., Hsieh, F. and Huff, H.E. 1994. Extrusion cooking of corn meal with soy fiber, salt, and sugar. *Cereal Chem* 71: 227–234.

Jin, Z., Hsieh, F. and Huff, H.E. 1995. Effects of soy fiber, salt, sugar, and screw speed on physical properties and microstructure of corn meal extrudate. *J Cereal Sci* 22: 185–194.

Johnson, E.J. 2000. The role of lycopene in health and disease. *Nutr Clinical Care* 3: 35–43.

Kammerer, D., Claus, A., Carle, R. and Schieber, A. 2004. Polyphenol screening of pomace from red and white grape varieties (Vitis Vinifera L.) by HPLC-DAD-MS/MS. *J Agric Food Chem* 52: 4360–4367.

Kaur, D., Sogi, D.S., Gary, S.K. and Bawa, A.S. 2005. Flotation-cum-sedimentation system for skin and seed separation from tomato pomace. *J Food Eng* 71: 341–344.

Khanal, R.C., Howard, L.R. and Prior, R.L. 2009. Procyanidin content of grape seed and pomace, and total anthocyanin content of grape pomace as affected by extrusion processing. *J Food Sci* 74: H174–H182.

Khanal, R.C., Howard, L.R., Brownmiller, C.R. and Prior, R.L. 2009. Influence of extrusion processing on procyanidin composition and total anthocyanin contents of blueberry pomace. *J Food Sci* 74: H52–H58.

Kirby, A.R., Ollett, A.L., Parker, R. and Smith, A.C. 1988. An experimental study of screw configuration effects in the twin-screw extrusion cooking of maize grits. *J Food Eng* 8: 247–272.

Kokini, J.L., Chang, C.N. and Lai, L.S. 1992. The role of rheological properties on extrudate expansion. In *Food extrusion science and technology*, eds. J.L. Kokini, C.-T. Ho and M.V. Karwe, 631–653. New York, N.Y.: Marcel Dekker Inc.

Koksel, H., Ryu, G.H., Basman, A., Demiralp, H. and Ng, P.K.W. 2004. Effects of extrusion variables on the properties of waxy hulless barley extrudates. *Nahrung* 48: 19–24.

Köksel, H., Ryu, G.H., Özboy-Özbaş, Ö., Basman, A. and Ng, P.K.W. 2003. Development of a bulgur-like product using extrusion cooking. *J Sci Food Agric* 83: 630–636.

Kris-Etherton, P.M., Hecker, K.D., Bonanome, A., Coval, S.M., Binkoski, A.E., Hilpert, K.F., Griel, A.E. and Etherton, T.D. 2002. Bioactive compounds in foods: their role in the prevention of cardiovascular disease and cancer. *American J Medicine* 113(9): 71–88.

Larrea, M.A., Chang, Y.K. and Martinez-Bustos, F. 2005a. Effect of some operational extrusion parameters on the constituents of orange pulp. *Food Chem* 89: 301–308.

Larrea, M.A., Chang, Y.K. and Martinez-Bustos, F. 2005b. Some functional properties of extruded orange pulp and its effect on the quality of cookies. *Lebens-Wiss Technol* 38: 213–220.

Laufenberg, G., Kunz, B. and Nystroem, M. 2003. Transformation of vegetable waste into value added products: (A) the upgrading concept; (B) practical implementations. *Bioresource Technol* 87: 167–198.

Launay, B. and Lisch, L.M. 1983. Twin-screw extrusion cooking of starches: Flow behaviour of starch pastes, expansion and mechanical properties of extrudates. *J Food Eng* 2: 259–280.

Le Meste, M., Champion, D., Roudaut, G., Blond, G. and Simatos, D. 2002. Glass transition and food technology: A critical appraisal. *J Food Sci* 67: 2444–2458.

Lee, E.Y., Ryu, G.-H. and Lim, S.-T. 1999. Effect of processing parameters on physical properties of corn starch extrudates expanded using supercritical CO_2 injection. *Cereal Chem* 76: 63–69.

Liu, Y., Hsieh, F., Heymann, H. and Huff, H.E. 2000. Effect of processing conditions on the physical and sensory properties of extruded oat-corn puff. *J Food Sci* 65: 1253–1259.

Llorach, R., Espín, J.C., Tomás-Barberán, F.A. and Ferreres, F. 2003. Valorization of cauliflower (Brassica oleracea L. var. botrytis) byproducts as a source of antioxidant phenolics. *Food Chem* 51: 2181–2187.

Lu, Y. and Foo, L.Y. 1997. Identification and quantification of major polyphenols in apple pomace. *Food Chem* 59: 187–194.

Lue, S., Hsieh, F. and Huff, H.E. 1991. Extrusion cooking of corn meal and sugar beet fiber: Effects on expansion properties, starch gelatinization, and dietary fiber content. *Cereal Chem* 68: 227–234.

Lue, S., Hsieh, F., Peng, I.C. and Huff, H.E. 1990. Expansion of corn extrudates containing dietary fiber: A microstructure study. *Lebens-Wiss Technol* 23: 165–173.

Martín-Carrón, N., Saura-Calixto, F. and Goñi, I. 2000. Effects of dietary fibre- and polyphenol-rich grape products on lipidaemia and nutritional parameters in rats. *J Sci Food Agric* 80: 1183–1188.

Masoodi, F.A. and Chauhan, G.S. 1998. Use of apple pomace as a source of dietary fiber in wheat bread. *J Food Process Pres* 22: 255–263.

Mendonça, S., Grossmann, M.V.E. and Verhé, R. 2000. Corn bran as a fibre source in expanded snacks. *Lebens-Wiss Technol* 33: 2–8.

Mercier, C. and Feillet, P. 1975. Modification of carbohydrate components by extrusion-cooking of cereal products. *Cereal Chem* 52: 283–297.

Meyer, A.S., Jepsen, S.M. and Sørensen, N.S. 1998. Ezymatic release of antioxidants for human low-density lipoprotein from grape pomace. *J Agric Food Chem* 46: 2439–2446.

Meyer, A.S., Yi, O., Pearson, D.A., Waterhouse, A. and Frankel, E. 1997. Inhibition of human low-density lipoprotein oxidation in relation to phenolic antioxidants in grapes. *J Agric Food Chem* 45: 1638–1643.

Mezreb, K., Goullieux, A., Ralainirina, R. and Queneudec, M. 2003. Application of image analysis to measure screw speed influence on physical properties of corn and wheat extrudates. *J Food Eng* 57: 145–152.

Moore, D., Sanei, A., Van Hecke, E. and Bouvier, J.M. 1990. Effect of ingredients on physical/structural properties of extrudates. *J Food Sci* 55: 1383–1402.

Moraru, C.I. and Kokini, J.L. 2003. Nucleation and expansion during extrusion and microwave heting of cereal foods. *Comp Reviews Food Sci Food Safety* 2: 120–138.

Moure, A., Cruz, J.M., Franco, D., Domínguez, J.M., Sinieiro, J., Domínguez, H., Núñez, M.J. and Parajó, J.C. 2001. Natural antioxidants from residual sources. *Food Chem* 72: 145–171.

Murphy, M.G., Skonberg, D.I., Camire, M.E., Dougherty, M.P., Bayer, R.C. and Briggs, J.L. 2003. Chemical composition and physical properties of extruded snacks containing crab-processing by-product. *J Sci Food Agric* 83: 1163–1167.

Nawirska, A. and Kwaśniewska, M. 2005. Dietary fibre fractions from fruit and vegetable processing waste. *Food Chem* 91: 221–225.

Ng, A., Lecain, S., Parker, M.L., Smith, A.C. and Waldron, K.W. 1999. Modification of cell-wall polymers of onion waste III. Effect of extrusion-cooking on cell-wall material of outer fleshy tissues. *Carbohydrate Polymers* 39: 341–349.

Obatolu, V.A., Skonberg, D.I., Camire, M.E. and Dougherty, M.P. 2005. Effect of moisture content and screw speed on the physical chemical properties of an extruded crab-based snack. *Food Sci Technol Int* 11: 121–127.

Onwulata, C.I., Konstance, R.P., Smith, P.W. and Holsinger, V.H. 2001. Co-extrusion of dietary fiber and milk proteins in expanded corn products. *Lebens-Wiss Technol* 34: 424–429.

Onwulata, C.I., Konstance, R.P., Strange, E.D., Smith, P.W. and Holsinger, V.H. 2000. High-fiber snacks extruded from triticale and wheat formulations. *Cereal Foods World* 45: 470–473.

Onwulata, C.I., Smith, P.W., Konstance, R.P. and Holsinger, V.H. 2001. Incorporation of whey products in extruded corn, potato or rice snacks. *Food Res Int* 34: 679–687.

Pelembe, L.A.M., Erasmus, C. and Taylor, J.R.N. 2002. Development of a protein-rich composite sorghum-cowpea instant porridge by extrusion cooking process. *Lebens-Wiss Technol* 35: 120–127.

Peschel, W., Śanchez-Rabaneda, F., Diekmann, W., Plescher, A., Gartzia, I., Jiménez, D., Lamuela-Raventós, R., Buxaderas, S. and Codina, C. 2006. An industrial approach in the search of natural antioxidants from vegetable and fruit wastes. *Food Chem* 97: 137–150.

Rampersad, R., Badrie, N. and Comissiong, E. 2003. Physico-chemical and sensory characteristics of flavored snacks from extruded cassava/pigeonpea flour. J Food Sci 68: 363–367.

Rao, A.V. and Agarwal, S. 2000. Role of antioxidant lycopene in cancer and heart disease. *J American College Nutr* 19: 563–569.

Rayas-Duarte, P., Majewska, K. and Doetkott, C. 1998. Effect of extrusion process parameters on the quality of buckwheat flour mixes. *Cereal Chem* 75: 338–345.

Roudaut, G., Dacremont, C., Vallés Pámies, B., Colas B. and Le Meste, M. 2002. Crispness: a critical review on sensory and material scienece approaches. *Trends Food Sci Technol* 13: 217–227.

Ryu, G.H. and Ng, P.K.W. 2001. Effects of selected process parameters on expansion and mechanical properties of wheat flour and whole cornmeal extrudates. *Starch* 53: 147–154.

Ryu, G.H. and Walker, C.E. 1995. The effects of extrusion conditions on the physical properties of wheat flour extrudates. *Starch* 47(1): 33–36.

Sacchetti, G., Pinnavaia, G.G., Guidolin, E. and Dalla Rosa, M. 2004. Effects of extrusion temperature and feed composition on the functional, physical and sensory properties of chestnut and rice flour-based snack-like products. *Food Res Int* 37: 527–534.

Sacchetti, G., Pittia, P. and Pinnavaia, G.G. 2005. The effect of extrusion temperature and drying-tempering on both the kinetics of hydration and the textural changes in extruded ready-to-eat breakfast cereals during soaking in semi-skimmed milk. *Int J Food Sci Technol* 40: 655–663.

Scalbert, A., Manach, C., Morand, C., Rémésy, C. and Jiménez, L. 2005. Dietary polyphenols and the prevention of diseases. *Critical Reviews Food Sci Nutr* 45: 287–306.

Schieber, A., Stintzing, F.C. and Carle, R. 2001. By-products of plant food processing as a source of functional compounds-recent developments. *Trends Food Sci Technol* 12: 401–413.

Sebio, L. and Chang, Y.K. 2000. Effects of selected process parameters in extrusion of yam flour (Dioscorea rotundata) on physicochemical properties of extrudates. *Nahrung* 44: 96–100.

Seker, M. 2005. Selected properties of native or modified maize starch/soy protein mixtures extruded at varying screw speed. *J Sci Food Agric* 85: 1161–1165.

Shrikhande, A.J. 2000. Wine by-products with health benefits. *Food Res Int* 33: 469–474.

Singh, B., Sekhon, K.S. and Singh, N. 2007. Effects of moisture, temperature and level of pea grits on extrusion behaviour and product characteristics of rice. *Food Chem* 100: 198–202.

Singh, J., Hoseney, R.C. and Faubion, J.M. 1994. Effect of dough properties on extrusion-formed and baked snacks. *Cereal Chem* 71: 417–422.

Singh, S., Gamlath, S. and Wakeling, L. 2007. Nutritional aspects of food extrusion: a review. *Int J Food Sci Technol* 42: 916–929.

Smith, A.C. 1992. Studies on physical structure of starch-based materials in the extrusion cooking process. In *Food extrusion science and technology*, eds. J. Kokini, C.-T. Ho and M.V. Karwe, 573–618. New York: Marcel Dekker Inc.

Stojceska, V., Ainsworth, P., Plunkett, A. and Ibanoglu, S. 2008. The recycling of brewer's processing by-product into ready-to-eat snacks using extrusion technology. *J Cereal Sci* 47: 469–479.

Stojceska, V., Ainsworth, P., Plunkett, A. and Ibanoglu, S. 2009. The effect of extrusion cooking using different water feed rates on the quality of ready-to-eat snacks made from food by-products. *Food Chem* 114: 226–232.

Stojceska, V., Ainsworth, P., Plunkett, A., Ibanoglu, E. and Ibanoglu, S. 2008. Cauliflower by-products as a new source of dietary fibre, antioxidants and proteins in cereal based ready-to-eat expanded snacks. *J Food Eng* 87: 554–563.

Suknark, K., Philips, R.D. and Chinnan, M.S. 1997. Physical properties of directly expanded extrudates formulated from partially defatted peanut flour and different types of starch. *Food Res Int* 30: 575–583.

Sun, Y. and Muthukumarappan, K. 2002. Changes in functionality of soy-based extrudates during single-screw extrusion processing. *Int J Food Prop* 5: 379–389.

Tahvonen, R., Hietanen, A., Sankelo, T., Korteniemi, V.M., Laakso, P. and Kallio, H. 1998. Black currant seeds as anutrient source in breakfast cereals produced by extrusion cooking. *Z Lebensm Unters Forsch A* 206: 360–363.

Thebaudin, J.Y., Lefebre, A.C., Harrington, M. and Bourgeois, C.M. 1997. Dietary fibers: Nutritional and technological interest. *Trends Food Sci Technol* 8: 41–48.

Tonucci, L.H., Holden, J.M., Beecher, G.R., Khachik, F., Davis, C.S. and Mulokozi, G. 1995. Carotenoid content of thermally processed tomato-based food products. *J Agric Food Chem* 43: 579–586.

Toor, R.K. and Savage, G.P. 2005. Antioxidant activity in different fractions of tomatoes. *Food Res Int* 38: 487–494.

Valiente, C., Arrigoni, E., Esteban, R.M. and Amado, R. 1995. Grape pomace as a potential food fiber. *J Food Sci* 60: 818–820.

Van der Burgt, M.C., Van der Woude, M.E. and Janssen, L.P.B.M. 1996. The influence of plasticizer on extruded thermoplastic starch. *J Vinyl Additive Technol* 2: 170–174.

Vandeven, J.C., Altan, A., Maskan, M. and McCarthy, K.L. 2007. Value-added extruded products: barley and fruit pomace blends. Book of Abstracts, Abstract No. 008–19. IFT Annual Meeting, Institute of Food Technologists, Chicago, IL.

Veronica, A.O., Olusola, O. and Adebowale, E.A. 2006. Qualities of extruded puffed snacks from maize/soybean mixture. *J Food Process Eng* 29: 149–161.

Vincent, J.F.V. 1998. The quantification of crispness. *J Sci Food Agric* 78: 162–168.

Wang, H.J. and Thomas, R.L. 1989. Direct use of apple pomace in bakery products. *J Food Sci* 54(3): 618–620, 639.

Wu, W., Huff, H.E. and Hsieh, F. 2007. Processing and properties of extruded flaxseed-corn puff. *J Food Process Pres* 31: 211–226.

Yağcı, S. and Göğüş, F. 2009. Development of extruded snack from food by-products: A response surface analysis. *J Food Process Eng* 32: 565–586.

Yağcı, S. and Göğüş, F. 2008. Response surface methodology for evaluation of physical and functional properties of extruded snack foods developed from food-by-products. *J Food Eng* 86: 122–132.

Yanniotis, S., Petraki, A. and Soumpasi, E. 2007. Effect of pectin and wheat fibers on quality attributes of extruded cornstarch. *J Food Eng* 80: 594–599.

Yao, N., Jannink, J.-L., Alavi, S. and White, P.J. 2006. Physical and sensory characteristics of extruded products made from two oat lines with different β-glucan concentrations. *Cereal Chem* 83: 692–699.

Yeh, A.I. and Jaw, Y.M. 1998. Modelling residence time distributions for a single screw extrusion process. *J Food Eng* 35: 211–232.

Yuliani, S., Torley, P.J., D'Arcy, B., Nicholson, T. and Bhandari, B. 2006. Effect of extrusion parameters on flavour retention, functional and physical properties of mixtures of starch and D-limonene encapsulated in milk protein. *Int J Food Sci Technol* 41: 83–94.

Zasypkin, D.V. and Lee, T.-C. 1998. Extrusion of soybean and wheat flour as affected by moisture content. *J Food Sci* 63: 1058–1061.

7 Extrusion of Snacks, Breakfast Cereals, and Confectioneries

Mahmut Seker

CONTENTS

INTRODUCTION

The extrusion process that is used for the processing of polymers in the plastic industry is also used for the processing of food materials with a screw rotating in an
extruder and a cutter at the end of a die in the food industry. Extrusion is a complex
process including several process variables that affect several measured feed and
process parameters, which determine the product quality detected with several factors. The complexity of extrusion is due to the presence of several process variables,
measured feed and process parameters, and quality factors and their multiple effects
on each other, so it is not easy to state the effect of one process variable or measured
feed and process parameter. Figure 7.1 shows the effect of one process variable or
measured feed and process parameter on another feed and process parameter or
quality factor in a simple way. Any parameter or factor is affected by one or more
process variable or measured feed and process parameter before the one directional
arrow showing the parameter or factor in Figure 7.1.

After cereals had been cooked with water and then expanded directly in the
extruder die with a flash of superheated water to produce snacks, the extruders were used for cooking cereals that were expanded during frying or baking.
Breakfast cereal can also be produced with the extruder. Not only the cereal processing industry but also the confectionary industry started to use the extruders.
The fortification of extruded products and coextrusion of cereals with cereals or
noncereal products enlarged the variety of products that can be produced with
extrusion. Research on the use of carbon dioxide (CO_2) for the expansion of extrudates, the improvement of the extruder with the design of the turbo-extruder, and
the improvement of postextrusion with applicator driers are advances in food
extrusion.

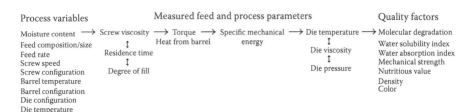

FIGURE 7.1   Process variables, measured feed and process parameters, and quality factors
affecting the extrusion process.

EXTRUDED SNACKS EXPANDED AT DIE

There is not a clear definition of snack foods, but snack food is known as a type of food that is eaten between main meals to reduce hungriness or to get its taste. These are candy, fruit, seed, and cereal based snacks. Cookies, popcorn, corn curls, chips, crackers, doughnuts, masa products, bars, pretzels, bagels, and crisp breads are types of cereal based snacks. Potato chips and baked crackers are known as the first generation snacks. Among those cereal based snack products, corn curls and crisp breads can be produced in the extruder. After the production of first generation snacks, corn curls were produced in a collet extruder and named as the second generation snacks. Corn curls are expanded directly upon exiting the collet extruder die, and dried by baking or frying so they are classified as baked and fried collets.

BAKED COLLETS

Corn grits with moisture content lower than 12% are used mostly in the baked collets after the degermination of corn. Removal of the oily part of corn with degermination is necessary due to the negative effect of oil on the conveying of feed in the extruder. The particle size of the corn grit affects the properties of the extrudates. Coarser corn grits increase the mechanical energy dissipation and then provide less dense extrudates than finer grits do, while the presence of finer grits such as flour affects extrusion negatively. Rice is also extruded with lower moisture content, approximately 8%, or with a longer screw for a desired expansion, but the higher cost of rice limits its usage. Potato solids can also be extruded with corn grits at higher moisture content, approximately 18% (Moore 1994).

Corn grits with low moisture are fed to an ingredient hooper, which is held full for choke feeding of the extruder. Gravity force causes corn grits to drop from the ingredient hooper above the extruder into the collet extruder including the die, barrel, and screw. The barrel and die of the collet extruder are not continuously heated with hot oil or an electrical heater, so it is heated before extrusion is started. The barrel of the extruder can be cooled in case of overheating of feed in the extruder. Corn grits with low moisture content are conveyed with screws in the barrel of the collet extruder and then processed to produce a baked collet as shown in Figure 7.2.

Corn grits are cooked with the dissipation of mechanical energy in the collet extruder that has a short screw (low length to diameter ratio) and small pitch angle. A shallow flight of screw, a small distance between the screw flight and barrel, or a grooved barrel are equipment variables that cause a high shear and the dissipation of mechanical energy during the conveying of corn grits in the collet extruder. The groves in the barrel are opposite to the screw that increases mixing during extrusion, and the hot barrel is maintained by the dissipation of mechanical energy in the collet extruder (Harper 1981).

On the other hand, low moisture content and high screw speed are process variables that cause a high shear and the dissipation of mechanical energy. Increasing the screw speed at a constant feed rate decreases the torque due to the reduced filled length of barrel and melt viscosity but increased the shear rate (Akdogan 1996; Garber, Hsieh, and Huff 1997). On the other hand, increasing screw speed increases

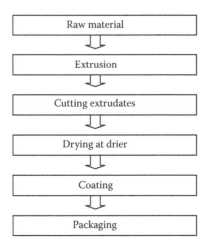

```
┌─────────────────────────────────┐
│          Raw material           │
└─────────────────────────────────┘
                ⇩
┌─────────────────────────────────┐
│            Extrusion            │
└─────────────────────────────────┘
                ⇩
┌─────────────────────────────────┐
│        Cutting extrudates        │
└─────────────────────────────────┘
                ⇩
┌─────────────────────────────────┐
│         Drying at drier          │
└─────────────────────────────────┘
                ⇩
┌─────────────────────────────────┐
│             Coating             │
└─────────────────────────────────┘
                ⇩
┌─────────────────────────────────┐
│            Packaging            │
└─────────────────────────────────┘
```

FIGURE 7.2   Flow diagram for production of baked collets.

specific mechanical energy (SME) due to higher shear and viscous dissipation at a constant feed rate (Lo, Moreira, and Castelll-Perez 1998). Increasing screw speed decreases torque but the effect of increased screw speed is not compensated with a decrease in torque, and the effect of screw speed dominates the effect of viscosity so increasing screw speed increases SME (Jin, Hsieh, and Huff 1994). Increasing screw speed increases the dissipation of SME at a constant feed rate, but it may not increase the SME at an increased flow rate due to a negative effect of an increased flow rate on SME (Seker 2005a).

The temperature of feed in the extruder increases due to the dissipation of mechanical energy. Starch is gelatinized and partially degraded, proteins are denatured, and water is superheated with a high pressure and temperature that are achieved by the dissipation of mechanical energy in the collet extruder. Increasing screw speed increases SME but decreases the residence time and degree of fill so increases of viscous dissipation are balanced with reduced barrel heating, so that the product temperature at die may not change significantly with changes in the screw speed (Chang and Halek 1991). Increasing screw speed reduces viscosity due to increased shear rate, molecular degradation, and SME, which in turn increases melt temperature (Lo, Moreira, and Castelll-Perez 1998). Since increasing screw speed decreases filled length of screw, less dough accumulates behind the die and then the pressure may decrease. Increases in SME and temperature but decreases in viscosity with increased screw speed also decrease the pressure at the die of the extruder (Garber, Hsieh, and Huff 1997; Hsieh et al. 1991).

Die pressure can reach 100 atm and the melt temperature at die exceeds 150°C depending on the process and equipment variables. Upon exiting a die superheated water flashes and expands the extrudates due to the difference in pressures inside and outside of the extruder. The overheating of feed in the extruder reduces the size of the collets. The extending of cooking and expansion of extrudates are also affected by the residence time of feed, the moisture of feed, and the size of the holes in the die of the collet extruder. Decreasing moisture content and size of the holes increase the extending of cooking and expansion of extrudates as increasing screw speed does (Moore 1994).

Extrudates are cut after exiting a die. The shape of a die is circular for corn but rectangular for potato based extrudates. Moisture content of the extrudates is reduced from 12% to 7% due to a flash of water after exiting a die, but the moisture content of the extrudates is still high for a microbial spoilage and the required crispy structure of the extrudates so moisture content of the extrudates are further reduced to 3% by drying in a rotary tumbling drier or perforated conveyer drier used as a baking oven in which the extrudates are moved with a conveyer while hot air at a temperature of approximately 150°C moves above or through the extrudate particles and removes moisture of the extrudates for 5–6 min. Residence time, temperature, and humidity in the drier depend on the moisture content of the extrudates. The residence time of the extrudates in the conveyer drier is adjusted by the speed of the conveyer while the extrudates are dried with air, the temperature of which is held as high as possible to hold more moisture and provide energy saving. The humidity of air is also controlled because air holds a definite amount of water vapor at a definite temperature so saturated air is not circulated.

After drying, the extrudates are coated with oil and flavor additives. Cheese, onion, and garlic are used as flavoring additives. Oil is mixed with flavoring additives in a hot blender and then pumped to be sprayed from a nozzle onto the extrudates in a tumbling drum. Salt is sprayed onto the extrudates upon coating with oil and flavoring additives (Harper 1981).

The coated extrudates are packed to prevent them from breaking in addition to reduce the migration of moisture, oxygen, and contaminating odor into the products. Polyethylene is used commonly as a packaging material due to its low cost, but polypropylene is also used alone or as a laminate with polyethylene as a packaging material.

Collet extruders are cheap, but extruding a variety of cereals, throughput, extruding cereals with higher moisture content, and control of the process and properties of extrudates are limited in a collet extruder. More expensive single-screw cooking extruders with greater throughput, capability to extrude a variety of cereals with higher moisture content, control of the process and properties of extrudates are also used for the production of baked collets.

FRIED COLLETS

Corn meals are moistened so that the moisture content of feed is approximately 18%, which is higher than the moisture content of feed of baked collets. Corn meals can be mixed with water in a blender as a batch system for homogenous distribution of moisture in feed. An uneven distribution of water in feed negatively affects the extrusion and the quality of extrudates. Increasing moisture content of feed decreases torque due to reduced melt viscosity (Harman and Harper 1973; Kirby et al. 1988). SME also decreases with an increase in moisture content of feed due to reduced torque and viscosity.

Moistened grits are introduced into the extruder with a feeder for uniform flow of a raw material in the extruder. Nonuniform flow of a raw material into the extruder causes unstable extrusion and extrudates of poor quality. Since the moisture content of feed is higher than the moisture content of a baked collet, fried collets can be produced with a collet extruder after the modification of the collet extruder to convey moistened feed. Fried collets are also produced in a single-screw extruder. Corn grits are conveyed in the single-screw extruder and then processed for the production of fried collets as shown at Figure 7.3. The length-to-diameter ratio of the single-screw extruders is less than 10 but higher than that of collet extruders. The extruder barrel is heated with an electrical or hot oil heater for the transfer of heat to feed from a barrel that is held at a high temperature depending on the parameters of process and the extruder. Heat is transferred from a barrel to feed by convection that is affected by the mixing of feed.

Upon introduction of raw materials into the extruder, the raw materials are conveyed in the extruder with pressure and screw at a definite residence time. The residence time of all the particles of feed materials is not similar in the extruder due to the friction between the barrel or screw wall and feed material so there is residence time distribution of particles in the extruder. There may be mixing of feed in the screw, and flow pattern is between plug and mixed flow depending on the screw configuration. A narrow spread of residence time distribution is desired for homogenous treatment of the particles in the extruder. Increasing screw speed may not change the flow pattern significantly, but it reduces mean residence time (MRT) and spread of residence time distribution (Seker 2005a).

Feed is cooked with energy coming from a hot barrel and the dissipation of mechanical energy in the extruder. Starch is gelatinized and partially degraded, proteins are denatured, and water is superheated during the cooking of feed in the extruder at high pressure. Product temperature may decrease with an increase in moisture as if SME decreases with an increase of moisture content (Garber, Hsieh, and Huff 1997) or the product temperature may not change with moisture (Della Valle, Tayep, and Melcion 1987). On the other hand, the apparent viscosity, of melt decreases with an increase of moisture content. Since increasing moisture content decreases SME, it decreases temperature that increases viscosity, but a viscosity increase is not observed with an increased moisture content of feed. The direct effect

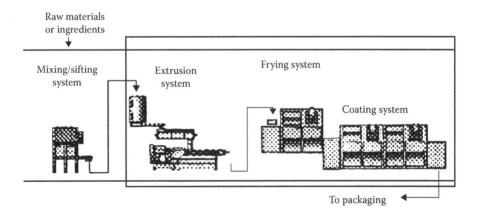

FIGURE 7.3 Flow diagram for production of fried collets.

of moisture content on viscosity with dilution is higher than its indirect effect on viscosity with SME and temperature (Lo, Moreira, and Castelll-Perez 1998). Die pressure decreases as moisture content increases due to reduced viscosity because less viscous material decreases pressure.

Upon conveying of raw materials with a screw in the barrel, they pass through a die. Increasing length and decreasing size of holes in a die increase residence time and heat received from the barrel of the extruder. Extended residence time with reduced hole size at the die can also increase the filled length of screws, torque, and then SME. Not only the length and size of holes in a die but also the number of die and whether the openings are singular or multiple affect shear in a die as if it affects the physical properties of extrudates.

Water in feed material turns into superheated form as a result of an increase in temperature and pressure of which extension depends on the variables of process. Extrudates expand to some degree with flashing off superheated water upon exiting a die due to lower pressure outside than inside of the extruder. Complete expansion is not achieved at the end of extrusion because feed have higher moisture and lower temperature at the exit than feed of baked collets have. After extrusion, extrudates are cut by a cutter with adjustable rotation speed that determines the length of extrudate at the end of extruder.

Moisture content of dough inside the extruder is approximately 18%, but it decreases after extrusion due to flashing off water. Moisture content of extrudates is further decreased to 3% to reduce microbial development and enhance the shelf-life of the extrudates. Moisture content of the extrudates is not reduced in an oven as in the case of baked collets, but it is reduced in a fryer at a high temperature of approximately 200°C for a short time. Frying not only reduces the moisture content but also expands the extrudates to complete their expansion (Moore 1994). Extrudates absorb fat during frying so the calories of fried corn collets increase with the oil absorbed during frying. Fried collets are more dense, crunchy, and expensive than baked collets. The cost of fried collets increases due to the cost of oil used for frying, which is higher than the cost of cereals.

Flavor additives and salt are incorporated into the extrudates with coating after the extrudates are fried. Hot extrudates with oil at their surface are coated with dry flavoring, seasoning, and salt additives (Harper 1981). Upon coating, the products are packed.

Twin-screw extruders can also be used to produce fried collets as if a collet extruder and single-screw extruder are used. A twin-screw extruder increases throughput and provides flexibility against the extrusion of various cereals, the cereals being coarse or fine flour, but it is the most expensive extruder.

FLAT OR CRISPY BREAD

Flour, water, and salt are used with or without leavening agents such as chemical agents of carbonates or biological starter culture of bacteria or yeast to obtain dough for the production of flat bread. The dough of flat bread can be proofed with or without fermentation. If it is fermented, it undergoes a shorter fermentation due to the use of weaker flour and dough development. After molding with rolling and sheeting of the dough, the dough is baked (Faridi 1988).

Crisp bread is produced with flour, water, and salt, which are kneaded while whipped with air. After molding, the dough is baked. The moisture content of crisp bread is further reduced by drying. The baking line in flat or crisp bread production is large and the drying cost is high due to the high moisture content of dough. As an alternative to conventional flat or crisp bread production, extrusion can also be used to produce flat or crisp bread with cooking in the extruder and expansion at the die instead of whipping or leavening with biological starter culture or chemical carbonates.

Wheat is mostly used as cereal in flat or crisp bread, but it is difficult to expand wheat in the single-screw extruder due to the presence of gluten as a protein having a strong network in wheat. The conveying of feed materials in the single-screw extruder occurs by friction that limits transport of feed materials like wheat in the single-screw extruder. But feed materials are transported with the interaction of one screw flight with the other screw in addition to friction in the twin-screw extruders that allow the conveying of feed materials like wheat as if conveying sticky and gummy feed materials. The conveying mechanism in the twin-screw extruder also provides more consistent extrusion of feed including fat than the conveying mechanism in the single-screw extruder does. Besides the advantages of twin-screw extruders, they are 60%–100% more expensive than single-screw extruders for a similar production capacity.

Wheat, salt, and water are mixed to obtain homogeneity depending on the mixing time, and then fed into the extruder with or without sugar, oil, milk powder, or casein. Feed materials are extruded with a moisture content lower than 15% at a temperature higher than 130°C in the twin-screw extruder and then processed to produce crisp bread as shown in Figure 7.4.

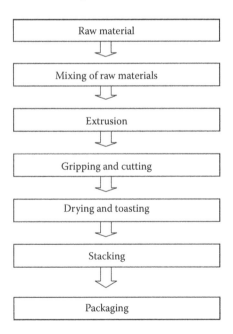

FIGURE 7.4 Flow diagram for the production of crispy bread.

Twin-screw extruders are not fed full, and the feed rate is one of the parameters affecting the degree of fill and residence time in the twin-screw extruder. Increasing feed rate increases the degree of fill and torque but decreases SME due to the greater effect of feed rate on SME than the effect of torque on SME (Lo, Moreira, and Castelll-Perez 1998; Hu, Hsieh, and Huff 1993). On the other hand, increasing feed rate decreases the residence time and spread of residence time distribution and provides more homogeneity (Ainsworth, Ibanoglu, and Hayes 1997; Vergnes, Barres, and Tayep 1992).

Feed material is conveyed in the extruder barrel that is maintained hot with an electrical heater or hot oil flowing in the jacket of the barrel. Feed materials and water are heated in the cooking section of the extruder with heat transferred from a barrel. Increasing barrel temperature may decrease torque due to reduced melt viscosity (Harman and Harper 1973; Kirby et al. 1988). SME also decreases with an increase in barrel temperature due to reduced torque and viscosity. The effect of increasing barrel temperature on MRT may not be observed at varying flow rate, but increasing barrel temperature slightly increases MRT at constant flow rate while it reduces SME (Davidson et al. 1983; Altomare and Ghossi 1986). Increasing barrel temperature causes opposite effects on the energy received from two sources of hot extruder barrel and the dissipation of mechanical energy due to the effect of barrel temperature on MRT and SME. Heat is transferred from a barrel of a twin-screw extruder by conduction and then transferred within feed by convection. Feed materials with water are not only cooked with heat transferred from the barrel in the extruder but also with the dissipation of mechanical energy in the kneading section of the extruder.

The modification of screws with cut flights in a screw provides better mixing and greater SME dissipation. Mixing elements increases MRT of raw materials in the extruder, filled length of screw, torque, and then SME. An increase in SME is not due to an increase in MRT alone because SME can be different at similar MRT in the extruder (Seker, Sadikoglu, and Hanna 2004). Not only flow pattern but also MRT affects the extension of mixing characterized with vessel dispersion number, when mixing in the twin-screw extruder is compared with mixing in the single-screw extruder (Seker et al. 2003b). Increasing mixing elements in the screw increases the spread of residence time distribution, which is reduced with an increase of screw speed so increasing number of mixing elements at higher screw speed homogenizes feed mixture and extends thermal treatment in the extruder (Seker 2004). The reduction of mechanical energy dissipation in a twin-screw extruder is compensated with the presence of reverse flight or lobe-shaped elements in the twin-screw. Barrel segments in addition to screw improve the capacity of the extruder with the configuration of a barrel.

The more positive conveying of two intermeshing twin-screws reduces changes in die pressure so that twin-screw extruders provide better process stability that prevents surging and changes of density, size, and color of product. The use of conical end sections in the twin-screw extruder provides two different flow ways that allow homogenous temperature and pressure before exiting a die.

High mechanical energy is dissipated into feed during extrusion due to the use of wheat as cereal. Proteins are denatured, starch is gelatinized, and water is superheated during extrusion. Product temperature may increase with an increase in barrel temperature (Della Valle, Tayep, and Melcion 1987; Chang and Halek 1991) and raising barrel temperature may decrease the viscosity of the material in the extruder (Lin and Armstrong 1990; Lo, Moreira, and Castelll-Perez 1998). Die pressure increases as barrel temperature decreases due to increased viscosity because viscous material increases pressure. Die pressure also rises with an increase of feed rate due to increased degree of fill of the barrel and buildup of dough mass behind the die (Hu, Hsieh, and Huff 1993).

Upon exiting a die, extrudates expand with water flash due to the pressure difference inside and outside of the extruder. Five percent of moisture in the extrudate can be removed during water flash at exiting a die, and a porous structure is obtained with the expansion of water. Moisture content, protein level, and type affect the expansion and porous structure of flat or crisp bread produced with the extrusion. The shape of the extrudate depends on the shape of die. The use of slit die produces ribbon shaped extrudate, but the use of round die produces finger shaped extrudate. The extruded flat or crisp breads are grabbed, pulled, and then cut. Extrudates are conveyed to an oven at a temperature of approximately 180°C to increase the flavor by toasting and to reduce the moisture in extrudates to 3% (Meuser and Wiedman 1989). Upon cooling, the extrudates are packaged.

Hollowed tubes or U-shaped extruded crisp breads can be produced with U-shaped die and then they can be filled with cheese to increase the nutritional value. Extruded crisp breads can also be enrobed with sweet products like chocolate to increase the types of products. Fibers are also incorporated into extruded flat or crisp bread to enhance the nutritional appeal of the product. Increasing fiber content increases extrudate density, which can be decreased by increasing starch content of feed. Low density and light structure of crisp bread draw the interest of the consumers as a healthy light product.

Feed material are mixed, kneaded, and cooked in one machine and then flat or crisp bread is produced with extrusion at shorter time and lower energy and cost. Extrusion also provides usage of whole grains rather than flour that is used in conventional flat or crisp bread production.

EXTRUDED SNACKS EXPANDED IN A FRIER OR OVEN

Extruded snacks can be cooked and shaped in the extruder and then expanded indirectly during drying in a fryer or oven rather than being expanded directly upon exiting a die of an extruder. Those snacks expanded indirectly are called third generation, semi or half products of snacks. Production processes of indirectly expanded snacks with extrusion are similar to the processes that are used for production of directly expanded snacks with extrusion, but they include some exceptions. A wide range of raw materials can be fed and cooked for production of indirectly expanded snacks in the preferred twin-screw extruder. Whole meal is used for the production of hard and crunchy structured extrudates, but starch is used for the production of soft and light structured extrudates. The mechanical energy used for the extrusion of whole meal is greater than the mechanical energy used for the extrusion of starch for a similar density of extrudates. Extruded snacks expanded in a fryer or oven can be produced as pellets or chips, depending on the shape of product, shaping process, moisture content, and degree of cooking of feed in the extruder.

EXTRUDED SNACK PELLETS EXPANDED IN A FRYER OR OVEN

A high level of starch (60%) is used for higher expansion and desired crunchiness and texture. Increasing the starch content above 60% in feed causes more expansion and results lighter and softer product. A high percentage of proteins can also be incorporated into the extruded snacks expanded in a fryer or oven. Shortenings, emulsifier, and vegetable oil are used as ingredients to reduce stickiness and provide more uniform cellular structure and expansion. The use of salt provides uniform moisture migration and equilibration after drying.

Dry ingredients of extruded snack pellets can be mixed in a blender of a batch system and then liquid ingredients like oil and water are sprayed in the batch system or injected into a preconditioner in a continuous system. The preconditioner can be used before extrusion for mixing, hydration, moisture equilibration, and flavor development. Water can be added in the form of liquid or steam that provides thermal energy. Since the production of steam as an energy source is cheaper than an electrical source, the use of a preconditioner reduces the heating cost of feed materials in the extruder. The use of a preconditioner also improves texture, increases flow rate, and reduces the cost of extrusion due to lowered wear. Feed materials are held at the preconditioner for several minutes at a temperature of approximately 100°C.

Raw materials with moisture content between 25% and 30%, which is higher than the moisture content in directly expanded snacks, are introduced into the extruder feeder. A flow diagram for the production of extruded snack pellets expanded in a fryer or oven is shown in Figure 7.5. Increasing moisture content

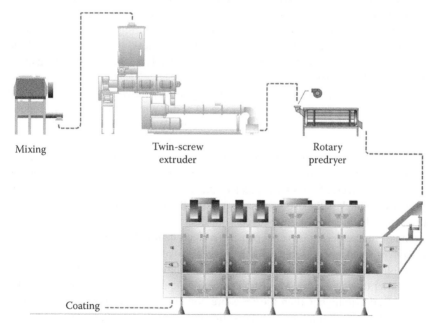

Mixing   Twin-screw   Rotary
                       extruder   predryer

Coating ------------

Finish drying and tempering

FIGURE 7.5 Flow diagram for production of extruded pellet expanded in a fryer or oven.

of feed reduces the viscosity of feed and dissipation of SME, which decrease the temperature and increase the viscosity at a die due to lowered viscous dissipation. The effect of moisture content on MRT of the particles in the extruder, which is an important parameter affecting the quality of the extrudate by its effect on molecular degradation and rheology of melt, depends on the extent of two opposite effects of moisture content on rheology of melt mentioned above (Seker 2005a).

Raw material with low density is compressed in the feeding zone of the extruder, and then further compressed in the kneading section with kneading screws including several flight or cut flight screws that increase residence time. Varying pitches also increase the residence time in the extruder. The density of feed increases and the granular structure starts to disappear while the degree of fill, temperature, and pressure increase. Particles start to agglomerate and form a melt, which is compressed and texturized in the cooking zone of the extruder in which shear rate, temperature, and pressure reach to maximum. Raw materials are cooked to gelatinize starch during extrusion with proper moisture content, temperature, and residence time in the extruder. Feed materials of extruded snack pellets are cooked at a barrel temperature between 90°C and 150°C in the cooking section of the extruder as in the case of direct expanded snack extrudates, but some part of the water in feed is vented to reduce moisture content and to cool and increase density of feed in the forming section of the extruder that is held at a temperature between 70°C and 90°C. The forming section of the extruder includes a screw with low shear configuration. Extruded snack pellets, which are cooked and formed in one extruder, are called half products. Cooking and forming of feed can also be conducted in two extruders, a cooking and then a forming extruder, to produce more complex shaped extrudate. Extruded snack pellets, which are cooked and formed in two extruders, cooking and forming extruders, are called third generation snacks (Harper 1981).

Upon cooking and then cooling feed during conveying with a screw in a barrel, cooled melt passes through a die. The use of a die with greater opening reduces pressure and expansion of extrudates leaving a die. Water of feed material is not in superheated form at the exit of the extruder as in the case of directly expanded snacks because temperature and pressure are lower at the exit of the extruder. Since temperature and pressure of melt of extruded snack pellets are lower than the temperature and pressure of feed of directly expanded snacks, water of melt does not expand the extrudates of snack pellets so that they conserve the shape of a die. Complex shaped extrudates, which cannot be produced with direct expansion due to the destruction of complex shape during direct expansion of feed material, are produced during the extrusion of snacks pellets expanded in a fryer or oven.

Extrudates are cut by a cutter with adjustable rotation speed to obtain their desired lengths after they exit a die. Moisture content of the dough of extruded snack pellets in the extruder is higher than the moisture content of dough of directly expanded snacks, and it is not reduced after exiting the die due to the absence of water flashing. Extrudates of snack pellets are dried with humid air below 95°C over a longer time (several hours) due to the dense structure of extrudates. Reducing the moisture content of feed below 12% decreases microbial development and enhances the shelf-life of the extrudates. Residence time, temperature, and humidity in the drier depend on the moisture content in the extrudates. Equilibration of water for 1 to 2 days after drying provides better expansion of snack pellets during frying (Moore 1994).

Extruded snack pellets with high density and transportation stability are sold to small snack producers that expand the pellets with an additional process of frying at 170°C–210°C and for 10–60 s. Heat is transferred from hot oil to cooler extrudate, the temperature of water in extrudates increases, water evaporates and then expands the extrudates during frying. Frying of extrudates not only dries and expands extrudates but also causes migration of oil to the extruded snacks pellets. Screw speed during extrusion and the temperature of melt in the extruder affect oil absorption of the extrudates during frying (Osman, Saai, and Jackson 2000). As an alternative to frying, extrudates can also be dried and expanded with rapid heat transfer in hot air impingement ovens or microwaves with high power for the production of extruded snack pellets with low oil content. Extruded snack pellets with high density are also expanded with frying or hot air impingement ovens at the factory in which they are produced.

Seasonings, flavor additives, and salt can be incorporated into the extrudates with coating after frying. Extrudates are covered with flavoring and seasoning additives in a drum. If they are dried and expanded by frying, then they are coated with dry flavoring and seasoning in a drum. Oil is sprayed onto the extrudates if they are dried and expanded in hot air impingement ovens or microwaves with high power, and then the extrudates are coated with seasoning and flavor additives. Extrudates are packaged after coating.

EXTRUDED SNACK CHIPS EXPANDED IN A FRYER OR OVEN

Potato chips are produced by slicing the potato and then frying the potato slices. Extruded snack chips, which resemble potato chips, are produced with various cereals as raw material. After the raw materials are mixed with ingredients in the blender, water is added to the mixture to increase the moisture content to approximately 35% and then the mixture is conveyed in the extruder and processed to produce snack chips that are expanded in a fryer or oven as shown in Figure 7.6. Moisture content of the raw materials of extruded snack chips is higher than the moisture content of raw materials of the extruded snack pellets. Raw materials are fed to the extruder as in the case of extruded snack pellets, but they are less cooked than pellets are while starch in feed materials is partially gelatinized by higher moisture and lower energy received by feed during extrusion (Moore 1994).

Extruded snack chips are in the form of sheets. Extrudates can exit a rectangular die or are passed through sheeting rolls to obtain chips. Upon cutting, they are not dried in an oven so they are fried to remove water, cook, and expand after extrusion. The oil content of the extrudates increases upon frying. As an alternative to frying, extruded snack chips can also be baked in an impingement oven to reduce moisture content and expand extrudates. Drying of extrudates in an impingement oven also reduces the oil content of the extrudates. Dried chips are coated with seasoning and flavor additives and then packaged.

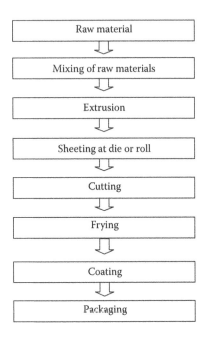

FIGURE 7.6   Flow diagram for production of extruded chips expanded in a fryer or oven.

PRODUCTION OF BREAKFAST CEREALS WITH EXTRUSION

Grains like corn, wheat, rice, or oats are cooked to produce breakfast cereals, which are consumed as a ready-to-eat meal with milk. Cooking cereals gelatinizes starch, converts cereals into edible forms, and provides development of flavor and texture. There are several types of breakfast cereals mainly flaked, expanded, and shredded types, which are produced with processes of cooking, tempering, drying, flaking, shredding, toasting, and coating.

FLAKED BREAKFAST CEREALS

Grain kernels of wheat, corn, rice, or oats being used for the production of flaked breakfast cereal are screened first to remove undesirable grains. After sizing and screening, whole grain kernels or part of whole grains as grits, and flavor additives like sugar, malt syrup, and salt, are cooked for the production of flaked breakfast cereals. Grains, which are cooked for gelatinization and fluidity of grains that prevent fracturing of grains during flaking, can be cooked with boiling water cookers that work at low temperatures of approximately 95°C, but it takes several hours to cook grains in boiling water cookers. Grains are not disrupted during cooking with hot water so the integrity of whole grain is maintained after cooking. Another type of cooking can occur with steam at a higher temperature of approximately 130°C, and in less time as an hour in a rotating steam cooker. Cooking time depends on the pressure inside the cooker and the moisture of grains that are increased with condensed steam during cooking. Moisture content of feed can reach 30%–50%, but the use of a superheated steam prevents the condensation of steam and moisture increase in grain. Cooking at higher temperature and pressure with steam reduces the cooking time to less than the cooking time in a boiling water cooker. Moisture content of cooked grains should not exceed 30%, depending on the grain type. The cooked grains are dumped and then delumped to separate the lumped grains into single grains (Fast 1990).

Grains are cooked with the heat of conduction and convection in boiling water or steam cookers. Extruders can also be used as an alternative for cooking grains in a shorter time with the conversion of mechanical energy, the degree of which depends on the type of extruder used for the production of flaked breakfast cereals. Cooking of grains at lower moisture content with extrusion also reduces drying load, requires less energy, and provides lower processing cost and flexibility for feed materials whether grits or not. Grain kernels or part of whole grains as grits are blended with flavor additives like sugar, malt syrup, and salt in the preconditioner. Preconditioning with steam before extrusion reduces mechanical energy dissipation and extruder wear. Moisture content of feed materials are increased to 26% and then processed to produce flaked breakfast cereal with extrusion as shown in Figure 7.7.

Moistened feed materials are cooked in the barrel at a temperature of approximately 140°C and converted into dough. Feed materials are formed after cooking during extrusion. Cooking and forming can be conducted in two extruders separately as cooking and forming extruders, and single-screw extruders are used for both cooking and forming. Cooking ruptures the granule, releases, and gelatinizes starch. Feed materials also undergo disruption with shear during conveying with a screw in the extruder, so not only cooking but also shearing disrupt and gelatinize starch during extrusion. After cooking, feed material is formed with low SME dissipation in a short forming extruder with L/D being smaller than 10 at low speed and deep flights of screw while heat is removed from feed by the circulation of cool water around the barrel (Harper 1981). Mass is compressed without excess shear in the forming extruder by adjusting the screw pitch and channel depth. High shear and low moisture degrade starch extensively and cause fast milk absorption and loss of the crispy structure during the consumption of breakfast cereals.

Twin-screw extruders can be used for lower shearing during the extrusion of flaked breakfast cereals. Cooking and forming can be conducted in one extruder in different sections of the twin-screw extruder. Feed material is cooked and steam is vented at the end of the cooking section of the extruder. Cooled feed is formed in the following section of the long twin-screw extruder with the L/D ratio being greater than 25. The forming section is held cooler than the cooking section of the twin-screw extruder, and cooked materials are processed at the screw speed of cooking during forming of cooked materials if cooking and forming are conducted in one extruder. If a forming extruder is used after cooking with a twin-screw extruder, then a shorter twin-screw extruder with L/D being shorter than 20 is used. Extrudate leaves a die of the forming extruder or the forming section of the extruder at a temperature below 100°C so there is not significant vaporization of water as extrudates exit a die of the forming extruder or forming section of the extruder as compact

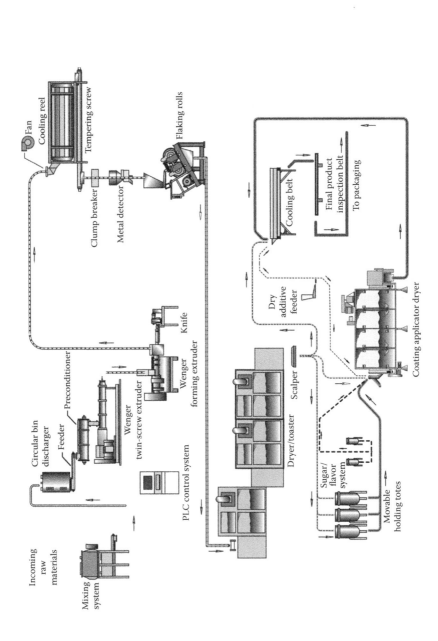

FIGURE 7.7 Flow diagram for production of flaked breakfast cereals. (Printed with permission of Wenger Manufacturing Inc., Sabetha, Kansas.)

pellets with a moisture content of approximately 23%. A die of extruder is designed in such a way that extrudates do not significantly expand upon exiting a die (Miller 1994).

Extruded pellets are cooled to a temperature of approximately 50°C in a rotating reel while air passes over the pellets and reduces the surface moisture in pellets. If moisture content in cooked pellets is high, then it can also be reduced to approximately 18% by drying at a higher temperature. Moisture in dried pellets is not uniform through the whole of the pellet. A dry surface due to lower moisture near the surface prevents proper flaking so dried and cooled products are tempered for the equilibration of moisture between the center and surface of the pellets in a tempering bin. The temperature of the pellets is held at approximately 50°C for proper flaking that prevents formation of clusters. A scalper or vibratory feeder is used to prevent clusters being fed into the flaking section.

The tempered grains are passed a small distance between a pair of counterrotating rolls at a temperature of approximately 45°C to produce thin flakes. The slight roughness of the roll surface and the hard or dry surface of the pellet provides friction and flow between the rolls without fracture of the pellet. Gelatinization, moisture, and strength of the pellet also provide flow between rolls and integrity of the pellet during passage through the rolls. High moisture causes sticky flakes that cannot leave the rolls easily. After the pellets pass through the rolls for flaking, the resulting flakes curl up and turn back to pellet shape so they have curl to flake shape (Fast 1990).

Crispness and blister of flakes are provided by drying and toasting of flakes. Flakes are held at a temperature of approximately 150°C to reduce the moisture content further and then held at temperatures of 200°C–300°C, depending on cereal type for toasting for several minutes (Miller 1994). Flakes are toasted in a rotary toasting oven with hot air or in a conveyorized oven with an electric heater or hot air. Products are cooled upon toasting.

Products can be coated with syrup and flavor additives. Sugar is moistened with a low percentage of water and then heated to high temperature under pressure. Hot and moistened sugar is pumped to be sprayed through a nozzle onto extrudates. Replacing some part of sucrose with invert sugars reduces the crystallization of sugars. The coated extrudates are re-dried after coating. Flavor additives are added after extrusion but it has drawbacks. If they are added before or during extrusion, their instability and volatility cause problems. The use of encapsulated flavors seems to be an alternative. Flavor substances must be encapsulated in temperature and shear resistant capsules (Yuliani et al. 2004). Dry additives like nuts and dried fruits can be added and then the final products are packaged to be transported to stores.

The reduction of starch degradation increases the bowl life of breakfast cereals and reduces an undesirable powdery mouthfeel of extruded pellets while the reduction of water absorption increases the shelf-life of breakfast cereals with enhancement of the crispy and crunchy structure. Starch with high amylose content causes harder and denser extrudate that absorb milk slowly, but starch with high amylopectin provides expanded extrudate. On the other hand, the modification of starch with cross-linking reduces both water absorption and solubility index of starch extrudates (Seker and Hanna 2005, 2006).

Expanded Breakfast Cereals

Grains are puffed to produce expanded breakfast cereal. Two types of puffing are used as conventional puffing: oven puffing and gun puffing. Rice or corn is expanded in an oven puffing system. Grains are mixed with sugar, salt, and malt and then cooked with steam to gelatinize the grains while the grains absorb moisture and their moisture content rise to approximately 30%. Cooked grains are dried to reduce the moisture content to approximately 20% and then tempered for several hours for the equilibration of moisture inside the grains. The tempered grains are passed through rolls without flaking to provide fissures required for expansion. After a second drying to 10% moisture and the equilibration of moisture inside the grain, cooked and dried grains are conveyed to an oven at a high temperature of approximately 300°C. A great temperature difference between the grain and environment causes rapid generation of steam inside the grain and then provides the expansion of grains by puffing. A crispy structure is achieved with the toasting of expanded product (Fast 1990).

Rice or wheat is used as grain for the production of another type of expanded breakfast cereal produced with gun puffing. Bran of rice or wheat is removed by pearling and then separated from the kernel with air. A saturated salt solution can also be sprayed onto wheat to toughen the bran being removed easily during the puffing of grain. Grain is introduced into a gun, which is preheated to a high temperature of approximately 250°C with hot flames of gas burners. The gun is closed after grain loading. Grains and their water are heated during rotation and by heating of gun with hot flames of gas burners while pressure reaches approximately 200 psi (1000000Pa). Opening a lid of the gun releases pressure, turns moisture of grains to steam, and then puffs grains. Pressure difference between the gun and atmosphere occurs as a lid is opened and grains are puffed. The puffed grains with moisture of approximately 6% are dried to 3% moisture and then packaged (Fast 1990).

Expanded breakfast cereals can also be produced with extrusion as in the case of expanded snack production, and a flow diagram for the production of expanded breakfast cereals with extrusion is shown in Figure 7.8. Raw materials can be grain or flour that increase nutritional value and flexibility of the raw material as the extruder is used for the production of expanded breakfast cereal. Changing the extruder and processes provides a variety of shaped and textured products. Cereals, salt and sugar are mixed in the blender and then fed into the preconditioner to increase the moisture content to approximately 20% and preheat the mixture. The mixture is then fed into the extruder. Expanded snacks are cooked by the extrusion of cereals and water without sugar, with energy coming from the dissipation of mechanical energy, but expanded breakfast cereals are produced with cereals, water, salt, and sugar that compete with starch for water. The reduction of water activity with salt and sugars negatively affects the gelatinization of starch. Lowered gelatinization requires extra energy that can be supplied with energy coming from a barrel. Increasing residence time in the extruder increases energy coming from the extruder barrel (Miller 1994). Amorphous feed is converted into melt dough and it is compressed with a screw including decreasing pitch and shear locks. The mixture is cooked with heat from SME dissipation and the barrel at a temperature of approximately 150°C. Starch undergoes gelatinization and some degree of degradation during extrusion for the

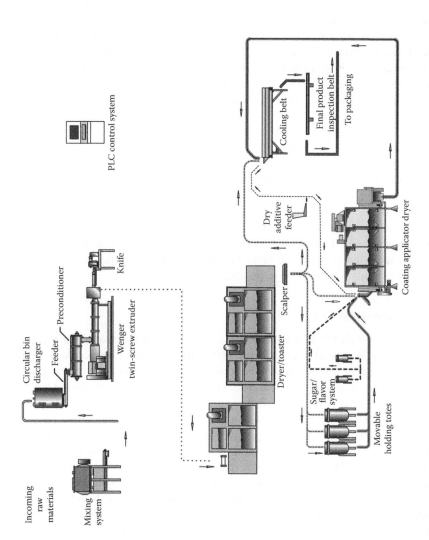

FIGURE 7.8 Flow diagram for production of expanded breakfast cereals. (Printed with permission of Wenger Manufacturing Inc., Sabetha, Kansas.)

expansion of extrudates but sectional expansion is not linearly correlated to starch degradation observed with SME dissipation (Seker, Sadikoglu, and Hanna 2004). A single-screw extruder can be used for the production of expanded breakfast cereals but replacement of a single-screw extruder with a twin-screw extruder provides an extruder configuration giving less starch damage and lower milk absorption.

Increases of temperature and pressure convert water into superheated water that flashes and expands extrudates as extrudate exit from the extruder die to the lower pressure of the atmosphere. Extrudates are shaped in the die according to the shape of die, but elasticity of melt also affects the shape of extrudate (Miller 1994). Extrudates are cut with a cutter after exiting a die. Moisture content of the extrudates at the die is approximately 15% for expanded breakfast cereals. Puffed breakfast cereals are less dense than flaked breakfast cereals but denser than puffed snack products.

Moisture content of the extrudates is further decreased to 3%–4% by the drying of extrudates, and then they are toasted. Extrudates can be coated or not. If extrudates are coated with syrup and flavor additives, then they are re-dried. Coating extrudates reduces absorption of milk with expanded breakfast cereal and provides a crispy structure. If dry additives are mixed with the extrudates, then the extrudates are packaged after the addition of dry additives.

The production of puffed breakfast cereals with extrusion shortens the process time, compared with the process time of oven puffing, provides continuous process compared with gun puffing, and gives flexibility for feed material being grain or flour and flexibility for the shape of products being varied.

SHREDDED BREAKFAST CEREALS

Wheat is used mostly for the production of shredded breakfast cereals. Wheat grains are cooked for 30 min at a temperature below 100°C in hot water cookers while absorbing 45%–50% water. Water in the cooker is removed after cooking and grains are conveyed to the cooling section. After cooling, the grains are tempered for the equilibration of water in grains and firming of grains. Cooked, cooled, and tempered grains are shredded by passing through a series of counterrotating paired rolls, of which one is smooth and the other is grooved. The temperatures of the rolls are held at 40°C, and one conveyer below several paired rolls collects shreds coming from several paired rolls. One shred comes from a pair of rolls and several shreds coming from several counterrotating paired rolls are layered in such a way that each layer is at the top of the other layer to form one piece on a conveyer below the rolls. A piece includes 10–20 layers. Cutting layered shreds with a dull cutter compresses the edges of layers and sticks the shreds. Shreds at 45% moisture are baked at a temperature of approximately 270°C for final moisture of 4%. Shrinkage and puffing of shredded products occurring during baking provides pieces thinner at the edges while thicker at the middle (Fast 1990).

Extruders can also be used to cook grains for a shorter time and at lower moisture content than a hot water cooker for the production of shredded breakfast cereals as shown in Figure 7.9. Usage of the extruder for the production of shredded breakfast cereals provides flexibility for type and form of grain being whole or flour. Grains

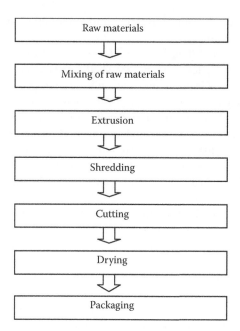

FIGURE 7.9 Flow diagram for production of shredded breakfast cereals.

are mixed with water to increase the moisture content to 25%–30% and then cooked in the extruder. After extrudates exit a die, they are shredded with shredding rolls. Since moisture of extrudates is lower, there may be shear in the rolls, so the rolls must be cooled. Shredded extrudates are cut and then baked. Shreds can also be formed and expanded in the extruder die so that both expanded and shredded product is obtained with extrusion. Shredded extrudates are cut while they expand so that a shredded product with the desired length is obtained. The production of shredded breakfast cereals with the extruder also reduces drying load in baking used in the conventional process (Miller 1994).

PRODUCTION OF CONFECTIONERIES WITH EXTRUSION

Liquorices, toffees, fudges, jellies, and chocolate are types of confectionery including sugar, cereal flour, starch, fat, cocoa, and milk powder as ingredients. Liquorices is manufactured by dissolving liquorices extract, sugar, starch or flour, gum Arabic or gelatin in water and then heating the mixture to a temperature of 135°C in a stirred kettle with a heating or cooling jacket or scrapped surface heat exchangers. The heated mixture is molded to produce liquorices and then dried. On the other hand, toffees are produced by boiling sugar, butter, and flour up to a temperature of 150°C in a kettle or heat exchanger to get a hard structure. The mixture is molded and then cooled.

There are also confectionery products that do not include starch or flour. Mixing and then heating sugar, butter, and milk to a temperature of 115°C in a kettle or heat exchanger produces fudge upon beating the mixture during cooling. Jelly beans, which are another confectionery not including flour, are produced by cooking corn syrup in a kettle or heat exchanger, pouring it to a starch tray, and then cooling and shelling with sugar panning.

In addition to sugar and flour based confectioneries, there are also cocoa based confectionery products. After removing the shell, cocoa beans are roasted and then crushed into cocoa nibs. Grinding the cocoa nibs provides cocoa liquor. Some part of cocoa liquor is separated into cocoa powder and cocoa butter that is used for chocolate production. After mixing and grinding of sugar, cocoa liquor, and cocoa butter, the mixture undergoes heating and mixing at a high temperature that is called conching. The conched mixture is tempered, molded, and then packaged.

There are not only dense confectioneries but also aerated confectioneries that are produced with aerating sugars, whipping agents, and stabilizing agents by beating the mixture to incorporate air.

Proteins undergo a Maillard reaction, degradation, and gelling, and sugars crystallize while starch gelatinizes and fats undergo polymorphism during the production of confectioneries. Confectioneries are produced by mixing, kneading, and cooking ingredients of related confectionery product at atmospheric pressure and temperature. Conducting several processes in one piece of equipment not only reduces the number of operations and equipments but also the space occupied by equipments in conventional confectionary technology. Conventional cooking, mixing, and kneading in the batch system of confectionery production can be replaced with continuous cooking, mixing, and kneading in the extruder.

EXTRUDERS OF CONFECTIONERIES

An extruder provides cooking, mixing, and kneading at lower moisture content, shorter time, and higher pressure. Cooking with lower moisture content reduces the cost of drying of confectionery manufacturing. Including venting and cooling zones in the extruder configuration reduces feed material temperature and moisture content. Inversion can be lowered while crystallization, gelatinization, gelling, and degradations can be better controlled during the extrusion of confectionery products. Confectioneries can also be aerated by injection of gas during extrusion to produce light product (Best 1994).

High viscosity and throughput cause pressure flow, so feed should be compressed near to a die rather than an inlet to reduce pressure flow and increase throughput. The presence of sugars and fats in confectionery products decreases torque and power. Since energy is provided from a barrel and mechanical energy dissipation, barrel length, degree of fill, and residence time should be increased to enhance energy received during extrusion. The use of an intermeshing twin-screw extruder is advantageous to reduce sticking and plugging problems.

MANUFACTURING OF CONFECTIONERY PRODUCTS WITH EXTRUSION

Liquorices can be manufactured by feeding, mixing, and cooking liquorices extract, sugar, starch, gum Arabic or gelatin, and water in the extruder. Extruding liquorices at low moisture reduce drying load as it reduces processing time. On the other hand, the production of toffees in the extruder with sugar, butter, and flour may not provide browning but reduces energy consumption of the toffee production processes.

The production of confectionery products like fudges, which do not include starch or flour but including sugar, butter and milk, with extrusion provides control of crystallization and cooling of product. Jelly beans, which are another confectionery product not including flour, are produced by extrusion that cook ingredients with low moisture so that drying load is reduced.

Cocoa beans are roasted at a temperature between 130°C and 150°C and then crushed into cocoa nibs after shell removal. Cocoa nibs can be ground and then cocoa liquor is pasteurized in the extruder. After mixing and grinding of sugar, cocoa liquor and cocoa butter, the mixture undergoes heating and mixing at a high temperature that is called conching in the extruder. The use of the extruder shortens chocolate production time. Continuous chocolate production with extrusion also eliminates the sedimentation of liquor appearing in the batch system (Best 1994).

Gums and chewing gums can also be produced by the extruder that reduces production time, energy consumption, and resting time for the production of gums and chewing gums with conventional production. Flavors of chewing gum can be better distributed and preserved with the extrusion process.

FORTIFICATION OF EXTRUDED SNACKS

Extruded snacks are produced with cereal based materials like corn, wheat, rice, or oats that include 2%–12% protein and 60%–70% starch. The fortification of extrudates with protein, meat, and vitamins increase their nutritious value but the fortification of extrudates with fiber, nuts, and antioxidant not only increases their nutritious value but also health aspects of extruded foods.

PROTEIN FORTIFICATION OF EXTRUDED SNACKS

Protein content of cereals and cereal based extruded snacks is low, and proteins of cereals and cereal based extruded snacks do not include amino acids like lysine and tryptophan. The incorporation of proteins into extruded snacks not only increases protein content and complements amino acid composition of extruded snacks but also affects physical properties of extrudates negatively. The addition of 10% whey protein (WP) to feed of extrudates does not change radial expansion and density of extrudates but increasing the WP level to 20% reduces radial expansion and increases the density of extrudates (Kim and Maga 1987). The increase of breaking strength with the addition of high level of WPs is ascribed to the interaction of proteins that reinforce cell walls and increase breaking strength (Onwulata et al. 1998). The extrudate expansion ratio also decreases as levels of nonfat dry milk or milk protein raffinates (MPR) are higher than 5% in extrudates while the breaking strength of extrudates increases with the incorporation of MPR into the feed of the extrudate (Singh et al. 1991).

Increased SME is desired for expanding cereal products, but SME is reduced as whey protein concentrate (WPC) and sweet whey solids (SWS) are incorporated into cereal based extrudates. Since increased SME is desired for the expansion of extrudates, reducing the moisture and adding reverse screw elements can be applied to increase SME and product expansion. The negative textural indicators associated with the inclusion of whey products can also be improved by adding wheat bran or fiber that improve SME along with product quality characteristics (Onwulata et al. 2001a). The effect of proteins added into feed on the expansion of extrudates is also ascribed to phase transition occurring with increased protein content of feed, but phase transition is not just one factor that can affect expansion of extrudates fortified with proteins. Melt elasticity is also considered as a factor that can affect the expansion of extrudates, and the type of starch being native or modified starch extruded with protein affects expansion of extrudates (Seker 2005b).

Legumes such as chickpea, white bean, pea, and lentil are also sources of protein and they can also be incorporated into extrudates as pure proteins are incorporated into extrudates, but the reduction of expansion and increase of density and breaking strength are observed with the incorporation of legumes like chickpea, white bean, pea, and lentil into extrudes (Singh, Sekhon, and Singh 2007; Bhattacharya and Prakash 1994; Lazou et al. 2007).

INCORPORATION OF MEATS INTO EXTRUDED SNACKS

Restructuring various types of meats is one of the interests of the meat industry. The disruption of muscle fibers negatively affects textural properties of meats so nonmeat materials are used to improve the textural properties of restructured meat products. Extrusion is considered for compression and texturization of meat particles. Fish, chicken, and beef are used as meat sources for restructured snack type extrudates. The incorporation of 6% tuna flesh into corn slightly increases expansion and brittleness while it decreases hardness of extrudates (Kim and Kim 2004). As the chicken meat content is increased in potato extrudates, fat and protein content increases while shear force and expansion of extrudates decrease due to the decrease in size and number of air cells in extrudates (IngJenq, Camire, and Bushway 1997). The incorporation of fat with raw beef into soy flour–corn starch mixture decreases expansion ratio and increases bulk density of extrudates (Park et al. 1993).

VITAMIN FORTIFICATION OF EXTRUDED SNACKS

Vitamins, being fat or water soluble, are one of the vital components of foods for continuity of healthy living. Natural foods include vitamins in small quantities, but thermal processing of foods causes losses of vitamins. Barrel temperature, screw speed, moisture of feed, and die diameter affect the retention of vitamins in feed during extrusion. B2, B6, B12, niacin, Ca-pantothenate, and biotin are stable against extrusion, but vitamin A, vitamin E, vitamin C, B1, and folic acid are sensitive to extrusion (Riaz, Asif, and Ali 2009). Since feed materials undergo thermal treatment during extrusion, vitamins are added to extrudates during coating after extrusion. Vitamins can also be added to feed mix in coated or crystal form, but the coated form of ascorbic acid, menadione, pyridoxine, and folic acid are more resistant to extrusion than the crystal form of them (Marchetti et al. 1999). Microencapsulation of vitamins also stabilizes vitamins, and incorporation of microencapsulated vitamins decreases vitamin losses during extrusion (Desai and Park 2005).

FIBER FORTIFICATION OF EXTRUDED SNACKS

Nutritional studies indicate that dietary fiber intake lowers the risk of serum cholesterol, coronary heart disease, cancer, blood pressure, obesity, and gastrointestinal disorders. Since foods supplemented with dietary fiber can lower the risk of the diseases mentioned above, the incorporation of dietary fiber into extruded snacks is desirable for healthy life. Sugar beet and fruits in addition to bran of wheat, oats, corn, and soybean include dietary fiber. The incorporation of fibers into extruded snacks increases bulk density and breaking strength of extrudates. Bran reduces the extensibility of cell wall and causes rupture of steam cells so that the formation of large cells and bubble expansion are prevented (Guy and Horne 1988). Increase of axial expansion is explained with the increase of extrudate temperature, and the reduction of radial expansion is ascribed to reduced elasticity and plasticity of dough while extrudates become darker as oat and wheat fibers (20%) are incorporated into extruded corn snack (Hsieh et al. 1989).

INCORPORATION OF NUTS INTO EXTRUDED SNACKS

Nuts like walnuts and hazelnuts are a source of protein and oil, and they draw attention due to their cholesterol reducing properties (Mukuddem-Petersen, Oosthuizen, and Jerling 2005). The incorporation of nuts into snacks improves the nutrition of snacks not only by reducing the cholesterol level but also by improving the protein content and amino acid composition of snacks. Increasing partially defatted hazelnut content or defatted peanut flour reduces the radial expansion index and hardness while it increases bulk density of extrudates (Yagci and Gogus 2009; Suknark et al. 1997). Extruded products with high lipid content show a considerable loss of fat during the extrusion and roasting processes. The presence of sugar esters in extrudate composition causes more oil to be retained than the presence of monoglycerides, diglycerides, and lecithin does during the extrusion of wheat flour and almond mixtures. Both single- and twin-screw extruders can be used for the incorporation of nuts into snacks during extrusion, but the twin-screw extruder is more efficient to extrude fatty feed than the single-screw extruder because it leads to a lower loss of fat during extrusion and a greater expansion of the products (De Pilli et al. 2004, 2007).

ANTIOXIDANT FORTIFICATION OF EXTRUDED SNACKS

The oil content of directly expanded snacks produced by baking may not be high but the oil content of third generation snacks and directly expanded snacks produced by frying is high. Oil is also used during coating snacks. Oil undergoes oxidation and reduces the quality of foods due to the undesirable taste of oxidized oils. The incorporation of antioxidant to extrudates prevents oxidation of oils in extrudates. Antioxidants not only prevent oxidation but also scavenge reactive oxygen species (ROS) and reduce cell proliferation in cancer lines (Kampa et al. 2000; Meyer et al. 2005). Antioxidants are also suggested for the preventation of cardiocirculatory diseases, therefore they have recently attracted considerable interest. Pure and synthetic antioxidants can be incorporated into extrudates but the metabolic effect of synthetic antioxidants is not known. In addition to pure antioxidants, foods including antioxidants can be added to feed materials that are extruded. Increasing the percentage of mulberry leaf as an antioxidant source from 5% to higher percentages in rice snacks decreases expansion (Charunuch et al. 2008), and the incorporation of blueberry concentrate in corn meal decreases expansion ratio and increases bulk density of extrudates while the antocyanins content of feed decreases during extrusion (Chaovanalikit et al. 2003).

PRODUCTION OF SNACKS WITH COEXTRUSION

The incorporation of ingredients mentioned above into an expanded snack provides a nutritional fortification of expanded snacks but negatively affects the expansion and breaking strength of extrudates. Those fortifying ingredients can be incorporated into an expanded extrudate with coextrusion in which fortification ingredients are filled inside a cereal based outer tube. It is also desired to have the extrudate including two materials with different texture, color, or shape. The extrudates with two different materials are also produced with coextrusion in which two different substances are extruded in different extruders, combined in a die of one extruder and then exit a die of extruder as shown in Figure 7.10, or one of the substances is extruded as a cereal based outer shell, the other is pumped to a die of an extruder to fill inside the cereal based outer shell to leave the system together as one part. Coextruded snacks can include a cereal based outer tube, and nut, protein, fiber, cream, chocolate, cheese or cereal based filling. In addition to a coextruded tube including filling inside the tube, a cereal based outer shell in a U-shape is also extruded. The U-shaped shell is filled and then rolled to cover the filling and to produce a coextruded snack.

The cereal based outer tube is cooked in the extruder. Inside of the outer tube is filled with cereal cooked in the other extruder or with noncereal based fillers, and

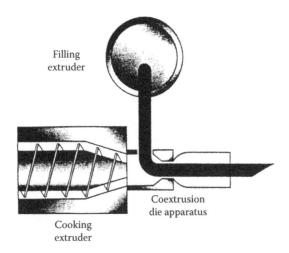

FIGURE 7.10   Die apparatus of extruder in coextrusion.

then exits a die as one piece of coextruded product. After cooling, the coextrudates are cut and then dried. The dried coextrudates are packaged and shipped to stores.

A common feed material problem encountered in coextrusion is the migration of moisture or oil from one substance to another in a coextruded product if there is a concentration gradient of moisture or oil in two substances coextruded into one piece. The other problems are incomplete filling and squeezing of filler.

Cereals store normal stress during their passage through the extruder and relax as they leave the extruder so extruded snacks swell as they exit the extruder. If material inside the tube of coextruded product is not cereal and filling occurs outside, then materials inside the outer tube do not store stress during their passage through the extruder and do not relax as they leave the extruder, therefore fillings do not swell and follow the outside tube of a coextruded product. That prevents full filling inside the outer tube of a coextruded snack. If filling occurs inside, there will be relaxation to inside so that it will squeeze the filler. Full filling can be provided by the controlled squeezing of filler. If the filler and outer tube of a coextruded snack contact and flow together, normal stresses can relax and provide full filling.

Rheological matching between filler and dough of the outer tube for a coextrusion process prevents undesired voids, and complete filling is obtained only for the materials exhibiting a stable network during the flow (Peressini et al. 2002).

The internal die swell relaxes within a short distance by changing geometrical, rheological, and production parameters (De Cindio et al. 2002). Reducing the distance between the filler nozzle and the extruder head outlet to zero may result in desirable product but the presence of die swell causes incomplete filling, which lowers the quality of the coextrudate. Internal die swell is controlled to have desired filling by adjusting the distance between the filler nozzle and the extruder head outlet with rheological properties of the outer tube and filling, and providing a short distance between the filler nozzle and extruder head (De Cindio et al. 2002).

EXTRUSION WITH SUPERCRITICAL CARBON DIOXIDE

Direct expanded extrudates are produced by gelatinization and partial degradation of starch. Increases of temperature and pressure of feed with a low moisture at high shear conditions flashes superheated water upon exiting a die to expand the extrudates. Extrusion with low moisture reduces drying cost but increases SME inputs and wear of the barrel and screw. Low moisture extrusion not only increases water absorption and solubility of extrudates that decrease the shelf-life of breakfast cereals with a loss of the crispy and crunchy structure but also increases the loss of vitamin and amino acids. Since water in feed expands the feed as its temperature is above 100°C and pressure is higher than atmospheric pressure, the expansion of extrudates including temperature sensitive ingredients requires the use of fluid expanding at a lower temperature. Expanded extrudates of foods are produced by the flashing off water but expanded plastic foams are produced by the injection of gas to plastic melt.

The injection of CO_2 into an extruder and the production of expanded extrudates by eliminating the effect of steam have prompted research on the use of CO_2 as a blowing agent as opposed to moisture flashing off to produce the expanded extrudates (Ferdinand, Clark, and Smith 1992). Expansion of carbon dioxide at a lower temperature than the expansion of water draws attention to the replacement of water with CO_2 for the expansion of extrudates at a lower temperature.

CARBON DIOXIDE AND SUPERCRITICAL CARBON DIOXIDE

Solid, liquid, or gas form of the substances depends on pressure and temperature so the form of a substance can be changed with pressure and temperature. Gases can be liquefied by decreasing temperature or increasing pressure. Fluids are termed as supercritical above critical temperature and pressure and are not liquefied by increasing pressure at temperature above the critical temperature as shown in Figure 7.11. Critical temperature and pressure of fluids vary from one fluid to another. Among the many fluids, CO_2 is safe and supercritical at the temperature and pressure of the extrusion process. CO_2 solubilizes in water, and solubility in water increases with a decrease in temperature. Measuring the equilibrium of CO_2 uptake by the starch–water system at elevated pressures up to 16 MPa at 50°C shows that the solubility of CO_2 increases linearly with pressure with maximum yield as 4.0 g CO_2/g starch–water system. Temperature and pressure affect the solubility of CO_2, but the solubility is not dependent on the degree of gelatinization (DG) of starch. On the other hand, the diffusion coefficient of CO_2 increases with the concentration of dissolved CO_2 but decreases with increasing DG in the range of 50%–100% (Chen and Rizvi 2006a).

Liquid CO_2 is held in a cool storage vessel to prevent vaporization while pressure of CO_2 is controlled at a compressor. Holding pressure of CO_2 above critical pressure prevents the flashing off CO_2 with increased solubility.

Upon adjusting its temperature during its passage through the heat exchanger, it is injected into the extruder barrel. CO_2 not only expands extrudates but also solubilizes heat sensitive vitamins and then provides the incorporation of vitamins with the injection of CO_2 to the extrusion line. CO_2 can be injected to the extruder both at sub or supercritical form and both forms expand melt of extrudates but lower solubility of subcritical CO_2 reduce the amount of CO_2 that can be incorporated into the melt

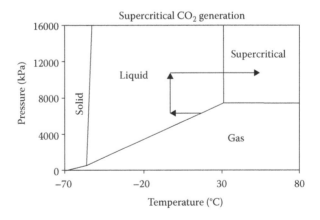

FIGURE 7.11 Generation of supercritical carbon dioxide.

of extrudate. An injection of CO_2 cannot provide the incorporation of heat sensitive vitamins into the melt of extrudate but an injection of supercritical CO_2 does.

INCORPORATION OF SUPERCRITICAL CARBON DIOXIDE INTO EXTRUDER AND ITS EFFECT ON DOUGH

Water is added to feed materials as a plasticizer and steam is added to heat feed materials for extrusion with supercritical carbon dioxide ($SC\text{-}CO_2$) but moisture content of feed expanded with $SC\text{-}CO_2$ is higher than moisture content of feed expanded with steam. A flow diagram for extrusion with $SC\text{-}CO_2$ is shown in Figure 7.12 (Mulvaney and Rizvi 1993). Feed with higher moisture content is introduced into the extruder. Heating feed not only cooks but also increases the gas holding

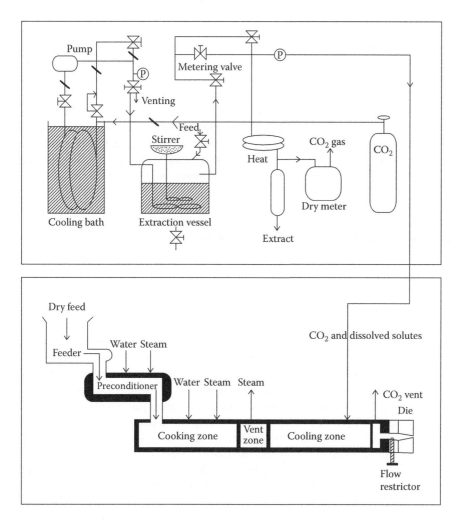

FIGURE 7.12   Supercritical extrusion system. (Printed with permission of *J. Food Science.*)

capacity of feed. After cooking feed with moderate shear in the extruder, steam is vented from feed to reduce the temperature of feed and prevent puffing with water flash-off. A cooling zone is also present in the extruder for a further decrease of feed temperature below 100°C. A longer extruder is used for extrusion with SC-CO$_2$ due to the presence of a cooling zone in the extruder. Pressure is increased gradually above supercritical pressure at the extruder. A reverse screw element is used before SC-CO$_2$ injection to prevent the backward flow of SC-CO$_2$ that is injected at lower pressure after the reverse screw element. A reverse screw element is also used before a die to mix SC-CO$_2$ with feed. The MRT in screw configurations to which SC-CO$_2$ is injected is higher than the MRT in conventional screw configurations due to the longer screw and extruder barrel of SC-CO$_2$ injected extrusion system (Schmid, Dolan, and Ng 2005). Ingredients like vitamins, which are injected with SC-CO$_2$, are deposited in feed as the density of SC-CO$_2$ decreases after the injection of SC-CO$_2$.

Screw, die and barrel configuration are arranged to allow conditions for the maintenance of SC-CO$_2$ and the prevention of back flow while a sufficient degree of fill is provided to eliminate the separation of melt phase to two phases being melt and gas phases. SC-CO$_2$ not only expands extrudates but also acts as a plasticizer for starch–water mixture at level of 0.45 g SC-CO$_2$/100 g sample and decreases the viscosity of the melt by an average of 14%. The reduction in viscosity depending on the DG and the amount of injected SC-CO$_2$ is explained with the mechanism of the free volume added to starch–water mixtures with SC-CO$_2$. Since gas retention capability, CO$_2$ diffusivity, and the pressure drop rates are affected with viscosity, SC-CO$_2$ may affect expansion and cellular characteristics of starch-based extrudates by its effect on viscosity (Chen and Rizvi 2006b).

Expansion with Supercritical Carbon Dioxide

SC-CO$_2$ is solubilized by shear provided with the mixing elements in the screw that break up SC-CO$_2$ into smaller bubbles after the injection of SC-CO$_2$. The bubble nucleation rate is maximized with the reduction of CO$_2$ solubility with thermal and pressure changes. Nuclei grow and then dissolved gas diffuses into the cells. The nucleated gas bubbles expand they extrudates after the extrudates exit a die. Since temperature at a die is lower than 100°C, extrudates puff due to the expansion of SC-CO$_2$. The growth time of nuclei in feed expanded with SC-CO$_2$ is longer than the growth time of nuclei in feed expanded with steam so growth time of nuclei is a factor controlling the cross-sectional expansion of extrudates that are expanded with SC-CO$_2$. Rapid initial cross-sectional and volumetric expansion of extrudates with SC-CO$_2$ cause collapse of extrudates so retarding the expansion of extrudates with SC-CO$_2$ provides the desired expansion and texture. The collapse of extrudates after expansion causes dense extrudate and high shear strength.

The incorporation of whey powder, corn fiber, and resistant starch as texturizing agents reduces collapse of extrudates and shear strength (Sokhey, Rizvi, and Mulvaney 1996). Incorporation of 4%–10% egg white and WPs into pregelatinized corn and potato starch extruded with SC-CO$_2$ increases the expansion ratio due to a reduced shrinkage of high-moisture extrudates and results in a product that has a uniform microcellular structure and cell density. The expansion ratio and bulk density of extrudates expanded with SC-CO_2 are comparable with the expansion ratio and bulk density of extrudates expanded with steam (Alavi et al. 1999). On the other hand, incorporation of WP alone acting as a diluent reduces the viscosity of melt, cross-sectional expansion ratio, and cell number density of extrudates expanded with SC-CO_2. Lower cell number density and the related decrease in expansion are attributed to a decrease in melt viscosity due to the addition of WP alone. Extrudates with 0.75% SC-CO_2 collapse during postextrusion drying at 85°C (Cho and Rizvi 2009). SC-CO_2 provides low to medium density extrudates (Mulvaney and Rizvi 1993).

The amount of injected SC-CO_2 also affects the expansion of extrudates. Since the pressure drop profile in a die controls the rate of expansion of extrudates with SC-CO_2 via nucleation and cell growth, the pressure drop profile is an important factor affecting the expansion of extrudates with SC-CO_2 (Cho and Rim 2008). Pressure not only affects expansion but also the water solubility index (WSI) that increases with increases of pressure when CO_2 is injected at 0.1–0.5 MPa (Jeong and Toledo 2004). Die pressure can be changed with CO_2 injection pressure, and increasing CO_2 injection pressure up to 2 MPa decreases die pressure but increases WSI, the water absorption index (WAI), and extrudate density due to the collapse of the foam. Increasing die pressure to 8.27 MPa increases density of corn meal extrudate puffed with SC-CO_2 at low (39 kg/h) and medium (49 kg/h) feed rate, but extrudate density is not affected with a further increase of die pressure. Increasing feed rate from 39 to 49 kg/h increases extrudate density but a further increase of feed rate to 57 kg/h decreases extrudate density at medium and high (>8.27 MPa) die pressures (Sokhey, Rizvi, and Mulvaney 1996).

The temperature of extrudates expanded with SC-CO_2 is approximately 100°C at a die. The density of extrudates decreases as temperature of extrudates increases from 95°C to 105°C, but a further increase in extrudate temperature does not affect bulk density (Sokhey, Rizvi, and Mulvaney 1996). On the other hand, decreasing moisture content of feed, which is extruded with SC-CO_2 at a die temperature of 80°C and screw speed of 300–400 rpm on a twin-screw laboratory-scale extruder, from 25% to 22% decreases bulk density of extrudates.

BENEFITS OF EXTRUSION WITH SUPERCRITICAL CARBON DIOXIDE

High moisture and low shear used during extrusion with SC-CO_2 as a blowing agent reduce starch degradation compared with the low moisture and high shear used during extrusion with water as a blowing agent. The reduction of starch degradation with SC-CO_2 lowers WSI (Lee, Ryu, and Lim 1999) that increases bowl life of breakfast cereals and reduces the undesirable powdery mouthfeel of extruded snacks.

Smaller cells with thick walls, which are observed at the extrudates expanded by SC-CO_2, are attributed to lower viscosity caused by SC-CO_2 (Mulvaney and Rizvi 1993). The extrudates expanded with SC-CO_2 have smaller and more uniform cells (Lee, Ryu, and Lim 1999). Steam, which expands the extrudates, condenses and collapses the cells if the transformation of melt of feed from a plastic to glassy state is not complete. But the use of SC-CO_2 reduces cell collapse due to reduced condensation and then provides a closed cell structure. The expansion of extrudates with CO_2 results in lower water absorption than the expansion of extrudates with steam does as extrudates are left to 100% relative humidity environment. Less porous and closed cell structures obtained with SC-CO_2 reduce the rate of water absorption, which increases the shelf-life of breakfast cereals with enhancement of the crispy and crunchy structure (Sokhey, Rizvi, and Mulvaney 1996; Lee, Ryu, and Lim 1999). The extrusion of wheat flour with SC-CO_2 at a die temperature of 80°C and screw speed of 300–400 rpm, and the extrusion of wheat flour without SC-CO_2 as a control at a die temperature of 150°C and screw speed of 200–300 rpm, also show that the extrudates obtained with SC-CO_2 result in lower WAI than control extrudates (Schmid, Dolan, and Ng 2005).

Openings on the surface of the extrudates expanded with water flash are a result of holes in the cell wall that are ruptured. The surface of extrudates expanded by SC-CO_2 is smoother than the surface of extrudates expanded by water flash because the temperature of extrudate at a die is not high to flash water (Mulvaney and Rizvi 1993; Lee, Ryu, and Lim 1999). Since coextruded products are manufactured by the extrusion of cereal tubes and pumping of filling material through the inner hollow of cereal tubes, the smooth surface of cereal tubes obtained with SC-CO_2 may provide better filling for the production of coextruded products in which filling material is introduced through the interior hollow surface of cereal tubes.

High moisture extrusion with SC-CO_2 also decreases vitamin and amino acid losses. Since SC-CO_2 expands below a temperature of 100°C and atmospheric pressure, temperature sensitive ingredients can be incorporated into melt of extrudate with lower losses of temperature sensitive ingredients due to lower temperature after their incorporation into melt.

The production of extrudates with SC-CO_2 provides the cells without opening and air passage into cells and reduces oxygen in cells. This prevents oxidative rancidity and then increases the shelf-life of extrudates.

The extrusion of corn meal and WP isolate with SC-CO_2 produces lighter extrudates due to a lowered Maillard browning reaction. Acidic conditions are provided with the use of SC-CO_2. Acidic conditions and higher moisture in feed reduce the Maillard reaction during extrusion with SC-CO_2 (Sokhey, Rizvi, and Mulvaney 1996).

Increasing moisture content of corn starch extruded in a corotating twin-screw extruder (24:1 L/D ratio, 31 mm screw diameter) with SC-CO_2 as a blowing agent reduces SME input (Lee, Ryu, and Lim 1999). Extrusion of high moisture feed with SC-CO_2 increases drying cost but decreases wear of barrel and screw due to a decrease in SME input during extrusion.

There are leavened confectionary and chocolate products in addition to plain confectionary and chocolate products. Since extruders are used for the production of confectionery and chocolate products, leavened confectionary and chocolate products can be manufactured with the injection of SC-CO_2 that leaven the products.

EXTRUSION WITH TURBO-EXTRUDER

Conventional extruders are used in the food industry producing extruded snack, breakfast cereals, and confectionary products. But lower equipment costs, lower maintenance costs with lower wear costs, higher throughput, better homogenization of ingredients in feed, better control of the process, easier dismantling and cleaning of the extruder, and the extrusion of feed with higher fat and sugar are targets of processors and equipment producers. A turbo-extruder as a special pump working on the basis of friction has been developed and patented by Schaaf Technologie GmbH in Germany to achieve some of the targets mentioned above (Heinz 1996).

TURBO-EXTRUDER AND FLOW OF MELT IN TURBO-EXTRUDER

A turbo-extruder has been designed to be used by itself or with a screw as a cooking extruder. Die and turbo units with or without a screw are components of a turbo-extruder as shown in Figure 7.13. The turbo unit, which is placed between the screw and die, includes a scrapper chamber consisting of at least two perforated plates and a scrapper-like propeller with arms also called a spatula. The scrapper surface can be designed as straight or curved, and it is thinner than the scrapper chamber and can be connected to other scrapper and screw with a shaft. The angle of scrapper to perforated plate is adjusted between 0° and 90°, and increasing the angle increases throughput but decreases pressure. The thickness, number, and size of holes in perforated plates can be varied to adjust resistance to flow, the shear the feed undergoes, energy dissipation, feed temperature, and homogenization. The roughness of the surface of a perforated plate causes friction. Large surface area of channels in a perforated plate can be used for effective heating and cooling of feed. The number of scrapper chambers can be adjusted according to desired melt temperature, residence time, and mechanical energy dissipation.

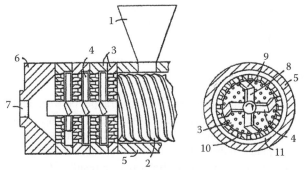

1. Hooper 2. Screw 3. Matrix of holes 4. Spatula
element 5. Barrel 6. Die 7. Nozzle 8. Arms 9. Offsets
10. Opening of the matrix of holes 11. Channel

FIGURE 7.13 Turbo-extruder and its components. (Printed with permission of Schaaf Technologie GmbH, Bad Camberg, Germany.)

Raw materials are conveyed with a screw to reach the first perforated plate in the turbo-extruder. Upon passing through and then exiting the first perforated plate, materials are scrapped by a rotating scrapper to the second perforated plate, and then pass through and exit the second perforated plate. Transportation occurs with a scrapper creating high pressure. The shape and distance of the arm of the scrapper from a perforated plate affect the pressure. Feed that passes through the second perforated plate move to a die. Heat is provided with mechanical energy input into the turbo-extruder. The continuous change in pressure by a scrapper is claimed to provide the homogenization of melt.

Water is not a required component in the plastic industry in which extruders are used but the extrusion of foods needs water in feed. Water affects cooking of extrudates via its effect on specific heat of raw material and energy dissipation in the extruder with friction. As extrudates exit a die, some of the superheated water in feed evaporates and then expand extrudates while some of the superheated water can condense. In the turbo-extrusion, feed is extruded with a scrapper through the first perforated plate into a scrapper chamber. Pressure is lower in front of a rotating scrapper element, but the pressure of feed coming to a scrapper is increased with a rotating scrapper element and then feed is passed through the first perforated plates and pressure drops behind the first perforated plates. The water of feed evaporates, expands dough, and then increases the viscosity of dough at a lower pressure after the passage of feed through the first perforated plate. Upon passing through the first perforated plate, feed contacts a scrapper that faces higher viscosity of dough that passed through the first perforated plate. After the scrapper creates pressure and condenses water vapor of feed, latent heat of vapor increases the temperature of dough and feed is passed through the second perforated plate. This process is repeated according to the number of scrapper chambers.

The viscosity of dough in the high pressure side of the scrapper chamber is lower than the viscosity of dough in the low pressure side. Dough in the high pressure side is forced to pass through a perforated plate. Dough is supposed to flow from the high pressure side back to the low pressure side of the scrapper but receives heat and gets hot when it passes through a perforated plate. Hot dough flows with the scrapper into a perforated plate rather than the low pressure side in the scrapper chamber, expands, and then moves in the direction of die.

BENEFITS OF TURBO-EXTRUSION

It is claimed that the movement of dough with a scrapper between high and low pressure sides in the turbo-extruder provides turbulence, homogenization of dough, and heat gain that prevent overheating. Expansion and compression of dough flowing through perforated plates in the turbo-extruder also improve heat transfer. Increasing screw speed increases the temperature linearly as a result of the increase in expansion and compression of dough so the temperature of melt is controlled better with screw speed during turbo-extrusion.

Residence time between 20–60 s in twin-screw extruders is achieved so that the process is called high temperature short time (HTST). The turbo-extrusion process is named as high temperature ultra short time (HTUST) with Schaaf Technologie

GmbH due to the reduction of residence time to 8 s in turbo-extrusion. Throughput can be increased with additional scrapper elements during turbo-extrusion.

It is claimed that turbo-extrusion provides better product stability, better product texture, and flexibility against raw materials like higher fat content (up to 10%) and sugar (up to 30%). The limitation of fat and sugar in turbo-extrusion does not depend on the transport capability of the turbo-extruder but depends on dough elasticity that affects expansion. As 3% fat is used in the extrusion of ingredients in the turbo-extruder, energy input and temperature are slightly affected. The use of a turbo-extruder provides linearity for motor current uptake in case of fat or sugar addition to ingredients.

The centered location of the extruder screw and perforated plate through which feed passes support radial forces generated during extrusion and prevent contact between the screw and extruder barrel so that reduced wear of related parts is claimed.

The wear of the scrapper is also low and production of a scrapper is economical. Easy assembling, disassembling, and cleaning of the scrapper chamber provide flexibility for its maintenance. It is claimed that using a conventional extruder, which was used before and underwent wearing, as a part of a turbo-extruder provides uniform shape and texture with increased capacity and reliable reproduction of operating parameters because wear in turbo-extruder does not affect the processes much compared with a conventional extruder.

In conventional extrusion, the presence of laminar flow starting from a screw to a die results in different flow behavior in outer and inner areas of a die opening so that nonuniform products exit a die. The presence of different shaped holes in a die also affects flow behavior that can provide uniform product thickness and shape for two simple shaped holes in a die but face a problem in the case of more than two different shaped holes in a die. A variation of throughput and raw material also affects the shape of extrudates. It is claimed that the turbo-extruder assists the production of multiple shapes from a die, and homogeneity and uniformness of energy distribution in the turbo-extruder improve flow in the extruder.

COATING APPLICATOR DRYER FOR POSTEXTRUSION PROCESS

Flavor ingredients are added to feed materials before extrusion, but the addition of heat sensitive vitamin and flavor ingredients to feed before extrusion cause loss of these ingredients due to high temperature and pressure of extrusion. Heat sensitive ingredients are incorporated to the extruded products by the coating of extrudates after extrusion. Coating can be one or two steps in the coater including a conveying belt or coating drum and spray solution tanks of vitamin, flavors, and sugar. Vitamin or flavor solutions are incorporated in the first step, and sugar solutions are incorporated in the second step of coating. Coating can be conducted with the passage of extrudates with a conveying-belt through a coater and spraying of coating solution from a nozzle of a spraying system to extrudates or tumbling extrudates in a coating drum and the spraying of a coating solution from a nozzle of a spraying system to extrudates. A coater drum is more effective than a conveying-belt system for the coating of extrudates. The tumbling of extrudates in a coater drum causes broken extrudates if the extrudates are fragile so a coater drum is not used for the coating of fragile extrudates. A conveying belt system is used instead of a coater drum for the coating of fragile extrudates. Moisture coming from coating agents in a coated extrudate is removed by drying of the coated extrudates in different equipment such as a drier at a temperature of approximately 150°C.

Wenger Manufacturing Inc. has developed and patented an applicator dryer in which extrudates are coated by the spraying of coating agents and dried with hot air passing through slots in the different sections of a perforated drum and producing a vortex action and gentle tumbling effect for the improvement of breakfast cereal coating (Singer 1992). An applicator dryer and the flow in it are shown in Figure 7.14. The coating and drying of extrudates several times in one single piece of equipment of an applicator dryer improve the coating of extrudates. Dry coatings like nut or fruit pieces can also be incorporated into extrudates after syrup coating. The use of applicator driers for both batch and continuous systems is another advantage of it.

Snack extrudates are coated by spraying oil to extrudates and then covering the extrudates with flavoring and seasoning, or by spraying premixed oil, flavoring and seasonings in the drum. The increase of calories in snacks with oil incorporated into extrudates during coating have required the use of a modified system. Third generation snack producers use microwaves or a hot air puffing system instead of frying. The use of gums as adhesive agents instead of fat and the development of the applicator dryer in which extrudates are coated with coating agents and dried with hot air

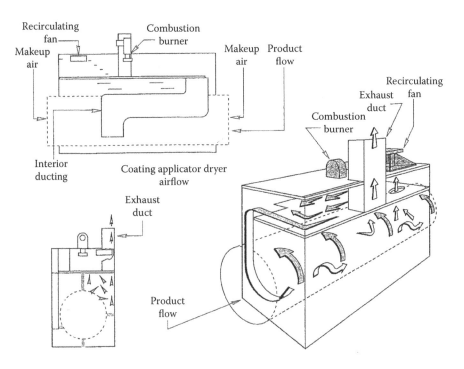

FIGURE 7.14 Flow diagram of coating applicator dryer. (Printed with permission of Wenger Manufacturing Inc., Sabetha, Kansas.)

passing through slots in the different sections of a perforated drum produces low calorie snack extrudates. Alternating coating and drying of extrudates several times in an applicator dryer provides the desired coating. Sweetener and spices in addition to color and flavor ingredients can also be added to extruded snacks with an applicator dryer.

ACKNOWLEDGMENTS

I thank Wenger Manufacturing Inc. for their permission to use some figures in this chapter and for their personnel communication for my questions at the Extrusion Process Center of the University of Nebraska-Lincoln. I appreciate Schaaf Technologie GmbH for providing information upon my requests and for their permission to use a figure in this chapter.

REFERENCES

Akdogan, H. 1996. Pressure, torque, and energy responses of a twin screw extruder at high moisture contents. *Food Res Int* 29: 423–429.

Alavi, S.H., Gogoi, B.K., Khan, M., et al. 1999. Structural properties of protein-stabilized starch-based supercritical fluid extrudates. *Food Res Int* 32: 107–118.

Altomare, R.E. and Ghossi, P. 1986. An analysis of residence time distribution patterns in twin-screw cooking extruder. *Biotechnology Progress* 2: 157–163.

Ainsworth, P., Ibanoglu, S. and Hayes, G.D. 1997. Influence of process variables on residence time distribution and flow patterns of tarhana in a twin-screw extruder. *J Food Eng* 32: 101–108.

Bhattacharya, S. and Prakash, M. 1994. Extrusion of blends of rice and chick pea flours. *J Food Eng* 21: 315–330.

Best, E.T. 1994. Confectionery Extrusion. In *The technology of extrusion cooking*, ed. N.D. Frame, 190–236. London: Blackie Academic & Professional.

Chang, K.L.B. and Halek, G.W. 1991. Analysis of shear and thermal history during co-rotating twin-screw extrusion. *J Food Sci* 56: 518–531.

Chaovanalikit, A., Dougherty, M.P., Camire, M.E., et al. 2003. Ascorbic acid fortification reduces anthocyanins in extruded blueberry-corn cereals. *J Food Sci* 68: 2136–2140.

Charunuch, C., Tangkanakul, P., Rungchang, S., et al. 2008. Application of mulberry (Morus alba) for supplementing antioxidant activity in extruded Thai rice snack. *Acta Horticulturae* 786: 137–145.

Chen, K.H.J. and Rizvi, S.S.H. 2006a. Rheology and expansion of starch-water-CO_2 mixtures with controlled gelatinization by supercritical fluid extrusion. *Int J Food Prop* 9: 863–868.

Chen, K.H.J. and Rizvi, S.S.H. 2006b. Measurement and prediction of solubility and diffusion coefficients of carbon dioxide in starch-water mixtures at elevated pressures. *J Polymer Sci, Part B-Polymer Physics* 44: 607–621.

Cho, K.Y. and Rim, S.S.H. 2008. The time-delayed expansion profile of supercritical fluid extrudates. *Food Res Int* 41: 31–42.

Cho, K.Y. and Rizvi, S.S.H. 2009. 3D microstructure of supercritical fluid extrudates I: Melt rheology and microstructure formation. *Food Res Int* 42: 595–602.

Davidson, V.J., Paton, D., Diosady, L.L., et al. 1983. Residence time distribution for wheat starch in a single screw extruder. *J Food Sci* 48: 1157–1161.

De Cindio, B., Gabriele, D., Pollini, C.M., et al. 2002. Filled snack production by coextrusion-cooking: 1. Rheological modeling of the process. *J Food Eng* 52: 67–74.

De Pilli, T., Severini, C., Baiano, A., et al. 2004. Study on extrusion and stability of fat meals. Comparison between single- and twin-screw extruders. *Sciences Des Aliments* 24: 307–322.

De Pilli, T., Carbone, B.F., Fiore, A.G., et al. 2007. Effect of some emulsifiers on the structure of extrudates with high content of fat. *J Food Eng* 79: 1351–1358.

Della Valle, G., Tayep, J. and Melcion, J.P. 1987. Relationship of extrusion variables with pressure and temperature during twin screw extrusion cooking of starch. *J Food Eng* 6: 423–444.

Desai, K.G.H. and Park, H.J. 2005. Recent developments in microencapsulation of food ingredients. *Drying Technol* 23: 1361–1394.

Faridi, H. 1988. Flat Bread. In *Wheat: Chemistry and technology*, ed. Y. Pomeranz, 471–488. St. Paul, MN: American Association of Cereal Chemists.

Fast, R.B. 1990. Manufacturing technology of ready to eat cereals. In *Breakfast cereals and how they are made*, eds. E.F. Caldwell and R.B. Fast, 15–41. St. Paul, MN: American Association of Cereal Chemists.

Ferdinand, J.M., Clark, S.A. and Smith, A.C. 1992. Structure formation in extrusion-cooked starch-sucrose mixtures by carbon-dioxide injection. *J Food Eng* 16: 283–291.

Garber, B.W., Hsieh, F. and Huff, H.E. 1997. Influence of particle size on the twin-screw extrusion of corn meal. *Cereal Chem* 74: 656–661.

Guy, R.C.E. and Horne, A.W. 1988. Extrusion and coextrusion of cereals. In *Food structure— its creation and evaluation*, eds. J.M.V. Blanshard and J.R. Mitchell, 331–349. London: Butterworths.

Harman, D.V. and Harper, J.M. 1973. Effect of extruder geometry on torque and flow. *Transactions of the ASAE* 1175–1178.

Harper, J.M. 1981. *Extrusion of foods*. Boca Raton, FL: CRC.

Heinz, S. 1996. Cooker-extruder apparatus and process for cooking-extrusion of biopolymers. US Patent 5567463.

Hsieh, F., Mulvaney, S.J., Huff, H.E., et al. 1989. Effect of dietary fiber and screw speed on some extrusion processing and product variables. *Lebensm-Wiss Technol* 22: 204–207.

Hsieh, F., Huff, H.E., Lue, S., et al. 1991. Twin-screw extrusion of sugar beet and corn meal. *Lebensm-Wiss Technol* 24: 495–500.

Hu, L., Hsieh, F. and Huff, H.E. 1993. Corn meal extrusion with emulsifier and soybean fiber. *Lebensm-Wiss Technol* 22: 204–207.

IngJenq, J., Camire, M.E. and Bushway, A.A. 1997. Properties of an extruded product prepared from potato flakes and chicken thigh meat. *Food Sci Technol Int* 3: 451–458.

Jeong, H.S. and Toledo, R.T. 2004. Twin-screw extrusion at low temperature with carbon dioxide injection to assist expansion: Extrudate characteristics. *J Food Eng* 63: 425–432.

Jin, Z., Hsieh, F. and Huff, H.E. 1994. Extrusion cooking of corn meal with soy fiber, salt and sugar. *Cereal Chem* 71: 227–234.

Kampa, M., Hatzoglou, A., Notas, G., et al. 2000. Wine antioxidant polyphenols inhibit the proliferation of human prostate cancer cell lines. *Nutrition and Cancer* 37: 223–233.

Kim, C.H. and Maga, J.A. 1987. Properties of extruded whey protein concentrate and cereal flour blends. *Lebensm-Wiss Technol* 20: 311–318.

Kim, H. and Kim, B. 2004. Characteristics of snack-like products from twin-screw extrusion of corn flour or wheat flour blended with tuna flesh. *Food Sci Biotech* 13: 219–224.

Kirby, A.R., Ollet, A.L., Parker, R., et al. 1988. An experimental study of screw configuration effects in the twin-screw extrusion cooking of maize grits. *J Food Eng* 8: 247–272.

Lazou, A.E., Michailidis, P.A., Thymi, S., et al. 2007. Structural properties of corn-legume based extrudates as a function of processing conditions and raw material characteristics. *Int J Food Prop* 10: 721–738.

Lee, F.Y., Ryu, G.H. and Lim, S.T. 1999. Effects of processing parameters on physical properties of corn starch extrudates expanded using supercritical CO_2 injection. *Cereal Chem* 76: 63–69.

Lin, J.K. and Armstrong, D.J. 1990. Process variables affecting residence time distributions of cereals in an intermeshing counterrotating twin screw extruder. *Transactions of ASAE* 33: 1971–1978.

Lo, T.E., Moreira, R.G. and Castelll-Perez, M.E. 1998. Effect of operating conditions on melt rheological characteristics during twin-screw food extrusion. *Transactions of ASAE* 41: 1721–1728.

Marchetti, M., Tossani, N., Marchetti, S., et al. 1999. Feeds. *Aquaculture Nutrition* 5: 115–120.

Meuser, F. and Wiedman, W. 1989. Extrusion plant design. In *Extrusion cooking*. eds. J.M. Harper, C. Mercier, and P. Linko, 91–155. St. Paul, MN: American Association of Cereal Chemists.

Meyer, F., Galan, P., Douville, P., et al. 2005. Antioxidant, vitamin, and mineral and prostate cancer preventation in the SU.VI.MAX trial. *I J Cancer* 116: 182–186.

Miller, R.C. 1994. Breakfast cereal extrusion technology. In *The technology of extrusion cooking*, ed. N.D. Frame, 73–109. London: Blackie Academic & Professional.

Moore, G. 1994. Snack food extrusion. In *The technology of extrusion cooking*, ed. N.D. Frame, 110–143. London: Blackie Academic & Professional.

Mukuddem-Petersen, J., Oosthuizen, W. and Jerling, J.C. 2005. A systematic review of the effects of nuts on blood lipid profiles in humans. *J Nutrition* 135: 2082–2089.

Mulvaney, S.J. and Rizvi, S.S.H. 1993. Extrusion processing with supercritical fluid. *Food Technol* 47: 75–82.

Onwulata, C.I., Konstance, R.P., Smith, P.W., et al. 1998. Physical properties of extruded products as affected by cheese whey. *J Food Sci* 63: 814–817.

Onwulata, C.I., Konstance, R.P., Smith, P.W., et al. 2001a. Coextrusion of dietary fiber and milk proteins in expanded corn products. *Lebensm-Wiss Technol* 34: 424–429.

Onwulata, C.I., Smith, P.W., Konstance, R.P., et al. 2001b. Incorporation of whey products in extruded corn, potato or rice snacks. *Food Res Int* 34: 679–687.

Osman, M.G., Saai, D. and Jackson, D.S. 2000. Oil absorption characteristics of a multigrain extrudate during frying. *Cereal Chem* 77: 101–104.

Park, J., Rhee, K.S., Kim, B.K., et al. 1993. Single screw extrusion of defatted soy flour, corn starch and raw beef blends. *J Food Sci* 58: 9–20.

Peressini, D., Sensidoni, A., Pollini, C.M., et al. 2002. Filled-snacks production by coextrusion-cooking. Part 3. A rheological-based method to compare filler processing properties. *J Food Eng* 54: 227–240.

Riaz, M., Asif, M. and Ali, R. 2009. Stability of vitamins during extrusion. *Critical Reviews Food Sci Nutr* 49: 361–368.

Schmid, A.H., Dolan, K.D. and Ng, P.K.W. 2005. Effect of extruding wheat flour at lower temperatures on physical attributes of extrudates and on thiamin loss when using carbon dioxide gas as a puffing agent. *Cereal Chem* 82: 305–313.

Seker, M., Sadikoglu, H., Ozdemir, M., et al. 2003a. Cross-linking of starch with reactive extrusion and expansion of extrudates. *Int J Food Prop* 6: 473–480.

Seker, M., Sadikoglu, H., Ozdemir, M., et al. 2003b. Phosphorus binding to starch during extrusion in both single- and twin-screw extruders with and without a mixing element. *J Food Eng* 59: 355–360.

Seker, M., Sadikoglu, H. and Hanna, M.A. 2004. Properties of cross-linked starch produced in a single screw extruder with and without a mixing element. *J Food Process Eng* 27: 47–63.

Seker, M. 2004. Distribution of the residence time in a single-screw extruder with differing numbers of mixing elements. *Int J Food Sci Technol* 39: 1053–1060.

Seker, M. 2005a. Residence time distributions of starch with high moisture content in a single-screw extruder. *J Food Eng* 67: 317–324.

Seker, M. 2005b. Selected properties of native or modified maize starch/soy protein mixtures extruded at varying screw speed. *J Sci Food Agric* 85: 1161–1165.

Seker, M. and Hanna, M.A. 2005. Cross-linking starch at various moisture contents by phosphate substitution in an extruder. *Carbohydrate Polymers* 59: 541–544.

Seker, M. and Hanna, M.A. 2006. Sodium hydroxide and trimetaphosphate levels affect properties of starch extrudates. *Industrial Crops Products* 23: 249–255

Singer, R.E. 1992. Methods and apparatus for combined product coating and drying. US Patent 5100683.

Singh, B., Sekhon, K.S. and Singh, N. 2007. Effects of moisture, temperature and level of pea grits on extrusion behavior and product characteristics of rice. *Food Chem* 100: 198–202.

Singh, R.K., Nielsen, S.S., Chambers, J.V., et al. 1991. Selected characteristics of extruded blends of milk protein raffinate or nonfat dry milk with corn flour. *J Food Process Preservation* 15: 285–302.

Sokhey, A.S., Rizvi, S.S.H. and Mulvaney, S.J. 1996. Application of supercritical fluid extrusion to cereal processing. *Cereal Foods World* 41: 29–34.

Suknark, K., Phillips, R.D. and Chinna, M.S. 1997. Physical properties of directly expanded extrudates formulated from partially defatted peanut flour and different types of starch. *Food Res Int* 30: 575–583.

Vergnes, B., Barres, C. and Tayep, J. 1992. Computation of residence time and energy distribution in reverse screw element of a twin-screw extrusion cooker. *J Food Eng* 16: 215–237.

Yagci, S. and Gogus, F. 2009. Selected physical properties of expanded extrudates from the blends of hazelnut flour-durum clear flour-rice. *Int J Food Prop* 12: 405–413.

Yuliani, S., Bhandari, B., Rutgers, R., et al. 2004. Application of microencapsulated flavor to extrusion product. *Food Reviews Int* 20: 163–185.

8 Extrusion Processing of Main Commercial Legume Pulses

Jose De J. Berrios

CONTENTS

CONTRIBUTION OF EXTRUSION COOKING TO FOOD PROCESSING

Extrusion cooking technology is considered a high-temperature-short-time, versatile, and modern food operation (Roos et al. 1998; Duizer, Campanella, and Barnes 1998). For the production of expanded food products, high pressure is also considered in the operation. Many companies use extruders in the production of human and animal foods and industrial products. Several benefits of extrusion cooking over traditional, batch cooking methods are as follows: the extruder acts as a complete processing plant where ingredients are mixed, cooked, formed, and sheared in one continuous process; a wide variety of products can be produced using one single machine by varying the ingredient composition in the mix and extrusion processing conditions; the extrusion process allows precise control over the cooking process, and changes can be made quickly during cooking; the high-temperature high pressure short-time cooking process eliminates micro-organisms, inactivates enzymes, and minimizes nutrient and flavor losses in the food being produced. Also, the final expanded, low moisture extruded food is considered a shelf-stable product. Additionally, extruders may be more cost-effective to operate than traditional cooking systems, because they perform multiple unit operations (e.g., mixing, blending, cooking, and forming) on one single machine which increases productivity and reduces production costs. Drum dryers, cooking vessels, and stirred reactors are among the main processing equipment that can sometimes be replaced by a single extrusion unit.

EXTRUDED FOOD PRODUCTS

Due to the processing flexibility offered by extrusion cooking technology, extrusion has become an important food operation used for the fabrication of value-added foods in the cereal and pet food industries, as well as in dairy, bakery, beverage, and confection industries.

The cereal food industry has used both single- and twin-screw extruders for the production of a large variety of products (breakfast cereals, snack foods, flat bread, flakes, instant foods, and many others) for more than 50 years. Most recently, extrusion processing technology has found application in the soy industry for the production of products such as texturized vegetable protein (TVP), meat analog type foods, and instant beverages.

Consumers find expanded, ready-to-eat (RTE) extruded snack products very attractive because of convenience, textural attributes, shelf stability, and enhanced flavor. Additionally, the nutritional appeal of high-protein, high-nutritional, low-caloric snacks is a value-added attribute of extruded products from plant origins such as legume seeds. Industrial production of snack foods from plant proteins has grown rapidly since the 1970s resulting from the interest of food manufacturers to produce a wide range of high-protein food products, using different sources of plant protein (Skierkowski, Gujska, and Khan 1990). Extrusion is used commercially to produce high value breakfast and snack foods based on cereals such as wheat or corn (Hazell and Johnson 1989; Jomduang and Mohamed 1994; Martinez, Figueroa, and Larios 1996; Lasekan et al. 1996). This processing method is not being commercially used for legume pulses, such as dry beans, lentils, chickpeas, and dry peas, due to the perception that legume pulses do not expand well in extrusion. The application of extrusion processing of pulses has been limited mainly to studies conducted on their flours, mixtures of pulses with other flours, and their high-starch and high-protein fractions.

ORIGIN AND DEFINITION OF LEGUMES AND LEGUME PULSES

The term "legume" originates from the Latin *legere*, to gather. Botanically, legume means the seed pod itself. However, the term is often applied both to the seed and to the plant or plant family. Grain legumes are crop plants belonging to the botanical family known as the Fabaceae, or alternatively the Leguminosae, which is one of the three largest families of flowering plants, comprising nearly 750 genera and

18,000–19,000 species. Legumes are further divided into three subfamilies. The first, the Papilionoids, is the largest of the three subfamilies and it displays the widest range of morphological diversity, including cultivated plants such as soybeans, common beans, peas, lentils, and chickpeas. The second subfamily, the Mimosoids, is made up of trees such as the acacia and mimosa. The third, the Caesalpinioids, comprises about 2000 species of large tropical trees that are still not well known.

Legumes are considered one of the oldest foods known to mankind. Archaeological findings indicate that legumes have formed part of the human diet for approximately 8000 years. A large number of species of legumes is cultivated in the semi-arid tropical countries of the world, where legumes have been an important staple for centuries. In more recent years, legumes have been grown extensively in many other parts of the word mainly to be exported to legume's consuming countries.

While legumes such as soybeans and peanuts, considered oilseeds, are mainly grown for oil extraction, other legumes such as dry beans are cultivated primarily for their edible grains (seeds). These legumes are harvested at maturity and marketed as a dry commodity called "pulses" by the industry and the trade.

The name "pulse" is derived from the Latin *puls* meaning thick soup or potage. Pulses are defined by the Food and Agricultural Organization of the United Nations (FAO) as annual leguminous crops yielding from one to twelve grains or seeds of variable size, shape, and color within a pod. Pulses are used for food and animal feed. The FAO use the term pulses for legume crops harvested solely for the dry grain. The term pulses exclude green beans and green peas, which are considered vegetable crops. The term also excludes crops which are mainly grown for oil extraction (oilseeds such as soybeans and peanuts), and crops which are used exclusively for sowing (clovers, alfalfa). The FAO recognizes 11 primary pulses (Table 8.1). Lentils (*Lens culinaris*), dry beans (*Phaseolus* spp., including several species now in *Vigna*), dry peas (*Pisum* spp.), chickpea (*Cicer arietinum*) also known as garbanzo and Bengal gram, are pulses of great economical importance worldwide (FAO 2004).

WORLD PRODUCTION OF LEGUME PULSES

Worldwide, legume pulses are second only to cereals as a source of human food and animal feed. Although there are more than 60 domesticated pulse species, only about 20 are most largely produced and consumed. Some main reasons for growing pulses are due to their recognized nutritional and health history, desirable functional properties of their ingredients/fractions, adaptability of different varieties, great variety of colors associated with freshness and health attributes, relation to emerging ethnic cuisine trends, and for environmental benefits that tie their ability to "fix" nitrogen in the soil leading to a more sustainable food production system by contributing to decrease in greenhouse gas production (carbon footprint).

Global production of pulses surpassed 40.6 million metric tons in 2007 (Figure 8.1), with dry beans, dry peas, chickpeas, and lentils representing approximately 46%, 26%, 20%, and 8% of the total worldwide production, respectively. The five largest producers of pulses were India, Canada, Brazil, China, and the U.S. with average production of about 11.0, 4.2, 3.2, 2.2, and 2.1 million metric tons per year, respectively (FAO 2007).

TABLE 8.1

Eleven Primary Pulses Recognized by the FAO

Dry beans (*Phaseolus* spp. including several species now in *Vigna*)

    Kidney bean, haricot bean, pinto bean, navy bean (*Phaseolus vulgaris*)

    Lima bean, butter bean (*Vigna lunatus*)

    Azuki bean, adzuki bean (*Vigna angularis*)

    Mung bean, golden gram, green gram (*Vigna radiata*)

    Black gram, Urad (*Vigna mungo*)

    Scarlet runner bean (*Phaseolus coccineus*)

    Rice bean (*Vigna umbellata*)

    Moth bean (*Vigna acontifolia*)

    Tepary bean (*Phaseolus acutifolius*)

Dry broad beans (*Vicia faba*)

    Horse bean (*Vicia faba equina*)

    Broad bean (*Vicia faba*)

    Field bean (*Vicia faba*)

Dry peas (*Pisum* spp.)

    Garden pea (*Pisum sativum* var. *sativum*)

    Protein pea (*Pisum sativun* var. *arvense*)

Chickpea, garbanzo, Bengal gram (*Cicer arietinum*)

Dry cowpea, black-eyed pea, black eye bean (*Vigna unguiculata*)

Pigeon pea, Toor, cajan pea, congo bean (*Cajanus cajan*)

Lentil (*Lens culinaris*)

Bambara groundnut, earth pea (*Vigna subterranea*)

Vetch, common vetch (*Vicia sativa*)

Lupins (*Lupinus* spp.)

Minor pulses include:

    Lablab, hyacinth bean (*Lablab purpureus*)

    Jack bean (*Canavalia ensiformis*), sword bean (*Canavalia gladiata*)

    Winged bean (*Psophocarpus teragonolobus*)

    Velvet bean, cowitch (*Mucuna pruriens* var. *utilis*)

    Yam bean (*Pachyrrizus erosus*)

Source: FAO 2004

In the U.S. the two main growing regions are the Northern Plains (Montana, North and South Dakota) and the Palouse region (eastern Washington, northern Idaho, and northeastern Oregon). In recent years, North Dakota has become the largest producer of pulse crops, followed by Montana. The Palouse region of the Pacific Northwest is responsible for most of the production of green peas and chickpeas. The production of peas, lentils, and chickpeas in the U.S. has doubled in the last 10 years. Approximately 75% of the U.S. pulses are exported to many pulse consuming countries around the world.

The Canadian pulse industry has become a major player in global pulse production and trade over the past 20 years. Over this time, Canada has emerged to become the world's largest exporter of lentils and peas, and a top five bean

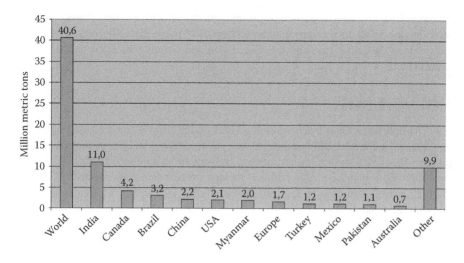

FIGURE 8.1   Largest producers of pulses in the world. (Data extracted from FAO, 2007. Food and Agriculture Organization of the United Nations, Available from FAOSTAT statistics database Agriculture, Rome, Italy.)

exporter. Canada had a record production of more than 4.8 million tons of pulses in 2005, with pulse production normally in the range of 4 to 4.5 million tons per year. The value of Canadian pulses exports alone exceeded $1 billion in 2006. Saskatchewan is the largest producer of peas, lentils, and chickpeas with a small bean industry, and Alberta produces beans under irrigation as well as peas, lentils, and chickpeas. Similar to the U.S., approximately 75% of Canadian pulse production is exported each year, normally to 150 countries around the world (Pulse Canada 2010).

NUTRITIONAL VALUE OF LEGUME PULSES

Legume pulses occupy a special place in the human diet, because they contain nearly three times more protein than cereals, and are considered the primary source of protein for vegetarians. The proximate composition of most legume pulses demonstrate that they contain approximately 10% moisture, 17%–30% crude protein, 0.8%–1.8% lipids, 2.2%–4.0% ash, and 60%–65% total carbohydrates, and supply 345–350 kcal/100g. The exception to the presented general proximate composition values are chickpeas (garbanzo beans), which contain significantly higher amounts of lipids (3.5%–5.5%) and therefore are of higher caloric value. The relative proportions of essential amino acids in pulse proteins are not well balanced for human dietary requirements as in meat, milk, or fish, since legumes are generally deficient in the essential amino acid methionine. Despite this limitation, pulse protein has high biological value, 0.8%–1.5% and contains high levels of the essential amino acid lysine, which is deficient in cereals. Therefore, in order to obtain a properly balanced or complete protein in the diet, the consumption of a combined food made from legumes and cereals is highly recommended. Many early civilizations developed around diets of pulses for protein, combined with a cereal crop to provide their nutritional dietary requirements. Examples of these diets are the use of beans and corn in the Americas, and the use of pita and humus (chickpea based) in the Middle East.

ANTI-NUTRIENTS

Food legumes are known to contain several compounds termed as anti-nutrients for humans, such as trypsin and chymotrypsin inhibitors, lectins, cyanogenic compounds, flatulence factors, lathyroghens, and saponins, among others (FAO 1977). However, most of these anti-nutritional factors are heat labile and destroyed during cooking, while others can be reduced, eliminated, or removed by various processing methods such as milling, soaking, germination, fermentation, extrusion, and other processing technologies.

ROLE OF LEGUME PULSES IN A HEALTHY DIET

Healthy eating is known to be important for proper growth and development. But more recently it has been recognized that healthy eating is a significant factor in reducing the risk of developing nutrition-related heath problems including obesity, heart disease, cancer, diabetes, hypertension (high blood pressure), osteoporosis, anemia, and some bowel disorders. In the U.S. and elsewhere, the rates of obesity, diabetes, and heart disease have reached what many healthcare professions have referred to as epidemic levels (Ludwig 2002; Ludwig and Ebbeling 2005; de Ferranti and Mozaffarian 2008). Diet-related health conditions are directly associated with very high medical cost and loss of productivity worldwide. In the U.S. diet-related health conditions cost society an estimated $950 billion annually in medical costs. Obesity is a growing health concern, about 61% of Americans are obese and obesity is the main cause of type 2 diabetes. Changes in physical activity and eating habits (in overweight or obese individuals) predicted weight loss, and weight loss in turn was associated with reduced diabetes risk (Wing et al. 2004). Therefore, obesity prevention and treatment are a priority for the World Health Organization, the European Union, nongovernmental organizations, and industry. Most recently (January 2010) in the U.S., the first lady and the Department of Health and Human Services announced an initiative to reduce overweight and obesity in children and adults through better nutrition and regular physical activity.

Pulses such as lentils, dry beans, and peas are a great fit for the healthy eating pattern. They occupy two main places on the USDA *Food Pyramid-Dietary Guidelines* as part of the high-protein and the vitamin-rich vegetable groups. Their consumption is also encouraged by Canada's *Food Guide to Healthy Eating*, as they are high in protein, complex carbohydrates, resistant starch, dietary fibers, B-vitamins, folate, anthocyanins, and significantly low in fat and sodium, and gluten-free. Based on all these important nutritional and health attributes conveyed by pulse components, pulses are considered natural functional foods that have great potential to benefit the consumer.

EXTRUSION COOKING OF PULSES

Extrusion technology had been used effectively to reduce the cooking time, improve the textural, nutritional, and sensorial characteristics of the final pulse and pulse-based extrudates. Extrusion processing variables such as temperature, feed moisture content and screw speed are the main important parameters inducing the above changes. Therefore, extrusion processing provides potential for using pulses and pulse-based mixes in the fabrication of value-added food products and ingredients with high nutritional and economic value.

Studies on the effect of extrusion processing on the physical, chemical, functional, and nutritional properties of pulses and pulse-based extrudates, as well as modification of their components and inactivation of undesirable compounds (antinutrients) by extrusion cooking will be reviewed.

EXTRUSION OF DRY BEANS (PHASEOLUS VULGARIS L.)

Berrios and Pan (2001) evaluated the effect of extrusion processing at different screw speeds on some nutritional and expansion parameters of black bean flours ground to different particle sizes. For this purpose, bean flours ground to particle sizes ranging from 0.85 to 2.28 mm in diameter were processed through a Werner & Pfleiderer 37 mm Continua twin-screw extruder run at screw speeds of 400, 450, and 500 rpm. The screw configuration, die temperature (160°C), feed rate, and water content of the feed (25 kg h^{-1} at 18% moisture, wb) were kept constant. Proximate analysis demonstrated that the black bean flour used in this study contained 9.87% moisture, 24.45% crude protein, 2.50% crude fat, 3.94% ash, and 59.24% total carbohydrates (calculated by difference). These values were similar to those reported for black bean flours by Koehler et al. (1987) and Berrios, Swanson, and Cheong (1999). Proximate composition of extruded black bean flours was not significantly affected ($p \leq .05$) by different particle sizes within the selected screw speeds of 400, 450, and 500 rpm used in this study. Therefore, the proximate composition average values of pin milled control and extruded black bean flours were used to illustrate the effect of the different screw speeds on nutrient contents of the flours (Figure 8.2). As observed from Figure 8.2, the percent fat and moisture of the control flour were significantly different ($p \leq .05$) than flours extruded under the different screw speeds. Percent moisture in the control flour represented the adjusted moisture content of the pre-extruded flour of 18% (wb). Based on this initial moisture content, the average moisture lost between the pre-extruded and extruded bean flours was 53.6%. In general, the proximate composition average values of extruded bean flours were not significantly ($p \leq .05$) affected by the different screw speeds or the particle sizes studied.

The percent torque and expansion ratio, within the different particle sizes evaluated, increased with an increase in screw speed (Table 8.2). Conversely, die pressure decreased under the same stated conditions. Greater expansion of extruded material is related to crispiness and is therefore considered a desirable attribute in the fabrication of snacks and RTE foods. The finer pin milled flours (<35 mm diameter) extruded at 500 rpm demonstrated the greater expansion in this study. Conway (1971) indicated that the particle size of ingredients fed to an extruder has been

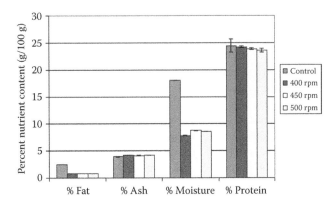

FIGURE 8.2 Proximate composition of control black bean flour and extruded black bean flours under different screw speeds.

shown to be important in single-screw extrusion. Additionally, Lai, Guetzlaff, and Hoseney (1989) and Mercier, Linko, and Harper (1989) reported that by increasing extruder screw speed, the expansion ratio of the extrudate material also increased. On the other hand, Chauban and Bains (1985) reported that the larger the particle sizes of wheat flour, the greater the expansion. The differences among the reported results may be due to the differences in raw materials and extruder type and processing conditions used in these studies.

Baking soda or sodium bicarbonate ($NaHCO_3$) is a leavening agent that has been widely used in the baking industry and found application in directly expanded

TABLE 8.2
Average Values of Expansion Ratio, Percent Torque, and Die Pressure of Black Bean Flours Extruded Under Different Particle Sizes and Screw Speeds

			Hammer-milled			
	Screw Speed	Pin-milled	0.85 mm	1.15 mm	1.53 mm	2.28 mm
Torque (%)	400 rpm	66.10 ± 0.74	72.40 ± 1.07	72.70 ± 0.67	69.50 ± 1.58	67.60 ± 1.07
	450 rpm	67.20 ± 0.79	71.50 ± 1.08	72.60 ± 1.17	70.20 ± 1.03	65.80 ± 0.92
	500 rpm	72.20 ± 0.79	75.70 ± 1.72	76.00 ± 1.25	72.50 ± 1.35	69.00 ± 1.25
Die pressure (psi)	400 rpm	184.00 ± 97.78	183.00 ± 75.14	223.51 ± 93.10	209.00 ± 98.60	129.00 ± 75.34
	450 rpm	148.00 ± 84.30	163.00 ± 70.40	151.00 ± 79.23	216.00 ± 95.36	107.80 ± 68.99
	500 rpm	130.00 ± 66.50	138.00 ± 55.34	149.00 ± 59.71	174.00 ± 71.99	107.00 ± 52.50
Expansion Ratio	400 rpm	6.29 ± 0.66	5.58 ± 0.75	4.99 ± 0.52	4.76 ± 0.47	4.75 ± 0.57
	450 rpm	6.33 ± 0.47	5.81 ± 0.81	5.08 ± 0.59	4.90 ± 0.30	4.71 ± 0.53
	500 rpm	6.74 ± 0.86	6.17 ± 0.62	5.52 ± 0.71	5.12 ± 0.49	5.08 ± 0.46

extruded cereal products to enhance their physical properties (Lai, Guetzlaff, and Hoseney 1989; Lajoie, Goldstein, and Geeding-Schild 1996; Parsons, Hsieh, and Huff 1996). Berrios et al. (2002) added $NaHCO_3$ in the range of 0.0% to 2.0% to black bean flours for the fabrication of expanded snack type products by twin-screw extrusion processing. A Leistritz Micro-18-GL co-rotating twin-screw extruder was used at a constant screw speed of 200 rpm. The temperature profile selected for this study was 23°C, 80°C, 100°C, 120°C, 140°C, and 160°C with the first temperature corresponding to the feed barrel section and the last to the die section. The die consisted of two circular openings 2 mm in diameter and the bean flour moisture adjusted to 20% (wb). They reported that the diameter measurements taken at the nod and at the area between the nods showed that these two regions increased proportionally with increased $NaHCO_3$ in the extrudates and that there was a significant difference between the control extrudate and those with added $NaHCO_3$ (Figure 8.3). And that the standard deviations within the extrudates with added $NaHCO_3$ were very small, within the range of 0.28–0.35 at the nod and from 0.26 to 0.36 at the area between the nods. They further indicated that the increase in extrudate diameter is reflected in the expansion ratio. Expansion ratio increased 1.6 fold at the nod and 1.4 fold at the area between the nods, between the control extrudate and extrudate with 0.1% $NaHCO_3$ addition. These differences increased to 2.0 and 1.8 fold for extrudates with 0.5% $NaHCO_3$ addition, at the same indicated areas. They stated that the difference in expansion distribution pattern on an extrudate would be important to consider in the fabrication of particular snack type products. Scanning electron micrographs of cross sections of black bean extrudates was used to illustrate the change in volume expansion with an increase of sodium bicarbonate ($NaHCO_3$) in the extrudate (Berrios et al. 2004).

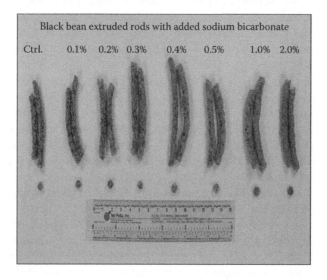

FIGURE 8.3 Expansion of black bean extrudates containing different concentration of sodium bicarbonate.

Scanning electron microscopy (SEM) observation of cross sections of black bean control extrudate demonstrated the presence of air cells of irregular size (Figure 8.4a, f). Also, it showed that control extrudate had the smallest overall expansion when compared to extrudates with added $NaHCO_3$. Moreover, the number of air cells increased and cell wall thickness decreased with an increase in $NaHCO_3$ addition. The authors indicated that the combined effect of the two factors mentioned above, caused the collapse of the cell walls and the appearance of large void spaces within the extrudates (Figure 8.4b, f). This was observed as an increase in volume of the extrudate, with an increase in the amount of $NaHCO_3$. Similarly, Lai, Guetzlaff, and Hoseney (1989) reported that the addition of sodium bicarbonate or sodium carbonate to wheat starch was shown to improve expansion but weakened the structure of the wheat starch extrudate. Berrios et al. (2004) concluded that the maximum expansion was obtained with 0.5% $NaHCO_3$ addition and that expansion characteristics of black bean extrudates corroborated SEM observations. Authors stated that extrusion conditions, which involved the use of heat and moisture, provided the necessary conditions for the release of CO_2 from $NaHCO_3$ during processing which caused changes in volume expansion of the extrudate.

Expansion and textural characteristics of extrudates of different types of dry beans have been the topic of several studies, using single- and twin-screw extruders. Avin, Kim, and Maga (1992) extruded red bean flours, adjusted to moisture content of 15% or 25% (wb), in a Brabender 19 mm laboratory single-screw extruder run at screw speeds of 80, 120, and 160 rpm and die temperatures of 90°C, 110°C, and 124°C. They reported that extrusion temperature significantly ($p < .01$) influenced torque, expansion, breaking strength, and water absorption index (WAI); feed moisture influenced torque and expansion; while screw speed only influenced extrusion yield. Edwards et al. (1994) reported extrusion characteristics of small white beans processed through a Werner & Pfleiderer 37 mm Continua twin-screw extruder, using three screw configurations of increased energy intensity. Extrudates were made over a moisture range of 15%–25% (wb) at a constant screw speed of 225 rpm, and die temperatures in the range of 114°C–177°C. As the energy intensity increased, expansion ratio, starting viscosity, and hot viscosity increased, while bulk density and ending viscosity decreased. They further indicated that all extrusion properties could be mathematically modeled as a function of feed moisture, die temperature, and specific mechanical energy (SME). Karanja et al. (1996) used a Brabender bench top single-screw extruder run at screw speeds of 49–107 rpm and five different die temperatures (100°C, 120°C, 140°C, 160°C, and 180°C) to evaluate expansion, WAI and water solubility index (WSI) of Canadian Wonder bean flours conditioned at 25% and 30% moisture (wb). Their results showed that expansion and WAI of the extrudate increased while WSI and bulk density decreased with increased extrusion temperature. Similar results were later reported by Martin-Cabrejas et al. (1999) for dry bean (cv. Horse-head) extrudates, processed in the same type of extruder, at die temperatures of 140°C, 160°C, 170°C, and 180°C. Balandran-Quintana et al. (1998) extruded pinto bean flours, adjusted to moisture content of 18%, 20%, and 22% (wb), using a Brabender 19 mm single-screw extruder run at screw speed (150, 200, and 250 rpm) and three different die temperatures (140°C, 160°C, and 180°C). They indicated that temperature and feed moisture conditions also significantly ($p < .05$)

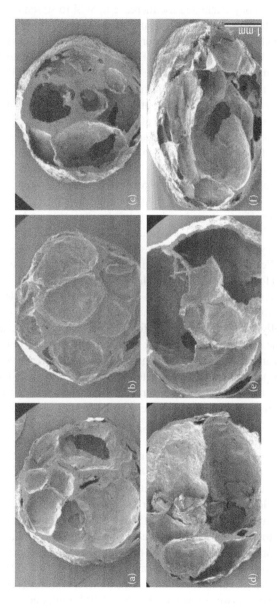

FIGURE 8.4 SEM cross sections of black bean extrudates with NaHCO$_3$ at the following levels: (a) 0%; (b) 0.1%; (c) 0.2%; (d) 0.3%; (e) 0.4%; and (f) 0.5%. Magnification bar = 1 mm.

affected bulk density, expansion, and WAI. Similar to Avin, Kim, and Maga (1992) and Karanja et al. (1996) they reported that expansion and WAI increased with increasing extrusion temperature. However, contrary to Karanja et al. (1996) and Martin-Cabrejas et al. (1999), they also reported an increase in WSI with increasing extrusion temperature. They attributed the increase in WSI to starch depolymerization at high extrusion temperatures. Their results confirmed those of Anderson et al. (1969) and Anderson (1982), who extruded corn and sorghum, and Gujska and Khan (1990), who extruded high starch fractions (HSF) of navy beans, pinto beans, and chickpeas. The lack of agreement among the different studies on WSI, as a consequence of high extrusion temperatures, requires future research clarification.

The effect of extrusion cooking on the nutritional value and digestibility of dry bean protein and starch and the inactivation of anti-nutritional components has been investigated by various researchers. Steel, Sgarbieri, and Jackix (1995) evaluated the effect of extrusion cooking on the inactivation of anti-nutrients and nutritional value of freshly harvested and hard-to-cook (HTC) brown Carioca SH bean cultivar. They used a Brabender bench top 19 mm single-screw extruder run at screw speed of 100 rpm, die temperature of 190°C, and feed moisture conditioned to 21.5% (wb). They reported that extrusion effectively reduced trypsin-chymotrypsin and hemagglutinin activities 88% and 95%, respectively. Previously, Edwards et al. (1994) reported a reduction of 85% of trypsin inhibitor activity on extrusion of small white bean flour extruded through a Werner & Pfleiderer 37 mm Continua twin-screw extruder with die temperature of 114°C–177°C. They further indicated changes in SME or die temperature caused large changes in trypsin inhibitor activity. Complete inactivation of the trypsin inhibitors is difficult to achieve, and residual activity between 10%–20% is normally found in processed foods (Anderson, Rackis, and Tallent 1979). To evaluate the nutritional value of the bean extrudates, their flours were fed to rats, and diet consumption, body weight gains, Protein Efficiency Ratio (PER), and other nutritional parameters were evaluated and compared to rats fed a control casein diet. They reported no statistical differences in intake between the diets, but the rate of growth was approximately double for rats in the casein diet compared to those fed the bean diet. The apparent digestibility and the net utilization of the extruded bean flour protein were significantly lower than that obtained with casein. Low digestibility of dry bean seed protein has been previously stated by different researchers (Sgarbieri and Whitaker 1982, 1989). However, the apparent biological value, which is an indicator of the degree of utilization of the absorbed amino acids, of the extruded bean flour protein was similar to that of the casein diet. The authors indicated that it was important to notice that the results of all nutritional parameters studied was similar for both freshly harvested and HTC bean extrudates. Karanja et al. (1996) used a Brabender bench top 19 mm single-screw extruder to evaluate different processing conditions for lectin inactivation capacity of HTC Canadian Wonder bean flours. They evaluated five different die temperatures (100°C, 120°C, 140°C, 160°C, and 180°C), two screw speeds (49 and 107 rpm), and two feed moistures (25% and 30%, wb). They indicated that at die temperature of 100°C, lectin activity was reduced at the lower screw speed. However, the greatest reduction on lectin activity was achieved at temperatures of 120°C and greater. Additionally, the authors indicated that lectin inactivation was higher with increased moisture in the flours. They concluded that lectin anti-nutrient activity decreased with increased extrusion temperature, increased moisture in the flour, and longer residence time. Using a similar type of extruder, die temperatures (140°C, 160°C, and 180°C), screw speeds (150, 200, and 250 rpm) and feed moisture content (18%, 20%, and 22%, wb), Balandran-Quintana et al. (1998) extruded pinto bean flours. Their results confirmed those reported by previous researchers (Edwards et al. 1994; Steel, Sgarbieri, and Jackix 1995; Karanja et al. 1996) that the extrusion temperatures and feed moistures used in the studies were highly effective for inactivating trypsin inhibitor activities. They also reported that extrusion processing increased the *in vitro* protein digestibility of the bean extrudate 8.3%, compared to the raw bean meal control, for all experimental conditions.

Alonso et al. (2000) studied the comparative effect of extrusion cooking, dehulling, soaking, and germination on protein and starch digestibility and reduction of anti-nutritional factors in kidney and faba beans. Extrusion processing was performed in a Clextral X-5 45 mm twin-screw extruder, operated at 100 rpm, 383 and 385 g/min feed rate at 25% moisture and die temperature of 152°C and 156°C. They reported that even though trypsin, chymotrypsin, α-amylase inhibitor activities of the two types of beans were decreased significantly by dehulling, soaking, and germination, haemagglutinating activity was not affected by the indicated conventional processing methods. However, extrusion cooking effectively inactivated all the antinutrients in the two beans under study, without altering protein content. Additionally, extrusion was the best method for improving protein and starch digestibility.

The effect of extrusion processing on some carbohydrate-oligosaccharide fractions (fructo-oligosaccharides and dietary fibers) of pulses has been studied over the last two decades, due to their role on product acceptability and health attributes. The effect of domestic and technological processes (soaking of pulse seeds in tap water or sodium bicarbonate solution, cooking of unsoaked or soaked seeds, and cooking under pressure or by autoclaving) leads to a larger decrease in α-galactosides content, according to a number of researchers (Reddy, Salunkhe, and Sharma 1980; Abdel-Gaward 1993, Sánchez-Mata, Cámara-Hurtado, and Díez-Marqués 1999; Han and Baik 2006). However, only a few studies have been done using extrusion cooking. Borejszo and Khan (1992) used a Wenger TX-52 twin-screw extruder to process pinto bean HSFs at die temperatures in the range of 110°C–163°C, screw speed of 300 rpm, and feed moisture of 18.8% (wb), to determine the effect of processing conditions on flatulence-causing sugar reduction. They reported that the levels of flatulence-causing sugar were lower in extruded compared to nonextruded samples with higher reduction at higher process temperature. Sucrose content decreased by 76% in samples extruded at 163°C, while raffinose and stachyose contents were reduced 47% to 60%, respectively. The authors also indicated that moisture content in the extruder barrel did not significantly affect sugars in the samples. Berrios and Pan (2001) processed black beans through a Werner & Pfleiderer 37 mm Continua twin-screw extruder run at screw speeds of 400, 450, and 500 rpm, die temperature of 160°C, and feed moisture of 18% (wb). They reported that the total oligosaccharides concentration of control and extruded black bean flours was not affected by difference in particle sizes, but was reduced significantly with an increase in screw speed (Figure 8.5). Berrios et al. (2002) processed black bean flours at 20%

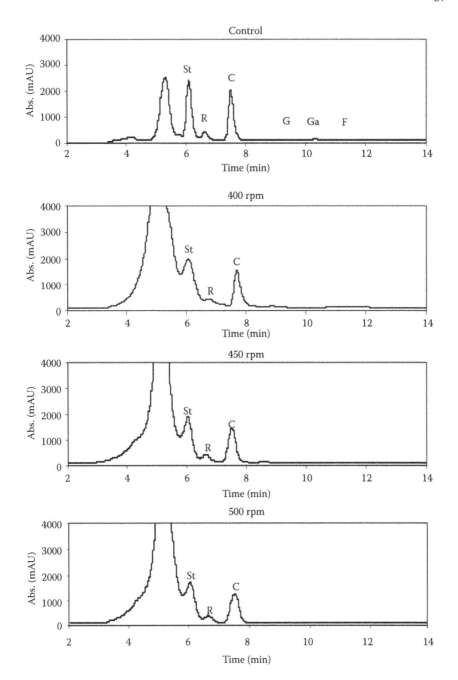

FIGURE 8.5 HPLC sugars profile in control black bean flours and black bean flours extruded under different screw speeds (St: stachyose, R: raffinose, C: cellobiose, G: glucose, Ga: galactose, F: fructose).

moisture, with sodium bicarbonate ($NaHCO_3$) added at levels from 0.0% to 2.0%, on a Leistritz 18 mm twin-screw operated at a screw speed of 200 rpm and die temperature of 160°C. The study demonstrated that extrusion conditions and $NaHCO_3$ addition reduced stachyose and sucrose in the extrudates. However, the individual free sugars studied were not significantly affected by the process or $NaHCO_3$ addition.

Contrary to Artz, Warren, and Villota (1990), who did not observe changes in fiber modification resulting from extrusion processing, Berrios et al. (2002) observed a decrease in insoluble fiber (IF) and an increase in soluble fiber (SF), which confirms that extrusion processing causes a redistribution of the IF to SF fractions, as has been previously reported by other researchers (Björck, Nyman, and Asp 1984; Lintas et al. 1995; Gualberto et al. 1997). This fiber fraction redistribution possibly resulted from hemicellulose depolymerization. Additionally, Martin-Cabrejas et al. (1999) reported that a considerable redistribution of IF to SF occurred as a result of extrusion, although total dietary fiber did not change. The pectic polysaccharides, arabinose, and uronic acids were the main sugars solubilized by the process. Most recently, Berrios et al. (2010) evaluated the changes in sugar fraction and dietary fiber composition of lentil, dry pea, and garbanzo flours as an effect of extrusion cooking for the potential fabrication of value-added, nutritious pulse-based snacks with low flatulence-promoting compounds. A Clextral EVOL HT32-H twin-screw extruder with co-rotating and closely intermeshing screws was used at screw speed of 500 rpm, die temperature of 160°C, and feed moisture of 17% (wb). They found that the extruded pulse flours exhibited consistently lower total available carbohydrates (TAC) values than the raw flours, and that the difference was highly significant ($p < .05$) in the case of chickpeas (Table 8.3). They indicated that the observed lower TAC content on the extruded pulse flours could be attributed to starch and other polysaccharides breaking down into lower molecular weight components. Additionally, they reported that extrusion processing did not significantly ($p < .05$) affect the total sugar content of dry pea and lentil flours. However, extrusion processing decreased the concentration of the raffinose family of oligosaccharides (raffinose and stachyose) in the extruded pulse flours. The oligosaccharides raffinose and stachyose are of particular importance since a number of researchers (Calloway and Murphy 1968; Reddy,

TABLE 8.3

Total Available Carbohydrates in Pulse's Flours (g/100 g)

	Lentils	Chickpeas	Dry Peas
Raw	62.490[a*, a**]	64.463[bc, ab]	65.708[a, b]
Extruded	61.426[a, b]	52.536[a, a]	63.921[a, b]

Values represent means of three replicate analyses.

[*] Means within a column followed by different letters are significantly different ($p < .05$).

[**] Means between columns followed by different letters are significantly different ($p < .05$).

Significant differences within and between columns are separated by a comma.

Salunkhe, and Sharma 1980; Fleming 1981) have demonstrated that these sugars are the principal causes of flatulence in humans and animals with monogastric digestive systems. Therefore, the observed reduction of these sugars, as a consequence of extrusion processing, indicates great potential for the fabrication of pulses-based snacks with reduced concentration of flatulence-producing factors.

EXTRUSION OF DRY PEAS (PISUM SATIVUM L.)

Dry peas are the second most important legume pulse in the world after dry beans. Dry peas are the only pulse from which protein has been extracted to produce protein concentrate and isolate at a commercial level. However, mainstream dry pea foods are made from whole and dry pea flours. Processing of dry peas into edible food has been limited mainly to domestic cooking. The use of extrusion cooking has been so far limited to basic studies on functionality, modification of nutritional component, and inactivation of anti-nutritional factors. Alonso et al. (2000) used a twin-screw extruder operated at 100 rpm, die temperatures of 148°C, and feed moisture of 25%, to evaluate the effect of processing on structural changes (mainly structural solubility and functional properties) of pea proteins. Extruded pea showed lower solubility than raw ones. This insolubilization of the protein was indicated to be caused by noncovalent interactions accompanied by disulphide bond formation promoted by the extrusion process. Della Valle, Quillien, and Gueguen (1994) have previously indicated that the decrease in protein solubility of extruded pea flour was attributed to the formation of noncovalent and disulphide bonds that could partially take place at the extruder die, by the influence of shear rate at the die. This report is in agreement with the hypothesis of Otun, Crawshaw, and Frazier (1986) who stated that protein alignment is favored by the existence of various flow conditions, such as shear rate and viscosity, which are necessary for the required texture of the product. Della Valle, Quillien, and Gueguen (1994) also reported that extrusion cooking significantly increased the water holding capacity (WHC) and the WSI of the extruded pea flour with respect to the raw flour. The greater WHC displayed by the extrudate was attributed to physical retention of water by capillary action in the new structure formed by the aggregation of proteins, confirming similar previous reports (Narayana and Narasinga Rao 1982; Gujska and Khan 1990) on other pulse flours. The authors further indicated that no changes in total nitrogen content occurred as a result of extrusion, but total and free sulphhydryl group and disulphide contents decreased largely due to extrusion cooking. Alonso, Grant, and Marzo (2001a) evaluated the effect of extrusion cooking and amino acid supplementation on the nutritional quality of the pea extrudates fed to rats. They indicated that growth, apparent nitrogen digestibility, biological value, and net protein utilization values for supplemented extruded pea (SEP) were similar to those values determined on a casein control diet. They also reported that SEP reduced serum total cholesterol, LDL, and cholesterol/HDL ratio; increased kidney and adrenal weights; and reduced liver weight. The authors concluded that extrusion treatment of pea flour preserved its hypocholesterolemic properties and improved its nutritional quality. The effect of extrusion on mineral bioavailability, carbohydrate composition, and inactivation of anti-nutritional factors in pea flour was studied by Alonso et al. (2001). They used a twin-screw extruder operated at screw speed of 100 rpm, die temperature of 150°C, and feed moisture adjusted to 25% in the extruder barrel. The study showed that the apparent absorption of Fe, Ca, and P significantly increased in extruded pea flour compared with unprocessed (raw) flours. Additionally, with amino acid supplemented pea diet, extrusion significantly increased the mineral (Ca, Mg, Fe, Cu, and P) apparent fecal absorption in rats. The study also showed that starch and nonstarch polysaccharides, stachyose, and inositol hexaphosphate, as well as tannin and lectin were significantly reduced as a consequence of extrusion cooking. Van der Poel, Stolp, and Van Zuilichem (1992) had previously reported a significant reduction on anti-nutritional factors (lectin and trypsin inhibitors) on the flours of two pea varieties processed by extrusion. Most recently, Berrios et al. (2010) indicated that dry pea showed the highest concentration of TAC, followed by chickpea and lentil and that extrusion processing did not significantly affect the TAC content of dry pea and lentil flours (Table 8.3). Moreover, extrusion processing decreased the concentration of raffinose in dry pea, chickpea, and lentil extrudates (Table 8.4). These studies demonstrated the overall benefit of extrusion cooking by increasing bioavailability of nutrients and decreasing undesirable compounds in the final extrudate.

EXTRUSION OF CHICKPEAS (CICER ARIETINUM L.)

Chickpeas are the third most important legume pulse in the world after dry beans and dry peas (Singh, Subrahmanyam, and Kumar 1991) and the most important legume crops in India. India produced 5.97 million tons of chickpea during 2008, which was about 75% of the world's production (FAO 2008). Chickpeas are considered one of the most nutritious of any dry edible pulse and with the least content of anti-nutritional factors (Chavan, Kadam, and Salunkhe 1986).

TABLE 8.4
Soluble Sugar Contents in Dry Pea Flour Samples (g/100 g)

	Ribose	Fructose	Glucose	Galactose	Sucrose	Maltose	Raffinose
Raw dry pea	0.521	0.124[a]	0.042	0.722	0.647[a]	0.191	1.564[b]
Extruded dry pea	ND	ND	ND	ND	1.299[b]	ND	0.816[a]
Raw lentil	0.305[b*]	0.085[a]	ND	ND	0.697[a]	0.039[a]	1.208[b]
Extruded lentil	0.136[a]	0.103[a]	0.033[a]	ND	0.601[a]	0.133[b]	0.222[a]
Raw chickpea	0.058[a*]	0.156[a]	0.028[a]	0.184[b]	0.794[b]	0.168[a]	0.605[a]
Extruded chickpea	0.062[a]	0.145[a]	0.041[b]	0.066[a]	1.003[b]	0.177[a]	0.754[ab]

Values represent means of three replicate analyses.
* Means within a column (under individual pair of pulse data) followed by different letters are significantly different ($p < .05$).
ND = not detected

The high content of starch in chickpeas of 40%–50% (Wang and Daun 2004; Huang et al. 2007) and high amylose content of 28.20%–52.82% (Singh et al. 2010) may favor high sheer processing, such as extrusion cooking, for the fabrication of directly expanded snack food products.

Gujska and Khan (1990) studied the effect of extrusion temperature on expansion and functional properties of the HSF of chickpea. They processed the HSF on a Wenger X-5 laboratory extruder at die temperatures in the range of 110°C–150°C. The best extrudate expansion index of 1.9 was obtained at 132°C. However, this expansion was considered the lowest, when compared to that of pinto bean and corn of 2.4 and 2.7, respectively. The lowest expansion index is attributed to the HSF of chickpea having higher protein and lipid contents than pinto bean and corn. Previous reports have indicated that expansion of cereals decreases with increased protein content (Faubion, Hoseney, and Seib 1982) and increased lipid content (Linko, Colomna, and Mercier 1981). Also, the authors reported that die temperature of 110°C showed the minimum expansion and the highest die temperature of 150°C resulted in burnt products. Oil emulsification capacity for chickpea extrudates, at the optimum die temperature of 132°C, was about 1.2 and 2.0 times higher than pinto bean and corn, respectively. This suggests that chickpea protein content and composition was the principal determining factor in response to these functional attributes. Under the same extrusion conditions, trypsin inhibitor activity was reduced by about 70%–85% and hemagglutinin was completely inactivated (Gujska and Khan 1990). Response surface analysis, which provides a means for optimization of formulation and process, was used by Bhattacharya and Prakash (1994) to develop an extruded snack from chickpea and rice flour blends. They extruded, using a single-screw extruder, three blends of rice–chickpea flours (100:0, 90:10, and 80:20) at three different die temperatures (100°C, 125°C, and 150°C). The authors indicated that incorporation of chickpeas into rice flour decreased torque and product expansion. It also increased bulk density and shear and breaking strength, indicating poor textural effects of chickpea flour inclusion. Similar results were reported by Shirani and Ganesharanee (2009) under a similar range of die temperatures. They added that the poor expansion and texture observed with an increase in inclusion of chickpea in the mix may be attributed to the high protein and dietary fiber content in chickpea compared to rice. This confirms what Faubion, Hoseney, and Seib (1982) reported for cereal snacks who indicated that higher protein in cereals decreased expansion of the final product. The above studies dealt with relatively simple raw material compositions. A more complex mixture, involving starch and protein fortification, may promote expansion and nutritional quality of a chickpea-based snack. The nutritional quality of a chickpea, based on iron bioavailability, was evaluated on extruded and conventional home-cooked chickpea by Poltronieri, Arêas, and Colli (2000). The results showed that extrusion cooking was as good as home-cooking with regard to preserving the bioavailability of iron, based on a standard rat study. Additionally, Cardoso-Santiago and Arêas (2001) extruded a mixture of chickpea–bovine lung (100:0, 95:5, and 90:10), at die temperature of 130°C and feed moisture of 13%, to develop snacks with improved iron and nutritional content. They reported that the extruded snacks contained a similar amount of total iron as the nonextruded samples and also, that the developed extruded chickpea snacks fortified with bovine lung were able to provide up to 30% of the iron RDA in a 30 g pack. The protein nutritional quality of the snacks, measured by the net protein ration (NPR) on standard animal bioassay, was similar to that of casein. Most recently, Moreira-Araujo, Araujo, and Arêas (2008) added corn to a mixture of chickpea and bovine lung to develop an acceptable, nutritious, high iron content extrudate to control iron-deficiency anemia in preschool Brazilian children. The fortified chickpea extrudate caused a significant drop in anemia incidence from 61.5% to 11.5% in the test group. The fortified chickpea snack was highly acceptable by the children and no undesirable effects were observed. Chickpea nutritional value has also been improved by defatting and dehulling techniques, among others. Optimization of extrusion cooking of defatted chickpea flour by Response Surface Methodology (RSM) demonstrated that expansion ration increased steadily with a decrease in feed moisture. Optimum product was obtained at die temperature of 132°C, feed moisture of 13%–14%, and screw speed of 200 rpm. The authors indicated that the defatted chickpea extruded snack was rated higher than a commercial corn snack (Batistuti, Barros, and Arêas 1991). Carrillo et al. (2000) evaluated the nutritional impact of extrusion of dehulled chickpeas on fresh and HTC chickpeas. A 19 mm single-screw extruder, operated at 189.5 rpm with die temperature set at 151°C, was used to process the flours. They found that the process improved *in vitro* protein digestibility significantly (9.3%–21.7%), apparent digestibility (6.5%–6.6%), true digestibility (5.8%–7.5%), protein efficiency ratio (27.2%–36.9%) and net protein ratio (5.2%–14.0%) of the fresh and HTC chickpea flours. Patil et al. (2005) achieved the highest protein digestibility in chickpea extruded at die temperature of 180°C and feed moisture of 28%. However, the expansion of the product was best at die temperature of 180°C and 20% moisture. Most recently, Meng et al. (2010) used RSM to evaluate the effects of extrusion conditions on system parameters and physical properties of a chickpea flour-based snack. Extrusion experiments were performed on a co-rotating Werner & Pfleiderer ZSK-57 twin-screw extruder. The screw speed, die temperature, and feed moisture were in the range of 250–320 rpm, 150°C–170°C, and 16%–18%, respectively. Their study demonstrated that product temperature and die pressure were affected by all three process variables, while motor torque and SME were only influenced by screw speed and barrel temperature. Desirable products, characterized by high expansion ratio and low bulk density and hardness, were obtained at low feed moisture, high screw speed, and medium to high barrel temperature. The expansion ratio of the extrudates was in the range of 3.06 and 4.99, similar to those reported for other pulses extruded under similar conditions (Patil et al. 2007; Berrios et al. 2002, 2004, 2010), and those reported for cereal-based extrudates (Ilo et al. 1996; Singh and Smith 1997). All these results indicate that nutritious snacks with desirable expansion and texture properties may be fabricated with chickpea flours and chickpea-based formulations, as well as from other pulses, by extrusion processing.

PRESENT AND FUTURE WORLDWIDE MARKET OF SNACK FOODS

The international market of snack foods suggest that the global volume sales of snack foods are expected to increase over the next three years by 9.8 million U.S. tons, due to the successful change in market attitude in favor of the consumer turning away from traditional snacks considered as salty and fatty type foods. Latin America, Asia Pacific and Eastern Europe are considered regions of the world that present the greatest opportunities for manufacturers of snack foods. However, the U.S. is the largest single market of cereal snacks, which represents approximately 42% of the total international market. Cereal bars represent the largest section of the cereal snack food market, equivalent to 70% of the total market. The snack fruit sector was worth approximately $25 million in 2005 compared to approximately $12 million in 1998, representing a 2.1 fold increase in market sales according to a recent U.S. Department of Agriculture (USDA) report. Trans-fat-free (TFF) snacks have also attracted the attention of consumers worldwide. Therefore, TFF snacks are expected to increase their sales as more consumers start becoming aware of heart disease and other health related problems associated with high fat consumption. Another important sector of the snack market that is continuously attracting the attention of the health conscious consumer is the gluten-free market. This is due to the fact that more than 7000 people every year are diagnosed as allergic to gluten, a disease known as celiac disease. There is great expectation that the market of gluten-free food products will increase largely in years to come and that it will revolutionize the cereal market. Main global food manufacturers, including General Mills are moving into the production of gluten-free snacks. In the U.S., one in 133 consumers are considered to be allergic to gluten. According to the USDA, the gluten-free market was expected to increase to approximately $1.7 billion by the year 2010 and is expected to increase in the coming years due to a reported estimated worldwide increase in the number of people suffering from gluten intolerance by a factor of 10.

Regarding children's food, a recent Canadian study indicates that nearly 90% of kid products, often marketed as "fun foods," did not meet established nutritional standards. What's more, 62% of the foods that researchers deemed to be of "poor nutritional quality" made positive nutritional claims on the package—such as being low-fat and containing essential nutrients. During 2006–2007, major food firms Kraft, Mars, PepsiCo, Dannon, Campbell, Bachman Company, Rudolph Foods Company, Shearer's Foods, and Ubiquity Brands supported the alliance, to assure reformulation of certain products, as well as to introduce new lines of healthier snacks for kids.

Regarding the general adult consumer, new reports indicate that on-the-go eating in the U.S. has markedly changed in the last decade due to consumer's job locations moving away from their main towns. Thirty-three percent of Americans regularly skip meals and rely on snack foods, in part fuelled by media reports about the health and weight-loss advantages of eating frequent small meals, on-the-go options from the food service industry.

Marketing companies are predicting that by 2011 commute times in Spain and Italy will increase by nearly 20 minutes and in France by 10 minutes. Other countries will see an average increase of between two and six minutes. These statistics indicate significant opportunities for industry to formulate snacks that offer time saving and cater to more indulgent tastes, while remaining convenient and healthy. The high cost of cereal grains in recent years and the consumer demand for healthy alternatives provide a great opportunity for the development of value-added legume pulse-based foods in a convenient form such as snacks.

AVAILABLE FOOD PRODUCTS FROM LEGUME PULSES

Legume pulses generally have a long cooking time. Therefore, to reduce day-to-day cooking time, pulses are processed into different products, which are later on used as food either by further cooking, frying or roasting. The methods, which are popularly followed in the developed nations, are those of canning the cooked pulses and aseptic packaging of the RTE precooked pulses like bean chili. However, in the developing countries pulses are processed in the form of a shelf stable product so that investment on packaging is minimal. They are processed into low moisture foods like baries, papads, leblebi, etc. Dr. Sethi (2008) covered effectively the most preferred foods in Indian cuisine. The following summary highlights the main commercial products based on legume pulses:

- Whole pulses: canned, micronized, split, and flaked
- Products made with whole pulses: soups, chilies, fried peas, lentils, and chickpeas, refried, beans, frozen entrees, snack chili mixes (hummus and others), roasted, and fried snacks
- Ground pulses: flours or powders (including naturally fermented flours)
- Products made with ground pulses: specialty bakery mixes, vegetarian foods (RTE frozen entrees and veggie burgers), pasta and noodles, extruded-fried snacks, and dips.

This demonstrates that the variety of pulse-based foods are very limited, compared to those made from soybeans, wheat, and other cereal grains. Additionally, most of these products do not match the on-the-go, sensible, nutritious, snacking desirable pattern. Food industries and food service sectors have a great opportunity of increasing their share of this growing snack market by adapting to new consumer attitudes and by developing legume pulse-based foods in a more convenient form, with indulgent tastes, while remaining safe and healthy.

DEVELOPMENT OF LEGUME PULSE-BASED EXTRUDED SNACK FOODS

Extrusion is used commercially to produce high value breakfast and snack foods based on cereals such as wheat or corn. However, this processing method is not being commercially used for legume pulse seeds due to the perception that they do not expand well in extrusion. The rise in consumer demand for convenience foods, health awareness and the role of pulses in meeting these demands motivated researcher Dr. Jose De J. Berrios of ARS' Western Regional Research Centre in Albany, California, Dr. Juming Tang and Dr. Barry Swanson at Washington State University to work on the development of unique, healthful, crunchy, great-tasting expanded snack-type foods from lentils, garbanzos, dry peas, and beans. Extrusion technology isn't new, but the scientists are the first to determine the processing speeds, heating temperatures, amounts of moisture, and formulations that create consistent, desirable textures, and tastes from every batch of pulse flour. The general process is outlined in the flowchart shown in Figure 8.6.

Several breakfast cereal type foods and snacks that come in a variety of shapes, from crisp bits and balls, to tubular puffs, were developed (Figure 8.7). These convenient food alternatives have already proven popular with 550 tasters at a lentil festival in Washington DC, who were encouraged to try the products in 2006.

Raw materials: pulse flours + food ingredients

↓

Homogenizing

↓

Grinding

↓

Mixing/blending

↓

Extrusion cooking

↓

Cooling/drying

↓

Coating

↓

Packaging

FIGURE 8.6   General process outlined for the fabrication of pulse and pulse-based snack foods.

The ARS presented festival-goers with barbeque, cheese, classic, and plain flavor snacks, as well as sweet and plain cereal foods. They were then asked to rate the product on a sliding scale from "like a lot" to "dislike a lot."

The combined result of flavor preferences demonstrated that the overall average liking percentage for the coated snack type lentil products was about 80%, independently of the type of coating used, and about 60% for the plain snack. The study also demonstrated that the overall average liking percentage for the breakfast cereal type product coated with brown sugar was above 80%, and above 60% for the plain one (Figure 8.8).

QUALITY OF THE PULSE-BASED EXTRUDED SNACK PRODUCT

The nutritional, healthy, tasty, attractive, and convenient, expanded, extruded legume pulse-based products with crunchy texture, in the form of snacks and breakfast cereal-type foods, are rich in protein and dietary fiber, very low in sodium and fat, cholesterol-free and gluten-free. The functional and superfood attributes of the high protein, low-calorie snacks could have a positive impact on reducing the risk of a variety of

FIGURE 8.7 High protein and fiber-rich extruded snacks and breakfast cereal-type foods made from lentils, dry beans, chickpeas, and dry peas based formulations by scientists from USDA-ARS and Washington State University.

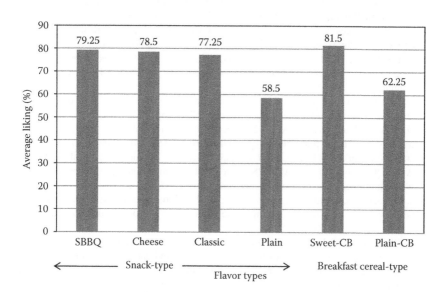

FIGURE 8.8 Overall liking of the consumers for different flavor coated lentil extruded products.

health problems such as obesity, heart disease, cancer and diabetes. A patent application, titled "Extruded Legumes," was summated to protect the value-added process and products developed by scientists from USDA ARS (Dr. Berrios) and Washington State University (Drs. Tang and Swanson) and the subject application was published on June 19, 2008 and was assigned publication number US-2008-0145483-A1.

Pulses are considered natural functional foods that have great potential to benefit the consumer. Therefore, making available pulse and pulse-based food products in convenient and diverse forms (RTE breakfast cereals, snack, and bar-type foods of different shapes and flavors, instant noodle-type, and beverages, among others) would open new market opportunities, create new jobs, help the sustainability of pulse crops worldwide and ensure that consumers enjoy a product that is unique, healthful, and easily distinguishable from traditional snack foods.

REFERENCES

Abdel-Gawad, A.S. 1993. Effect of domestic processing on oligosaccharide content of some dry legume seeds. *Food Chem* 46(1): 25–31.

Alonso, R., Grant, G. and Marzo, F. 2001. Thermal treatment improves nutritional quality of pea weeds (Pisum Sativum L.) without reducing their hypocholesterolemic properties. Nutr Res 21(7): 1067–1077.

Alonso, R., Orue, E., Zabalza, M.J., Grant, G. and Marzo, F. 2000. Effect of extrusion cooking on structure and functional properties of pea and kidney bean proteins. *J Sci Food Agric* 80(3): 397–403.

Alonso, R., Rubio, L.A., Muzquiz, M. and Marzo, F. 2001. The effect of extrusion cooking on mineral bioavailability in pea and kidney bean seed meals. *Animal Feed Sci Technol* 94(12): 1–13.

Anderson, R.A. 1982. Water absorption and solubility and amylograph characteristics on roll-cooked small grain products. *Cereal Chem* 59: 265–269.

Anderson, R.A., Conway, H.F., Pfeifer, V.F. and Griffin, E.J., Jr., 1969. Roll and extrusion-cooking of grain sorghum grits. *Cereal Sci Today* 14: 372–375, 381.

Anderson, R.L., Rackis, J.J. and Tallent, W.H. 1979. Biologically active substances in soy products. In *Soy protein and human nutrition*, eds. H.L. Wilcke, D.T. Hopkins and D.H. Waggle, 209–233. New York: Academic Press.

Artz, W.E., Warren, C. and Villota, R. 1990. Twin-screw extrusion modification of a corn fiber and corn starch extruded blend. *J Food Sci* 55: 746–750.

Avin, D., Kim, C.H. and Maga, J.A. 1992. Effect of extrusion variables on the physical characteristics of red bean (phaseolus vulgaris) flour extrudates. *J Food Proc Pres* 16: 327–335.

Balandran-Quintana, R.R., Barbosa-Cánovas, G.V., Zazueta-Morales, J.J., Anzaldúa-Morales, A. and Quintero-Ramos, A. 1998. Functional and nutritional properties of extruded whole pinto bean meal (Phaseolus vulgaris L.). *J Food Sci* 63: 113–116.

Batistuti, J.P., Barros, R.M.C. and Arêas, J.A.G. 1991. Optimization of extrusion cooking process for chickpea (Cicer arietinum L.) defatted flour by response surface methodology. *J Food Sci* 56(6): 1695–1698.

Berrios, J.D.J. and Pan, J. 2001. Evaluation of extruded of extruded black bean (Phaseolus vulgaris L.) processed under different screw speeds and particle sizes, Institute of Food Technologists: Chicago, IL Annual Meeting of the Institute of Food Technologists, New Orleans, LA, June 23–27 (Abstract 15D-9 # 8704), 30, 2001.

Berrios, J.D.J., Camara, M., Torija, M.E., and Alonso, M. 2002. Effect of extrusion cooking and sodium bicarbonate addition on the carbohydrate composition of black bean flours. *J Food Proc Pres* 26(2): 113–128.

Berrios, J.D.J., Morales, P., Camara, M., and Sanchez-Mata, M.C. 2010. Carbohydrate composition of raw and extruded pulse flours. *Food Res Int* 43(2): 531–536.

Berrios, J.D.J., Swanson, B.G. and Cheong, W.A. 1999. Physico-chemical characterization of stored black beans (Phaseolus vulgaris L.). *Food Res Int* 32(10): 669–676.

Berrios, J.D.J., Wood, D.F., Whitehand, L., and Pan, J. 2004. Sodium Bicarbonate and the microstructure, expansion and color of extruded black beans. *J Food Proc Pres* 28(5): 321–335.

Bhattacharya, S. and Parakash, M. 1984. Extrusion of blends of rice and chickpea flours: A response surface analysis. *J Food Eng* 21(3): 315–330.

Björck, I., Nyman, M. and Asp, N-G. 1984. Extrusion cooking and dietary fiber: Effects of dietary fiber content and on degradation in the rat intestinal tract. *Cereal Chem* 61(2): 174–179.

Borejszo, Z.B. and Khan, K.H. 1992. Reduction of flatulence-causing sugars by high temperature extrusion of pinto bean high starch fractions. *J Food Sci* 57(3): 771–777

Calloway, D.H. and Murphy, E.L. 1968. The use of expired air to measure intestinal gas formation. *Annals of the New York Academy of Sciences* 150, 82.

Canada Food Guide. Available from: http://www.hc-sc.gc.ca/fn-an/food-guide-aliment/index-eng.php (Accessed 25 September 2010)

Cardoso-Santiago, R.A. and Arêas, J.A.G. 2001. Nutritional evaluation of snacks obtained from chickpea and bovine lung blends. *Food Chem* 74(1): 35- 40.

Carrillo, J.M., Moreno, C.R., Rodelo, E.A. Trejo, A.C. and Escobedo, R.M., 2000. Physicochemical and Nutritional Characteristics of extruded flours from fresh and hardened chickpeas (Cicer arietinum L). *Lebensm Wiss Technol* 33(2): 117–123.

Chauban, G.S. and Bains, G.S. 1985. Effect of granularity on the characteristics of extruded rice snack. *Int J Food Sci Tech* 20(3): 305–309.

Chavan, J.K., Kadam, S.S. and Salunkhe, D.K. 1986. Biochemistry and technology of chickpea (Cicer arietinum L.) seeds. *Crit Rev Food Sci Nutr* 25(2): 107–132.

Conway, H.F. 1971. Extrusion cooking of cereals and soybeans. *Food Prod Dev* 5(2): 14–17, 27–29.

de Ferranti, S. and Mozaffarian, D. 2008. The perfect storm: Obesity, adipocyte dysfunction, and metabolic consequences. *Clin Chem* 54(6): 945–955.

Della Valle, G., Quillien, L. and Gueguen, J. 1994. Relationships between processing conditions and starch and protein modifications during extrusion cooking of pea flour. *J Sci Food Agric* 64(4): 509–517.

Duizer, L.M., Campanella, O.H. and Barnes, G.R.G. 1998. Sensory, instrumental and acoustic characteristics of extruded snack food products. *J Texture Stud* 29(4): 397–411.

Edwards, R.H., Robert, B., Mossman, A.P., Gray, G.M. and Whitehand, L. 1994. Twin-Screw Extrusion Cooking of Small White Beans (Phaseolus vulgaris). *Lebensm Wiss Technol* 27(5): 472–481.

FAO, 1977. Food and Agriculture Organization of the United Nations, Legumes Agriculture, Rome, Italy.

FAO, 2004. Food and Agriculture Organization of the United Nations, Available from FAOSTAT statistics database Agriculture, Rome, Italy.

FAO, 2007. Food and Agriculture Organization of the United Nations, Available from FAOSTAT statistics database Agriculture, Rome, Italy.

FAO, 2008. Food and Agriculture Organization of the United Nations, Available from FAOSTAT statistics database Agriculture, Rome, Italy.

Faubion, J.M., Hoseney, R.C. and Seib, P.A. 1982. Functionality of grain components in extrusion. *Cereal Food World* 27: 212–216.

Fleming, S.E. 1981. A study of relationships between flatus potential and carbohydrate distribution in legume seeds. *J Food Sci* 46(3): 794–798.

Gualberto, D.G., Bergman, C.J., Kazemzadeh, M. and Weber, C.W. 1997. Effect of extrusion processing on the soluble and insoluble fiber and phytic acid contents of cereal brans. *Plant Foods for Hum Nutr* 51(3): 187–198.

Gujska, E. and Khan, K. 1990. Effect of temperature on properties of extrudates from high starch fractions of navy, pinto and garbanzo beans. *J Food Sci* 55(2): 466–469.

Han, I.H. and Baik, B.-K. 2006. Oligosaccharide content and composition of legumes and their reduction by soaking, cooking, ultrasound, and high hydrostatic pressure. *Cereal Chem* 83(4): 428–433.

Hazell, T. and Johnson, I.T. 1989. Influence of food processing on iron availability in vitro from extruded maize-based snack foods. *J Sci Food Agric* 46(2): 365–374.

Huang, J., Schols, H.A., van Soest, J.J.G., Jin, Z., Sulmann, E. and Voragen, A.G.J. 2007. Physicochemical properties and amylopectin chain profiles of cowpea, chickpea and yellow pea starches. *Food Chem* 101(4): 1338–1345.

Ilo, S., Tomschik, U., Berghofer, E. and Mundigler, N. 1996. The effect of extrusion operating conditions on the apparent viscosity and the properties of extrudates in twin-screw extrusion cooking of maize grits. *Lebensm Wiss Technol* 29(7): 593–598.

Jomduang, S. and Mohamed, S. 1994. Effect of amylose/amylopectin content, milling methods, particle size, sugar, salt and oil on the puffed product characteristics of a traditional Thai rice-based snack food (Khao Kriap Waue). *J Sci Food Agric* 65(1): 85–93.

Karanja, C., Njeri-Maina, G., Martin-Cabrejas, M., Esteban, R.M., Grant, G., Pusztai, A., Georget, D.M.R., Parker, M.L., Smith, A.C. and Waldron, K.W. 1996. Extrusion-induced modifications to physical and antinutritional properties of hard-to-cook beans. In *Agrifood quality: An interdisciplinary approach*, ed. G.R. Fenwick, 279–283. Cambridge: Royal Society of Chemistry

Koehler, H.H., Chang, C.H., Scheier, G. and Burke, D.W. 1987. Nutrient composition, protein quality, and sensory properties of thirty-six cultivars of dry beans (Phaseolus vulgaris L.). *J Food Sci* 52(5): 1335–1340.

Lai, C.S., Guetzlaff, S. and Hoseney, R. 1989. Role of sodium bicarbonate and trapped air in extrusion. *Cereal Chem* 66(2): 69–73.

Lajoie, M.S., Goldstein, P.K. and Geeding-Schild, D. 1996. Use of bicarbonates in extrusion processing of ready-to-eat cereals. *Cereal Food World* 41(6): 448–451.

Lasekan, O.O., Lasekan, W., Idowu, M.A. and Ojo, O.A. 1996. Effect of extrusion cooking conditions on the nutritional value, storage stability and sensory characteristics of maize-based snack food. *J Cereal Sci* 24(1): 79–85.

Linko, P., Colomna, P. and Mercier, C. 1981. High temperature short-time extrusion cooking. In *Advances in cereal science and technology*, ed. Y. Pomeranz, 145. St. Paul, MN: American Association of Cereal Chemists.

Lintas, C., Cappeloni, M., Montalbano, S. and Gambelli, L. 1995. Dietary fiber in legumes: Effect of processing. *Eur J Clin Nutr* 49(3): S298–302.

Ludwig, D.S. 2002. Physiological mechanisms relating to obesity, diabetes, and cardiovascular disease. *JAMA* 287: 2414–2423.

Ludwig, D.S. and Ebbeling, C.B. 2005. Overweight Children and Adolescents. *The New Engl J Med* 353: 1070–1071.

Martin-Cabrejas, M.A., Jaime, L., Karanja, C., Downie, A.J., Parker, M.L., Lopez-Andreu, F.J., Maina, G., Esteban, R.M., Smith, A.C. and Waldron, K.W. 1999. Modifications to physicochemical and nutritional properties of hard-to-cook beans (Pheseolus vulgaris L.) by extrusion cooking. *J Agric Food Chem* 47(3): 1174–1182.

Martinez, B.F., Figueroa, J.D.C. and Larios, S.A. 1996. High lysine extruded products of quality protein maize. *J Sci Food Agric* 71(2): 151–155.

Meng, X., Threinen, D., Hansen, M. and Driedge, D. 2010. Effects of extrusion conditions on system parameters and physical properties of a chickpea flour-based snack. *Food Res Int* 43(2): 650–658.

Mercier, C.C., Linko, P. and Harper, J.M. 1989. Extrusion cooking of starch and starchy products. In *Extrusion cooking*, eds. C. Mercier, P. Linko and J.M. Harper, 263. St. Paul, MN: American Association of Cereal Chemists.

Moreira-Araujo, R.S.R., Araujo, M.A.M. and Arêas, J.A.G. 2008. Fortified food made by extrusion of a mixture of chickpea, corn and bovine lung controls iron-deficiency anaemia in preschool children. *Food Chem* 107(1): 158–164.

Narayana, K. and Narasinga Rao, M.S. 1982. Functional properties of raw and heat processed winged bean (Psophocarpus tetragonolobus) flour. *J Food Sci* 47: 1534–1538.

Ohio State University Extension Fact Sheet. Available from: http://ohioline.osu.edu/hyg-fact/5000/5553.html (Accessed 25 September 2010).

Otun, E.L., Crawshaw, A., and Frazier, P.J. 1986. Flow behavior and structure of proteins and starches during extrusion cooking. In *Fundamentals of dough rheology*, eds. H. Faridi and J.M. Faubion, 37–53. St. Paul, MN: American Association of Cereal Chemists.

Parsons, M.H., Hsieh, F., and Huff, H.E. 1996. Extrusion cooking of corn meal with sodium bicarbonate and sodium aluminum phosphate. *J Food Proc Pres* 20(3): 221–234.

Patil, R.T., Berrios, J. De J., Tang, J. and Swanson, B.G. 2007. Evaluation of the methods for expansion properties of legume extrudates. *Appl Eng Agric* 23(6): 777–783.

Patil, R.T., Berrios, J. De J., Tang, J., Pan, J. and Swanson, B.G. 2005. Empirical modeling of extrusion cooking of chickpea flour for process scale up, Paper No. 056068 for presentation at the ASAE Annual International Meeting at Tampa, Fl.

Poltronieri, F., Arêas, J.A.G. and Colli, C. 2000. Extrusion and iron bioavailability in chickpea (Cicer arietinum L.). *Food Chem* 70(2): 175–180.

Pulse Canada, 2010. Available from: http://www.pulsecanada.com/pulse-industry. Accessed 25 September 2010.

Reddy, N.R., Salunkhe, D.K. and Sharma, R.P. 1980. Flatulence in rats following ingestion of cooked and germinated black gram and a fermented product of black gram and rice blend. *J Food Sci* 45(5): 1161–1164.

Roos, Y.H., Roininen, K., Jouppila, K. and Tuorila, H. 1998. Glass transition and water plasticization effects on crispness of a snack food extrudate. *Int J Food Prop* 1(2): 163–180.

Sánchez-Mata, M.C., Cámara-Hurtado, M.M. and Díez-Marqués, C. 1999. Effect of domestic processes and water hardness on soluble sugars content of chickpeas (Cicer arietinum L.). *Food Chem* 65(3): 331–338.

Dr. Sethi. Leguminous Convenience (TFPJ, June–July, 2008, Rs 75)

Sgarbieri, V.C. and Whitaker, J.R. 1982. Physical, chemical and nutritional properties of common bean (Phaseolus vulgaris L.) proteins. A review. In *Advances in food research*, ed. C.O. Chichester, 28, 93–165, New York: Academic Press.

Sgarbieri, V.C. and Whitaker, J.R. 1989. Composition and nutritional value of bean (Phaseolus vulgaris L.). *World Rev Nut Diet* 60: 132–198.

Shirani, G. and Ganesharanee, R. 2009. Extruded products with fenugreek (trigonella foenumgraecium) chickpea and rice: Physical properties, sensory acceptability and glycaemic index. *J Food Eng* 90(1): 44–52.

Singh, N. and Smith, A.C. 1997. A comparison of wheat starch, whole wheat meal and oat flour in the extrusion cooking process. *J Food Eng* 34(1): 15–32.

Singh, N., Kaur, S., Isono, N. and Noda, T. 2010. Genotypic diversity in physico-chemical, pasting and gel textural properties of chickpea (Cicer arietinum L.). *Food Chem* 122(1): 65–73.

Singh, U., Subrahmanyam, N. and Kumar, J. 1991. Cooking quality and nutritional attributes of some nely developed cultivars of chickpea (Cicer arietinum). *J Sci Food Agric* 55(1): 37–46.

Skierkowski, K., Gujska, E. and Khan, K. 1990. Instrumental and sensory evaluation of textural properties of extrudates from blends of high starch/high protein fractions of dry beans. *J Food Sci* 55(4): 1081–1083

Steel, C.J., Sgarbieri, V.C. and Jackix, M.H. 1995. Use of extrusion technology to overcome undesirable properties of hard-to-cook dry beans (phaseolus vulgaris l.). *J Agric Food Chem* 43(9): 2487–2492.

USDA Food Pyramid. Available from: http://www.mypyramid.gov/pyramid/index.html. Accessed 25 September 2010.

van der Poel, A.F.B., Stolp, W. and Van Zuilichem, D.J. 1992. Twin-screw extrusion of two pea varieties: Effects of temperature and moisture level on antinutritional factors and protein dispersibility. *J Sci Food Agric* 58(1): 83–87.

Wang, N. and Daun, J.K. 2004. Effect of variety and crude protein content on nutrients and certain antinutrients in field peas (Pisum sativum). *J Sci Food Agric* 84(9): 1021–1029.

Wing, R.R., Hamman, R.F., Bray, G.A., Delahanty, L., Edelstein, S.L., Hill, J.O., Horton, E.S., et al. 2004. Achieving weight and activity goals among diabetes prevention program lifestyle participants. *Obes Res* 12(9):1426–1434.

9 Extrusion of Pet Foods and Aquatic Feeds

Kasiviswanathan Muthukumarappan, Kurt A. Rosentrater, Chirag Shukla, Nehru Chevanan, Murali Rai, Sankaranandh Kannadhason, Chinnadurai Karunanithy, Ferouz Ayadi, and Parisa Fallahi

CONTENTS

PET FOOD AND AQUATIC FEED SOURCE

Distillers dried grains with solubles (DDGS) is the dried residue remaining after the starch fraction of corn is fermented using selected yeasts and enzymes to produce ethanol and carbon dioxide. After complete fermentation, the alcohol is removed by distillation and the remaining fermentation residues are dried. The starch in corn is consumed in the fermentation process, while other nutrients such as protein are concentrated as much as threefold during production of DDGS (William, Horn, and Rugala 1977). Reddy, Stoker, and Roger (1993) incorporated DDGS on a 50–50 weight basis with wheat flour in preparation of noodles and baked foods. It is also known that up to 20% DDGS in a product significantly improved the flavor of extruded product for companion animals (Wampler and Gould 1984). The utilization of DDGS as a source of protein, fat, and fiber needs further exploration in preparation of pet foods and aquaculture feeds. Being a rich source of protein, DDGS can open newer and extremely profitable avenues for the pet food industry if proven beneficial in terms of health and foods' physical qualities.

DDGS is one of the co-products produced by the dry milling ethanol industry. Approximately 98% of DDGS in North America is produced by the ethanol producing industry and the rest comes from the alcohol beverage industry or other sources. According to the National Corn Growers Association (NCGA 2009), the U.S. produced approximately 12.1 billion bushels of corn in 2008. It was estimated that 30% (3.7 billion bushels) of the corn had been utilized for ethanol production, and 31.25 million tons of DDGS was produced in the U.S. in 2008. Assuming $90/ton DDGS, its value was $2.8 billion, potentially making ethanol production a profitable value-added venture. The future profitability of ethanol production via dry milling is hinged upon establishing localized marketing channels for DDGS.

COMPOSITION

DDGS contains 25%–30% crude protein and 8%–12% fat. Moisture is often kept under 12% but depends on the utility and customization of the product. DDGS may contain 42.2% insoluble fiber and only 0.7% soluble fiber. This amount of insoluble fiber is higher than the amount found in corn (4.7%) and soybean meal (13.2%), and higher than two common fiber sources occasionally used in swine diets, sugar beet pulp (3.9%) and soy hulls (5.5%). Whitney et al. (2002) believed insoluble fiber may prevent pathogenic organisms from attaching to the gut or serve as a nutrient source for some of the beneficial bacteria in the gut. If DDGS is effective in reducing losses due to ileitis, this would be yet another added value of DDGS. Drying DDGS decreases protein degradability while increasing the bypass protein value. Bypass protein is the protein that escapes digestion in the rumen, and, as a result, is digested in the intestinal tract, similar to that of urea. DDGS is high in phosphorus and potassium, but low in calcium. Some additional characteristics of DDGS are low starch, highly digestible fiber, and protein that are partially protected from rumen breakdown. A light colored DDGS is an indication of high amino acid digestibility and provides an excellent energy and available phosphorus source (Buchheit 2002).

It has been observed that fat, ash, and protein contents of DDGS fractions increase with decreasing particle size (Wu and Stringfellow 1982). However, neutral detergent fiber value decreases with decreasing particle size and is respectively low at high moisture. Fractions from initial high-moisture DDGS fractions lose moisture but those from initial low-moisture DDGS gain moisture after milling and screening (Wu and Stringfellow 1982). To avoid differences in protein, fat, ash, and moisture, DDGS was used as such. The particle size was greater than 40-mesh. Using DDGS without milling eliminated protein shift. Protein shift is a calculated value to compare protein displacement and equals the sum of the amount of protein shifted into the high protein fractions and out of the low protein fractions as a percentage of the total protein present in the starting material (Gracza 1959). Milling was also discouraged as results showed an increase in leucine, methionine + cystine, and phenylalanine + tyrosine, and a decrease in lysine and tryptophane with decreasing particle size of DDGS. Decreased lysine content may considerably imbalance amino acid composition in DDGS, which has already shown concerns. The high fiber content of DDGS, however, limits the amount that can be put in blended foods (Wu and Stringfellow 1986). The utilization of DDGS in dry extruded food in our research is a base to evaluate the viability of DDGS emphasizing the functional properties of the final product. Nutritional analysis leaves scope for further development of this new product.

BENEFITS

DDGS provides the following advantages: it is a great source of rumen bypass protein for dairy and beef cattle; it is easily available; it has high energy due to lipid contents from corn; it is cheap; it is an excellent source of essential minerals potassium and phosphorus; it has enhanced palatability, and it is an excellent source of linoleic acid, methionine, cystine and vitamin E and can be eaten by humans provided the production of DDGS follows generally recognised as safe (GRAS) standards. There are many potential uses for DDGS (Shurson 2002; Kohler and Kohler 2001; Morrison 2001). DDGS is an excellent feedstuff for broilers (Scott 1970) when used at 5% or less, as it is assumed that DDGS provides unidentified growth factor activity (Jensen 1981). This was one area that investigated inclusion of DDGS in snack foods. Lysine (Scott 1970), tryptophane and arginine (Wu and Stringfellow 1982) are amino acids that DDGS is deficient in. Despite such deficiencies, studies have indicated that DDGS can be used effectively as a protein source in dairy rations, and yield lactation responses equal to or exceeding soybean meal when the amount in the total ration does not exceed 19%. The seasonal variation in composition of DDGS has partially limited its use in commercial foods, but with improved technology and processes, better compositional consistency is now achievable. Expecting to compensate for the deficiency of lysine, meat and bone meal was used. Desired feeding qualities of DDGS include low rumen degradability of protein (Klopfenstein and Stock 1981; Zhang and Bhatnagar 2000), low amount of soluble carbohydrate, relatively high fiber and fat content and unidentified nutritional factors (Poos 1981).

EXTRUSION PROCESSING

Extrusion processing is as much an art as it is a science. Extrusion has been the most widely used technology for snack and animal food manufacturing. Both single-screw and twin-screw extruders have welcomed newer ingredients. Researchers have studied ways to optimize processing parameters by using a variety of ingredients in different proportions and extruding under various conditions. Of the heterogeneous composition of products, water is one of the major components used during extrusion processing.

Early single-screw extruders were mainly used to produce breakfast cereal products and snack foods with considerable commercial success, all of which have a relatively simple feed material system composed of starch and protein with low or medium moisture content. In recent years, blending various ingredients in preparation of food has gained interest. With the benefits of increased sources of nutrition, blending many ingredients has posed a challenge of process optimization due to varied product heterogeneity. In extrusion cooking, two important features of an extruder are its ability to perform mixing, heating, shearing, and other required operations in the barrel of the extruder simultaneously, and to bring in desired texture to final products exiting from the die (Rakosky 1987; Lusas and Khee 1987; Harper 1989).

In general, extrusion and associated processes involve grinding and mixing the ingredients, cooking and forming through extrusion, followed by drying. Like other heat treatments, extrusion provides numerous benefits, such as making food healthier, safer, tastier, and more shelf-stable (Zhang and Bhatnagar 2000). Originally run with no drying step, more recent extrusion installations have included modest drying to permit operation at high moistures. This has increased flexibility by providing an additional control variable, and has improved product quality and reduced excess wear found at low moistures (Miller 1984). Extrusion temperature, pressure and sometimes the product moisture are further increased to the proper levels for cooking and expansion of the product to desired bulk density. Physical parameters of products have always been of prime interest to extrusion scientists. Color, product density, water absorption and solubility, expansion, texture and product rheology are among the widely studied parameters. Chemical composition is focused during product scaling and commercialization.

Extrusion is rapid, flexible, and relatively inexpensive when compared to most other food processing operations. It exposes the raw material to high temperature, pressure, and shear-force to mix and cause physical and chemical changes, which constitutes cooking and gives extruded snacks a variety of shapes and flavors. Extruded snacks are frequently low in nutritional density and high in fat and starch. Twin-screw extruders, on the other hand, provide more versatility and are successfully used to process more complicated material systems. Snacks have long been a part of the American diet and are rapidly increasing in popularity. Extruders take the privilege of dominating the technology of producing most snack foods. In developed countries where protein-energy is deficient, snacks are widely consumed (Omueti and Morton 1996). Extrusion cooking, a dominant technology in pet food manufacturing, is a widely used technique in snacks and the confectionary industry as well. Technological developments in expansion of breakfast cereals and snack items, in particular, were hindered by the difficulty of achieving and then maintaining steady conditions for extrusion with single-screw extruders. A prime target of consumer advocates and nutritionists is to effectively improve the quality and nutrition of snack foods (Matz 1984).

Extruders have made their position in the food industry, which seems to be irreplaceable. The variety of extruded products ranges from pasta, ready-to-eat breakfast cereal, and snack food, to confectionery, pet foods, and aquaculture feeds (Harper 1979; Harper and Jansen 1985). Its most important commercial application is high-shear extrusion cooking, a process used to produce dry expanded foods in a wide range of forms (Enterline 1984). Of the many operations of an extruder, mixing, heating, and shearing are among the most important to bring about desired texture in the final product exiting from the die (Harper 1989; Lusas and Khee 1987; Rakosky 1987). Preconditioning, extrusion, cutting, and drying are primary unit operations in most industrial extrusion processing. Extrusion of heterogeneous ingredients makes as good a use of art as it does science. Controlled expansion is a desirable feature in extrusion processing. Several groups have evaluated corn and wheat DDGS in extruded foods and all have reported a loss in extrudate expansion when more than 20% DDGS was included in the formulation, especially at higher temperatures (Anderson et al. 1981; Breen, Seyam, and Bansik 1977; Satterlee et al.

1976). It has also been observed that high moisture level permits gelatinization at lower temperatures (Harper 1979). As a primary step towards optimization of extrusion parameters, Shukla (2003) studied the properties of DDGS when extruded as a heterogeneous blend consisting additionally of corn meal, meat and bone meal, and pork in a single-screw extruder with varying levels of DDGS and extruder die temperatures and focusing towards pet food.

DEVELOPMENT OF PET FOODS USING A SINGLE-SCREW EXTRUDER

Shukla (2003) developed pet foods using a Brabender Plasticorder Single-screw Extruder Model 2003H/C (PL2000 controller) with three temperature zones, 20:1 *L/D* ratio, 3:1 compression ratio, uniform pitch-single flight screw design having 0.15″ flute depth, variable speed up to 225 rpm powered by 7½ HP motor, maximum torque of 24,000 meter-gram and ¾″ barrel diameter (South Hackensack, NJ). Die diameter was 2.7 mm. The extruder was equipped with a variable speed drive and a tachometer. The barrel was equipped with two electrically heated, compressed air cooled collars controlled by thermostats. The extended die plate was electrically heated and controlled by a thermostat but did not have a cooling system. The extruder was operated at a constant speed of 120 rpm. The exterior of the single-screw extruder is shown in Figures 9.1 through 9.3. The extruder was set to 120°C at the feeding zone, 130°C at the compression zone, and at respective treatment temperature for the die zone.

The composition of ingredients used in this study is shown in Table 9.1. Results showed that an increase in temperature caused a decrease in moisture content at all levels of DDGS as shown in Figure 9.4. Processing 10% and 15% DDGS (65% and 62% corn meal, respectively) at 80°C resulted in significantly high moisture (14.05% at 10% DDGS and 12.81% at 15% DDGS) extrudates than it did at 100°C and 120°C (10.53% and 9.92%, respectively, at 10% DDGS, and 11.35% and 11.30%, respectively, at 15% DDGS).

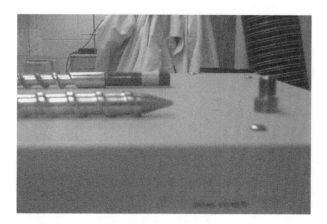

FIGURE 9.1   Extruder screw front (zoomed) and rear end.

FIGURE 9.2 Side view of the extruder shows thermostats and other elements end.

The effect of increased temperature became evident at 20% and 25% DDGS level. At these levels of DDGS, processing formulation at 120°C die temperature caused a significant reduction in moisture than processing at 80°C and 100°C. Lowest moisture was observed at 120°C die temperature for 10% and 25% DDGS. Moisture remained fairly constant among 15% and 20% DDGS extrudates. This interesting outcome could be a result of the start of optimum interaction between free moisture, corn meal, and DDGS at 15% and 20% moisture causing better gelatinization. Contrary to 15% and 20% DDGS levels, 10% and 25% DDGS based extrudates might not have reached the beginning of the gelatinization stage, and exposure to a higher temperature (120°C) at the die might have prompted more moisture removal. During production of low moisture foods, a 10% or 25% DDGS level could result in the least moisture when processed at 120°C. Processing either 10% or 25% DDGS

FIGURE 9.3 Front view of the extruder shows controls and die mounting zone.

TABLE 9.1
Ingredients (%) Used in Formulations for Pet Foods

Ingredient	Control	Formula 1	Formula 2	Formula 3	Formula 4
Pork	25	20	15	10	7
CM	65	65	62	59	55
DDGS	0	10	15	20	25
MBM	5	5	5	5	5
Water	5	0	3	6	8

CM Corn meal; *MBM* Meat and bone meal

at 80°C should result in high moisture extrudates. Based on these results, processors may select proper process control to selectively optimize their products to desired qualities. Again moisture control may also affect various other parameters of interest, which is discussed in subsequent sections.

The effect of DDGS on moisture content showed an interesting trend. Extrudates obtained from processing with 15% DDGS at 80°C were not significantly different from extrudates having 20% DDGS. Similarly, extrudates having 15% and 20% DDGS showed no significant difference from one another at 100°C and 120°C die temperature. This result may help producers to select a range of DDGS levels if they were to consider optimum gelatinization. Formulations having 10% and 25% DDGS processed at 80°C retained significantly higher moisture (14.05% and 13.87%) than did those having 15% and 20% DDGS formulations (12.81% and 12.79%). At 100°C processing temperature, moisture showed an increasing trend from 10% to 25% DDGS as moisture increased from 10.53% to 13.6%. It was also observed that at 80°C and 100°C moisture increased when processing 25% DDGS formulation (13.87% and 13.6% moisture, respectively). Corn meal might have interacted with free moisture but less residence time in the barrel might not have helped in moisture

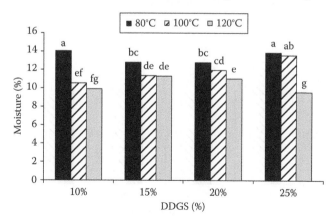

FIGURE 9.4 Effect of die temperature and DDGS on moisture content of extrudates. (Bars with the similar letter(s) on top are not significantly different ($p < .05$).)

expulsion, thereby extrudates resulted in higher moisture. At 120°C, both 10% and 25% levels of DDGS resulted in extrudate moisture of 9.92% and 9.58%, respectively, which was the lowest observed. The high die temperature processing might have brought about a rapid expulsion of moisture despite reduced retention time and probably, reduced gelatinization.

EFFECT OF DDGS AND EXTRUDER DIE TEMPERATURE ON $L*$ VALUE OF EXTRUDATES

In color measurement of extrudates, $L*$ value is an important aspect. It indicates the luminosity or brightness. A higher $L*$ value represents bright extrudate color. The study revealed that with an increase in die temperature from 80°C to 120°C at each level of DDGS from 15% to 25%, the $L*$ value either decreased or remained unchanged as a general trend, as shown in Figure 9.5. On the other hand, a significant increase in $L*$ value from 44.08 to 50.58 was observed when the temperature increased from 80°C to 100°C with 10% DDGS level, but a decrease in $L*$ from 50.58 to 45.83 with a change in die temperature from 100°C to 120°C was not significant. Processing 10% DDGS level at 100°C shows maximum brightness of 50.58 $L*$ value. This may be due to optimized gelatinization and a relatively improved expansion. An improved expansion causes thinning and stretching of cell structure. This imposed a shiny sheen on the extrudates causing improved luminosity expressed as $L*$ value.

An increase in $L*$ value was also evident at 120°C die temperature for 10% DDGS (45.83 $L*$) and 100°C die temperature and 15% DDGS combination. The capability of starch from corn meal to optimally bind water and expand despite the short residence time could be a reason for this behavior. No significant differences were observed in brightness at 20% and 25% DDGS level for 100°C and 120°C die temperature, respectively. The interior of the extrudate, when visually studied, showed brighter color than the exterior. It is assumed that higher die temperatures tend to cause localized burning on the exterior of the extrudate.

Processing 20% and 25% DDGS at higher die temperatures may result in a darker extrudate resulting in an $L*$ value in the range of 36.17 to 39.52. At 80°C die

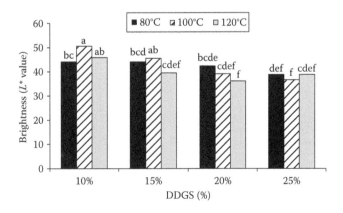

FIGURE 9.5 Effect of die temperature and DDGS on luminosity/brightness ($L*$) of extrudates. (Bars with the similar letter(s) on top are not significantly different ($p < .05$).)

temperature, a darker extrudate resulted with an $L*$ value of 38.76 when the highest level (25%) of DDGS was used. DDGS itself is reddish brown in color and its increased addition could have darkened the extrudate.

A 10% and 15% DDGS formulation showed improved brightness (50.58 and 45.53, respectively) but a significantly darker extrudate was resulted by processing 20% and 25% DDGS at 100°C (39.14 and 36.61). Apruzzesse, Balke, and Diosady (2000) explained how the extrusion process can break starch to lower molecular weight macromolecules due to thermal and mechanical forces. Davidson et al. (1984) explained that the largest branched molecules of amylopectin would break down through mechanical forces, due to shearing by the extruder. The forces expose starch to the hot barrel. In cases where moisture is not added during extrusion, severe burning can cause the extrudate to darken. Moisture removal from matrices during extrusion was explained by hydrogen bond theory. Some of the molecular changes of starch, which take place during extrusion, can be explained in terms of hydrogen bonding. During heating, the hydrogen bond network within each starch granule becomes disrupted, thus allowing water molecules to occupy new spaces within amylopectin molecules. When this matrix is opened up, heat penetrates the matrix, thereby reducing moisture in the extrudate and possible burning.

At a higher level of die temperature (120°C), increased DDGS levels (15%, 20%, and 25%) showed consistently lower $L*$ values of 39.52, 36.17, and 38.78, respectively. From the above results, a processor may be able to decide on the right amount of DDGS and die processing temperature to use to produce extrudates of desired brightness.

Effect of DDGS and Extruder Die Temperature on Apparent Bulk Density of Extrudates

Apparent bulk density (BD) is an important factor for the packaging section of the food industry, and this section would find these results interesting. BD, also known as packaging bulk density in Asian countries, is an important parameter of study for expanded foods. A higher BD represents less usage of packaging material and also means a higher price tag for a smaller package thereby affecting customer psychology. BD remains fairly unchanged at all die temperatures and DDGS levels. A combined effect of positive and negative forces affecting expansion may be balanced thereby resulting in a constant bulk density. This may prompt producers to utilize maximum amount of DDGS provided other physical and chemical properties are in acceptable limits. The presence of pork and meat and bone meal in corn meal-DDGS blend can significantly suppress the effects of DDGS and die temperature on food. It should act as a suggestion on whether to use a simpler blend or complex blend to suit their product specifications at specific production conditions. As heterogeneity increased in formulations, the same processing parameters resulted in a different physical structure of food.

Effect of DDGS and Extruder Die Temperature on Radial Expansion of Extrudates

The snack food industry looks for taste, nutritional density, and product texture. Out of the several different product textural properties, radial expansion is the most

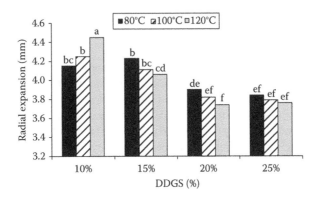

FIGURE 9.6 Effect of die temperature and DDGS on radial expansion of extrudates. (Columns with the similar letter(s) on top are not significantly different ($p < .05$).)

important one to both consumers and producers. In short, radial expansion is the most interesting parameter in expanded food preparation.

The highest radial expansion of 4.45 mm was obtained by processing 10% DDGS at 120°C as seen in Figure 9.6. As noted by mentioned researchers, an optimum barrel temperature of 153°C causes maximum starch expansion. In our case, the barrel temperature was constant. Elevating the extruder die temperature to 120°C brought about a maximum expansion of food. Again, 10% DDGS extrudate had 65% of corn meal, which was the maximum among all formulations. The increased amount of corn meal present in the formulation might have caused an increase in radial expansion.

At each DDGS level, with an increase in die temperature, radial expansion decreased, except for 10% DDGS level where expansion increased. Protein is suspected to suppress expansion (Camire, Camire, and Krumhar 1990), and DDGS being rich in protein could be an explanation to reduction of radial expansion. At 10% DDGS level, an increase in radial expansion because of an increase in die temperature seemed to be more effective than the effect of DDGS in suppressing expansion. This could be a reason why 10% DDGS showed an opposite trend. At 80°C, no significant difference was observed when the DDGS level was increased from 10% to 15% as radial expansion varied in the narrow range of 4.15 to 4.23 mm. However, a significant decrease in radial expansion was observed when the level of DDGS was raised from 15% to 20%, causing reduction in expansion from 4.23 to 3.90 mm. Again, 20% DDGS extruded at 100°C and 120°C die temperature, and 25% DDGS extruded at all die temperatures did not cause significant reduction in radial expansion.

As discussed earlier, the highest expansion was observed at 120°C and 10% DDGS level, and the lowest expansion was observed at 20% DDGS (100°C and 120°C) and 25% DDGS (all temperatures). Apart from an increase in expansion with an increase in die temperature at 10% DDGS, other levels of DDGS indicated either a decrease or no change in radial expansion. Processors may possibly exploit this capability of DDGS to reduce expansion by introducing DDGS in foods where expansion needs to be controlled at standard processing parameters. It can also be suggested that the theory of increased expansion due to an increase in temperature holds true for the 10% DDGS level only for the single-screw extruder used in the study.

EFFECT OF DDGS AND EXTRUDER DIE TEMPERATURE ON PEAK LOAD OF EXTRUDATES

Peak load is a measure of resistance of the extrudate to fracture. In other words, it indicates how much load food could sustain before fracturing. It was observed that the peak load remains fairly constant and did not necessarily show a trend with varying DDGS level or die temperature. Maximum peak load of 1.53 kg was observed when 10% DDGS was extruded at 100°C and 120°C. As a general observation, 10% and 15% DDGS extrudates (Avg: 1.51 kg) were stronger than 20% and 25% DDGS extrudates (Avg: 1.34 kg). Extruding longer strands of extrudate from a larger die could explain peak load better. From the results, it could be suggested that the addition of higher DDGS (20% and 25%) produced weaker extrudate. It could be said theoretically that increasing DDGS might cause slipping of product as DDGS and pork might release lipids more easily than other ingredients. Product might thus be exposed to less amount of shear resulting in a less compact structure. It can be suggested that peak load could be increased by permitting more residence time and optimized gelatinization conditions thereby binding DDGS inside the matrix and strengthening the extrudate.

DEVELOPMENT OF AQUATIC FEEDS USING SINGLE-SCREW AND TWIN-SCREW EXTRUDERS

Single-screw extruders are widely used to manufacture aquatic feeds due to low capital investment and operating costs. In fact, the cost of installing a twin-screw extruder can be up to two to three times as much as for a single-screw extruder. A thorough comparison between single- and twin-screw extruders can be found in Harper (1981). There have been a large number of studies on the use of DDGS in various fish diets over the years, but until recently there was no information on the actual processing of DDGS into aquatic feeds. After 2007, several studies have appeared in the literature, which examined the effects of both processing conditions as well as feed ingredient compositions on resulting extrudate (i.e., pellet) properties, and which aimed to maximize the quantity of DDGS in various aquatic feed blends while maintaining product quality. These studies are summarized in Table 9.2. Most of this work examined processing using single-screw extruders; two studies were conducted in a commercial scale twin-screw extruder. Overall, the studies found that the high levels of fiber in the DDGS were problematic, especially as high concentrations of DDGS were used. Various binding materials appeared to help, however, several processing conditions, including moisture, also impacted pellet quality. Each study will be briefly discussed below, with primary emphases placed on unit density and pellet durability (as these are important to viable aquatic feeds).

An initial study was conducted by Chevanan, Rosentrater, and Muthukumarappan (2008). They formulated three isocaloric (3.5 kcal/g) blends, with the net protein adjusted to 28% (wb), with 20%, 30%, and 40% (wb) DDGS; the balance of each

TABLE 9.2

Key Properties in Work Published to Date on Processing of DDGS into Aquatic Feeds[a]

DDGS (%)	Binder	Properties		Citation
		Unit Density (g/cm³)	Durability (%)	
Single-Screw Extrusion				
20	None	0.96	89.00	Chevanan, Rosentrater, and Muthukumarappan 2008
30		0.93	65.00	
40		0.93	56.00	
20	Whey protein	1.05	94.04	Chevanan, Muthukumarappan, and Rosentrater 2009
30		1.07	94.02	
40		1.06	93.52	
20	Cassava Starch	0.78	81.6	Kannadhason, Muthukumarappan, and Rosentrater 2009a
30		0.88	84.2	
40		0.86	86.1	
20	Corn Starch	0.90	85.3	Kannadhason, Muthukumarappan, and Rosentrater 2009a
30		0.94	75.7	
40		0.91	62.7	
20	Potato Starch	0.79	81.8	Kannadhason, Muthukumarappan, and Rosentrater 2009a
30		0.88	85.3	
40		0.90	87.4	
20	Corn Starch	1.03	70.9	Rosentrater, Muthukumarappan, and Kannadhason 2009a
25		1.00	90.8	
30		1.02	69.7	
20	Tapioca Starch	0.94	90.2	Kannadhason, Muthukumarappan, and Rosentrater 2009b
25		0.93	95.8	

TABLE 9.2   (Continued)

Key Properties in Work Published to Date on Processing of DDGS into Aquatic Feeds[a]

DDGS (%)	Binder	Properties		Citation
		Unit Density (g/cm³)	Durability (%)	
30		0.99	84.1	
20	Potato Starch	0.85	89.1	Rosentrater, Muthukumarappan, and Kannadhason 2009b
25		0.97	96.4	
30		0.93	82.4	
Twin-Screw Extrusion				
20	Whey protein	0.24	97.76	Chevanan, Rosentrater, and Muthukumarappan 2007
40		0.34	97.60	
60		0.61	96.85	
0–27.5	Whey protein	0.58–0.99	88.4–97.0	Kannadhason et al. 2010

[a]   Unit density and durability values reported in this table are mean values from each of the studies listed.

consisted of soy flour, corn flour, fish meal, mineral and vitamin mix. Blends were conditioned to 15%, 20%, and 25% (wb) moisture content prior to processing, and no binders were used. The blends were extruded using a single-screw (compression ratio of 3:1) extruder (barrel length of 317.5 mm; length-to-diameter ratio of 20:1), with a die length/diameter of 13 mm/2.7 mm (thus $L/D = 4.81$), using screw speeds of 100, 130, and 160 rpm, and processing temperatures of 90°C, 90°C, and 120°C (corresponding to the feed, transition, and die sections, respectively). The extruder used for this study is shown in Figure 9.1. All extrudates had unit densities less than 1.0 g.cm³, which indicated that they all floated. Unfortunately, the durability values were low, and ranged from 56% to 89%, that indicates that they fell apart relatively easily. They found that increasing the DDGS content from 20% to 40% resulted in a 37.1% and 3.1% decrease in extrudate durability and unit density, respectively. An increase in the screw speed from 100 to 160 rpm resulted in a 20.3% increase in durability. An increase in the moisture content from 15% to 25% (wb) resulted in a 28.2% increase in durability, but an 8.3% decrease in unit density.

A follow-up study was then conducted by Chevanan, Muthukumarappan, and Rosentrater (2009), using the same blend formulations and extruder setup. In this study, however, 5% whey protein was added to the ingredient blends to improve binding. Compared to the previous research, the durability and unit density of the resultant extrudates were found to increase substantially, which was specifically due to the addition of the whey protein. All durability values were greater than 86% (indicating cohesive products which held together well), but unit densities were all greater than 1.05 g/cm³ (which indicates that the extrudates did not float). Overall, increasing the moisture content of the blends from 15% to 25% resulted in an increase of 8.7% in durability. This study demonstrated that ingredient moisture content and screw speed are critical considerations when producing extrudates with feed blends containing DDGS; but further work was necessary to produce floating feeds.

Chevanan, Rosentrater, and Muthukumarappan (2010) extended this work and investigated the effects of DDGS level (20%, 30%, and 40%), blend moisture content (15%, 20%, and 25%, wb), barrel temperature profile (90°C-100°C-100°C, 90°C-130°C-130°C, and 90°C-160°C-160°C), and screw speed (80, 100, 120, 140, and 160 rpm) on various extrusion processing parameters. They used the same balance of ingredients (with 5% whey protein as a binder), and the same single-screw extruder, with a 2.7 mm-diameter circular die (*L*/*D* of 4.81). Processing parameters measured included mass flow rate, net torque required turning the screw, specific mechanical energy consumed during processing, apparent viscosity, and temperature and pressure of the dough melt inside the barrel and die. For all blends, as the temperature profile increased, mass flow rate exhibited a slight decrease, die pressure decreased, and apparent viscosity exhibited a slight decrease as well. The net torque requirement, specific mechanical energy consumption, and apparent viscosity decreased as screw speed increased, but mass flow rate increased. As moisture content increased, die pressure decreased. At higher temperatures in the barrel and die, the viscosity of the dough was lower, leading to lower torque and specific mechanical energy requirements. An increase in the DDGS content, on the other hand, resulted in a higher mass flow rate and decreased pressure inside the die. Again, they found that processing temperature and moisture content levels are critical for processing of DDGS-based ingredient blends.

Experimental results from the previous studies were then combined by Chevanan, Muthukumarappan, and Rosentrater (2007) and developed various multiple linear regression and neural network models to explain the effects of DDGS inclusion, blend moisture content, extruder temperature profile, screw speed, and die dimension on resulting extrusion processing parameters and extrudate properties. In general, the regression and neural network models predicted the extrusion processing parameters with better accuracy than they did the extrudate properties. Regression models for extrudate properties, resulted in R^2 values ranging from 0.29 to 0.84. Regression models using three (moisture, temperature, and die *L*/*D*), and six (moisture, temperature, die length, die diameter, die *L*/*D*, and screw speed) input variables predicted extrusion processing parameters with R^2 values of 0.56 to 0.97 and 0.75 to 0.97, respectively. But, neural network models (using 3, 5, and 6 input variables) had better performance and predicted extrusion processing parameters with R^2 values of 0.82 to 0.98, 0.86 to 0.99 and 0.90 to 0.99, respectively. With the regression modeling, even though increasing the number of input variables from three to six resulted in better R^2 values, no decrease in the coefficient of variation between the measured and predicted variables were observed. On the other hand, the neural network models developed with six input variables resulted in more accurate predictions with reduced coefficients of variation and standard errors. Because of the ability to produce accurate results with reduced variation and standard error, neural network modeling has greater potential for developing robust models for extrusion processing than does regression modeling. Thus, more work needs to be done to develop this tool.

To further extend the understanding of the aquatic feeds development, Kannadhason, Muthukumarappan, and Rosentrater (2009a) examined multiple protein levels and multiple starch sources, in addition to multiple DDGS levels. Feeds blends were formulated with DDGS (20%, 30%, and 40% wb), the balance of which consisted of ingredients similar to those used in prior studies, except cassava, corn, and potato starches were used as binders (instead of whey), and three protein levels were used (28%, 30%, and 32% wb). All blends had a moisture content of 20% wb. The same single-screw extruder was used, with a 3.0 mm diameter circular die, using a temperature profile of 90°C-120°C-120°C, and screw speed of 130 rpm. All extrudates had durability values greater than 62%, and most were higher than 81%. Unit density values ranged from 0.78 to 0.94 g/cm³. For all three starch sources, increasing the DDGS level resulted in a significant increase in sinking velocity. Moreover, as DDGS and protein levels increased, unit density and pellet durability also increased for cassava and potato starch extrudates. For the cassava starch blends, a 20% level of DDGS and 28% level of protein exhibited better expansion and floatability. But the corn starch blends at 40% DDGS and 32% protein exhibited better durability.

More in depth studies were conducted by Rosentrater, Muthukumarappan, and Kannadhason (2009a) and prepared ingredient blends using three levels each of DDGS (20%, 25%, and 30% db), protein (30%, 32.5%, and 35% db), and feed moisture content (25%, 35%, and 45% db), along with appropriate quantities of corn starch, soybean meal, fish meal, whey, vitamin, and mineral mix. The blends were extruded with the same single-screw extruder, using three screw speeds (100, 150, and 200 rpm) and three extruder barrel temperature profiles (80°C-90°C-100°C, 80°C-100°C-125°C, and 80°C-125°C-150°C), and a 2.9 mm diameter circular die (*L/D* of 3.19). Extrudate unit density values ranged from 0.87 to 1.28 g/cm³ (thus some extrudates floated while others sank), whereas durability ranged from 14.3% to 99.3%. As moisture content increased, unit density increased by 16.7%. But as protein increased, it decreased by 1.4%. Increasing the DDGS levels from 20% to 25% db, protein content from 30% to 32.5% db, feed moisture content from 25% to 35% db, processing temperature from 100°C to 125°C, and screw speed from 100 to 150 rpm increased the durability values by 28.1%, 18.1%, 31.8%, 6.6%, and 32.2%, respectively; all of these decreased as these independent variables increased to their highest levels, however.

Furthermore, Kannadhason, Muthukumarappan, and Rosentrater (2009b), examined similar blends (except that tapioca starch was used as a binder instead of corn starch), using the same extruder and the same processing conditions as Rosentrater, Muthukumarappan, and Kannadhason (2009a). Extrudate unit densities ranged from 0.63 to 1.24 g/cm³, while durability values ranged from 49.2% to 98.4%. Thus, compared to the corn starch experiments, the extrudates from this study had somewhat better durability. Increasing the DDGS levels from 20% to 30% db, protein content from 30% to 35% db, feed moisture content from 25% to 45% db, and processing temperature from 100°C to 150°C decreased the resulting durability values by 7.5%, 16.2%, 17.2%, and 16.6%, respectively.

Rosentrater, Muthukumarappan, and Kannadhason (2009b) used the same extruder, the same processing conditions, and similar feed blends except that potato starch was used as a binder instead of corn starch as Rosentrater, Muthukumarappan, and Kannadhason (2009a). In this study, unit densities ranged from 0.60 to 1.23 g/cm³, while durability values ranged from 53.9% to 98.6%. Thus, these pellets were somewhat better in terms of structural cohesion compared to the corn starch/DDGS extrudates than that of Rosentrater, Muthukumarappan, and Kannadhason (2009a). Again, some extrudates floated while others did not. An increase in the DDGS levels from 20% to 30% (db), protein content from 30% to 35% (db), feed moisture content from 25% to 45% (db), and processing temperature from 100°C to 150°C decreased the durability values by 7.5%, 10.7%, 4.0%, and 16.8%, respectively, but in a curvilinear fashion.

Small scale single-screw extrusion processing is a very effective tool for proof of concept and quantifying general behaviors, but extrusion at pilot and commercial scales must also be done in order to truly replicate the production of DDGS-based feeds which can be used in commercial aquaculture settings. Accordingly, feed blends containing DDGS at three levels (20%, 40%, and 60% wb) were formulated with soy flour, corn flour, fish meal, vitamin mix, and mineral mix to achieve a net protein content of 28% by Chevanan, Rosentrater, and Muthukumarappan (2007) that was similar to the blends used by Chevanan, Muthukumarappan, and Rosentrater (2009). These blends were extruded in a Wenger TX-52 twin-screw extruder (see Figure 9.7), which had two 52 mm diameter screws (each of which consisted of 25 individual screw segments), a barrel length-to-diameter ratio of 25.5:1, and dual 3.175 mm diameter circular dies. During processing, two levels of blend moisture content were attained (15% and 19%, wb), and two screw speeds were used (350 and 420 rpm). Unit density values for the resulting extrudates ranged from 0.24 to 0.61 g/cm³ (the low values were a consequence of the high expansion, which ranged from 5% to 66%), and thus all pellets floated in water—which was one of the goals of the study. Durability values ranged from 95% to 99%; hence these pellets were highly resistant to breakage during storage and transport—another goal of the study. Increasing the DDGS content from 20% to 60% resulted in a 36.7% decrease in the radial expansion, leading to a 159% increase in the unit density; but even the 60% DDGS pellets were better than those produced by small scale single-screw extruder. Thus, all pellets were of very high quality, for all DDGS levels and for all processing conditions used. Consequently, it was determined that DDGS could be included at a rate of up to 60% (using these formulations) and viable pelleted floating feeds could be produced.

In a recent study, Kannadhason et al. (2010) prepared six isocaloric (3.65 kcal/g), isonitrogenous (35% dry-basis; db protein), ingredient blends with 0%, 17.5%, 20%, 22.5%, 25%, and 27.5% DDGS and other ingredients (soybean meal, corn, fish meal, whey, soybean oil, vitamin, and mineral mix). The blends were moisture balanced to 15% db, then extruded in the same twin-screw extruder used by Chevanan, Rosentrater, and Muthukumarappan (2007) using a 2 mm die at 190 rpm, and a 3 mm die at 348 rpm. The 0% DDGS had the lowest unit density and the extrudates that contained 20% and 27.5% DDGS had the highest durability values. Results reported by Chevanan, Rosentrater, and Muthukumarappan (2007) for extruded tilapia blends which incorporated DDGS, showed a great increase in unit density values (159%) with change in DDGS levels from 20% to 60%. In contrast, no significant differences in unit density values were noticed by

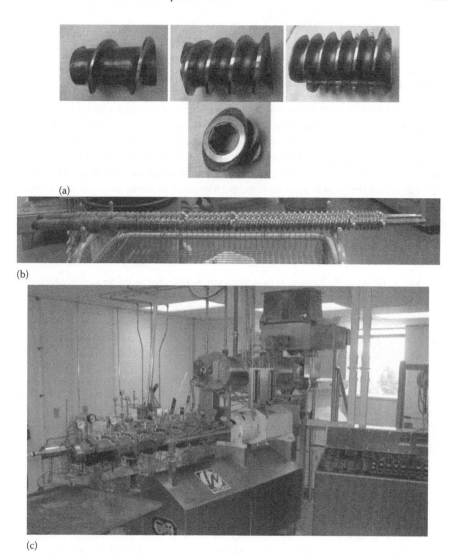

FIGURE 9.7 An example of a twin-screw pilot-scale extruder: (a) various screw elements; (b) the screw is formed by attaching several elements together in various configurations; (c) the entire extruder during operation, note the feed hopper and preconditioner are also shown.

Rosentrater, Muthukumarappan, and Kannadhason (2009a, 2009b) for a change in DDGS levels. This contradiction was probably due to differences in the feed compositions used in the studies. Their results are similar to the findings of Chevanan, Rosentrater, and Muthukumarappan (2007) and Rosentrater, Muthukumarappan, and Kannadhason (2009a, 2009b). All conditions led to fairly high pellet durability values, and were thus resistive to the destructive forces commonly encountered by feed materials during handling and storing, and are thus important to maintaining the quality and value of the feed product.

CONCLUSIONS

The body of literature on extrusion processing of pet foods and aquatic feed blends containing DDGS is growing. The studies done to date emphasize the importance of the balance between ingredient composition (which is a function of the target pet and fish species), the physical properties of the raw feed blends, the type of extruder, and processing conditions which are used. It is also important to understand how all of these can affect the growth response and performance in actual dog and fish—which is an area that still needs more attention in order to most effectively use DDGS as a pet food and aquatic feed ingredient.

REFERENCES

Anderson, Y., Hedlund, B., Jonsson, L., and Svensson, S. 1981. Extrusion cooking of a high-fiber cereal product with crisp bread characters. *Cereal Chem* 58: 370–374.

Apruzzesse, F., Balke, S.T., and Diosady, L.L. 2000. Inline color and composition monitoring in the extrusion cooking process. *Food Res Int* 33: 621–628.

Breen, M.D., Seyam, A.A., and Bansik, O.J. 1977. The effect of mill by-products and soy protein on the physical characteristics of expanded snack products. *Cereal Chem* 54: 728–736.

Buchheit, J.K. 2002. Distiller's dried grains with soluble. In *Alternative agriculture, rural enterprise and alternative agriculture development initiative report*. Report 11, June 2002.

Camire, M.E., Camire, A., and Krumhar, K. 1990. Chemical and nutritional changes in food during extrusion. *CRC Crit Rev Food Sci Nutr* 29: 35–57.

Chevanan, N., Muthukumarappan, K., and Rosentrater, K.A. 2009. Extrusion studies of aquaculture feed using distillers dried grains with solubles and whey. *Food Bioprocess Technol* 2: 177–185.

Chevanan, N., Rosentrater, K.A., and Muthukumarappan, K. 2008. Effect of DDGS, moisture content, and screw speed on the physical properties of extrudates in single screw extrusion. *Cereal Chem* 85(2): 132–139.

Chevanan, N., Rosentrater, K.A., and Muthukumarappan, K. 2010. Effects of processing conditions on single screw extrusion of feed ingredients containing DDGS. *Food Bioprocess Technol* 3: 111–120.

Chevanan, N., Muthukumarappan, K., and Rosentrater, K.A. 2007. Neural network and regression modeling of extrusion processing parameters and properties of extrudates containing DDGS. *Transactions of the ASABE* 50(5): 1765–1778.

Chevanan, N., Rosentrater, K.A., and Muthukumarappan, K. 2007. Twin screw extrusion processing of feed blends containing distillers dried grains with solubles. *Cereal Chem* 84(5): 428–436.

Davidson, V.J., Paton, D., Diosady, L.L., and Rubin, L.J. 1984. Degradation of wheat starch in a single screw extruder: characteristics of extruded starch polymers. *J Food Sci* 49: 453–458.

Enterline, W.R. 1984. The production of extruded pet foods. Sprout–Waldron Division, Koopers Co., Inc. 1984.

Gracza, R. 1959. The subsieve-size fractions of soft wheat flour produced by air classification. *Cereal Chem* 36: 465.

Harper, J.M. 1979. Food Extrusion. *CRC Crit Rev Food Sci Nutr* 11: 155–215.

Harper, J.M. 1981. *Extrusion of Foods, Volume I*. Boca Raton, FL: CRC Press, Inc.

Harper, J.M. 1989. Food extruders and their applications. In *Extrusion cooking*, eds. C. Mercier, P. Linko, and J.M. Harper, 1–15. St. Paul, USA: American Association of Cereal Chemists, Inc.

Harper, J.M., and Jansen, G.R. 1985. Production of nutritious precooked foods in developing countries by low-cost extrusion technology. *Food Reviews Int* 1(1): 27–97.

Jensen, L.S. 1981. Value of distillers dried grains with solubles in poultry feeds. In *Proc Dist Feed Res Council*, 87–93.

Kannadhason, S., Muthukumarappan, K., Rosentrater, K.A., and Brown, M. 2010. Twin Screw Extrusion of DDGS-Based Aquaculture Feeds. *J World Aquaculture Society* 41: 1–15.

Kannadhason, S., Muthukumarappan, K., and Rosentrater, K.A. 2009a. Effect of starch sources and protein content on extruded aquaculture feed containing DDGS. *Food Bioprocess Technol*. DOI: 10.1007/s11947-008-0177-4.

Kannadhason, S., Muthukumarappan, K., and Rosentrater, K.A. 2009b. Effects of ingredients and extrusion parameters on aquafeeds containing DDGS and tapioca starch. *J Aquaculture Feed Sci Nutr* 1(1): 6–21.

Klopfenstein, T., and Stock, R. 1981. Distillers grains for ruminants. *Feedstuffs* 53(41): 26.

Kohler, C.C., and Kohler, S. 2001. *The feasibility of locating a tilapia production facility in conjunction with a dry mill distillery owned and operated by Golden Triangle Energy Cooperative in Craig, Missouri*. Rural Enterprise and Alternative Agricultural Development Initiative Report (11). June 2002. Prepared by Joshua K. Buchheit. Southern Illinois University, Carbondale, publication. http://www.siu.edu/~readi/grains/factsheets/distillerdriedgrainswsolubles.pdf.

Lusas, E.W., and Khee, K.C. 1987. Extrusion processing as applied to snack foods and breakfast cereals. In *Cereals and legumes in food supply*, eds. J. DuPont, and E.M. Osman, 163–172. Ames, Iowa: Iowa State University Press.

Matz, S.A. 1984. Snack Food Technology, Second Edition, AVI Publishing Company, Inc.

Miller, R.C. 1984. Low moisture extrusion: Effects of cooking moisture on product characteristics. *J Food Sci* 50: 249.

Morrison, E.M. 2001. *A fertilizer made from distiller's grain and fish may protect Minnesota's liquid assets*. Farming for Energy: Ethanol Co-Products. Ag Innovation News. Vol 10, #3.

National Corn Growers Association. 2009. The World of Corn 2009. www.ncga.com/09world/main/index.html.

Omueti, O., and Morton, I.D. 1996. Development by extrusion of soyabari snack sticks: A nutritionally improved soya-maize product based on the nigerian snack (kokoro). *Int J Food Sci Nutr* 47(1): 5–13.

Poos, M.I. 1981. Ruminant nutritional utilization of alcohol production byproducts. *Feedstuffs* 53(26): 18.

Rakosky, J. 1987. Soy products in food service. In *Cereals and legumes in food supply*, eds. J. DuPont, and E.M. Osman, Ames, 277–288. Iowa: Iowa State University Press.

Reddy, J.A., Stoker, R. 1993. Source: http://www.uspto.gov. United States Patent. Patent number 5225228, 6 July 1993.

Rosentrater, K.A., Muthukumarappan, K., and Kannadhason, S. 2009a. Effects of ingredients and extrusion parameters on aquafeeds containing DDGS and corn starch. *J Aquaculture Feed Sci Nutr* 1(2): 44–60.

Rosentrater, K.A., Muthukumarappan, K., and Kannadhason, S. 2009b. Effects of ingredients and extrusion parameters on aquafeeds containing DDGS and potato starch. *J Aquaculture Feed Sci Nutr* 1(1): 22–38.

Satterlee, L.D., Vavak, D.M., Abdul-Kadr, R., and Kendrick, J.G. 1976. The chemical, functional and nutritional characterization of protein concentrates from distiller's grains. *Cereal Chem* 53: 739–749.

Scott, M.L. 1970. Twenty-five years of research on distillers feeds for broilers. In *Proc Dist Feed Res Council*, 19–24.

Shurson, J. 2002. Increasing the utilization of DDGS in livestock and poultry production systems. Powerpoint presentation: http://www.ddgs.umn.edu/pps-poultry/mnagexpo-presentations. pps. University of Minnesota Ag Expo 2002, Jackpot Junction, Morton, MN.

Shukla, C. 2003. Properties of extruded corn distillers dried grains with solubles (ddgs) using single screw extruder. Unpublished MS Thesis, South Dakota State University, Brookings, SD.

Wu, Y.V., and Stringfellow, A.C. 1982. Corn distillers' dried grains with solubles and corn distillers' dried grains: Dry fractionation and composition. *J Food Sci* 47: 1155–1180.

Wu, Y.V., and Stringfellow, A.C. 1986. Simple dry fractionation of corn distillers' dried grains and corn distillers' dried grains with soluble. *Cereal Chem* 63(1): 60–61.

Wampler, D.J., and Gould, W.A. 1984. Utilization of distillers' spent grains in extrusion processed doughs. *J Food Sci* 49: 1321–1322.

Whitney, M.H., Shurson, G.C., Guedes R.M., and Gebhart, C.J. 2002. The relationship between distiller's dried grains with solubles (DDGS) and ileitis in swine. 63nd Minnesota Nutrition Conference, Eagan, MN. September 17–18, 2002.

William, M.A., Horn, R.E., and Rugala, R.P. 1977. Extrusion: An in-depth look at a versatile process. *Food Eng* 49: 99.

Zhang, P., and Bhatnagar, S. 2000. Effect of extrusion processing on digestibility and bioavailibility of nutrients in pet foods. Ralston Purina Co. Supplement to Compendium on Continuing Education for the Practicing Veterinarian. Sep. 2000. 22 (9A).

10 Industrial Application of Extrusion for Development of Snack Products Including Co-Injection and Pellet Technologies

Jorge C. Morales-Alvarez* and Mohan Rao*

CONTENTS

---

* This paper reflects the views of the authors, and does not represent the official views of PepsiCo, or any of its affiliated companies.

INTRODUCTION

Product innovation in extruded snack foods and most ready-to-eat breakfast cereals relies greatly on formulation and product shape. While these remain critical factors in product development, substantial product definition and novelty can be attained if defined color application patterns or internal flavors are added into the extruded products. This chapter will cover the co-injection of food substances into an extruder die with the objective of creating defined colored patterns, adding internal flavors, or achieving other food injection applications into cereal-based extruded products. A die assembly capable of achieving the co-injection of liquid food substance at the die to develop color patterns in direct expanded product while maintaining extrusion process stability will be presented in detail.

This chapter will also deal with development of half products or third generation snacks (pellets). Focus will be given to processes for producing corn masa- and potato-based pellets and the identification of critical process parameters for providing precise control in hydration, as well as gelatinization levels, overcoming variations in the characteristics of starting raw materials. These constitute key aspects to maintaining target product design and desirable pellet product quality.

CO-INJECTION IN FOOD EXTRUSION

The objective of co-injection in most food extrusion applications is to coat, variegate, or incorporate ornamental patterns into the extrudates, normally using food additives exhibiting substantially different flow behavior properties than the main extruding dough. The result is an extruded product with ornamental patterns and designs on its surface and throughout its cross section (Figure 10.1).

In this context, we consider co-injection as a special case of co-extrusion technology. In their most general interpretation, both co-injection and co-extrusion deal with the process and flow behavior of multiple layered materials through a common flow channel. We distinguish them only on the basis of the much higher viscosity of the complementary food materials introduced in co-extrusion, such as fillings, pastes, and secondary extruded dough (Millauer 1999; Moore 1999). Furthermore, beyond the obvious differences in the design of resultant products, from the flow behavior perspective, we make an important distinction between co-injection and co-filling technologies. In the latter, the introduction of complementary food substances is relatively independent of the pressure at the die; they are introduced through an

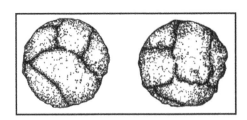

FIGURE 10.1   Co-injected pattern on direct expanded ball, simulating a sport-ball. U.S. Patent D 463,088 S.

independent flow channel and incorporated into the main extruding dough at the lowest pressure drop, near the exit of the die (Figure 10.2).

The methods by which liquid food additives such as food dyes and other similar liquid food substances are supplied and injected into extruder dies have remained unchanged over the years (Benson 1958a, 1958b; Fries and Kaufman 1967; McCulloch 1984; Parsons and Chatel 2003). Benson (1958a, 1958b) describes a method to achieve colored patterns in ready-to-eat cereal flake products using gravimetric force to supply food color substances. Fries and Kaufman (1967) present a technology designed to incorporate fluid food fillings into shaping pieces of dough for the production of baked products. This technology requires injecting the filling materials into a piston extruder die operating at low pressure relative to that required for direct expansion. McCulloch (1984) describes a method and apparatus to incorporate food additives throughout an expanded food product at attendant pressure, however, pattern definition, in this case, is not critical to the process.

The relatively high pressure, often exceeding 5000 kPa (725 psi), encountered in most direct expanded snacks and ready-to-eat breakfast cereal extrusion operations presents a challenge to the direct injection of food substances into the die. These challenges increase if, in addition, one demands:

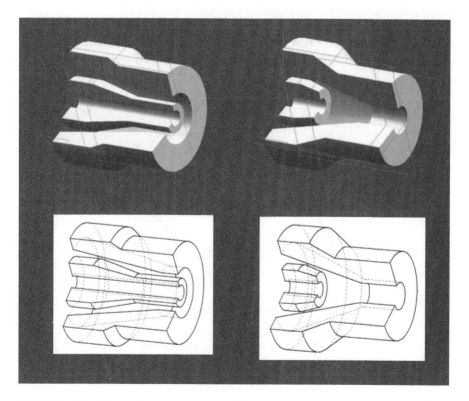

FIGURE 10.2 Schematic representation of co-filling and co-injection general designs. Observe how co-filling is independent of internal pressure conditions at the die.

1. The development of clear injection patterns on the final expanded product
2. Good injection performance at a variety of pressures
3. No clogging or blockage of the injection device and die orifice
4. Ease of cleaning and maintenance.

The stability of the process is paramount to all these considerations and directly connected to the aforementioned high pressure regimes at the extruder die. Therefore, an understanding of the flow behavior of multiple layered materials through a common flow channel becomes critical to the development of a technology capable of co-injecting food substance at the die while maintaining extrusion process stability.

Multiple Layer Extrusion Flow through a Common Channel

Flow Behavior Analysis

To offer a quantitative interpretation of multi-layer flow phenomena, let us consider the case of symmetrical three-layer flow through a cylindrical channel, with the core layer being surrounded by two identical surface layers showing equal volumetric flow rate (Figure 10.3). This will help to understand key criteria followed in the design and development of the technology subject of this chapter.

Based on the law of conservation of momentum (equation of motion), and considering only the direction of flow (Figure 10.3), the Navier–Stokes equation in cylindrical coordinates for a Newtonian fluid with constant density and viscosity is:

$$\rho\left(\frac{\partial v_z}{\partial t} + vr\frac{\partial v_z}{\partial r} + \frac{v_\theta}{r} + vz\frac{\partial v_z}{\partial z}\right) = -\frac{\partial P}{\partial z} + \rho g_z + \eta\left[\frac{1}{r}\frac{\partial}{\partial r}\left(r\frac{\partial v_z}{\partial r}\right) + \frac{1}{r^2}\frac{\partial^2 v_z}{\partial \theta^2} + \frac{\partial^2 v_z}{\partial z^2}\right]$$

$$(10.1)$$

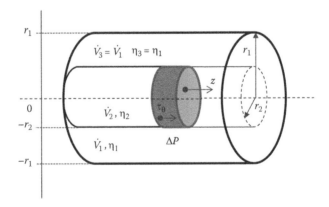

FIGURE 10.3 Symmetrical three-layer flow through a cylindrical channel with core layer ($\dot{V}_2$) surrounded by two identical surface layers with equal volumetric flow rate ($\dot{V}_1 = \dot{V}_3$).

Assuming steady state and fully developed flow: steady state, $\partial v_z/\partial t = 0$; fully developed flow, $\partial v_z/\partial z = 0$; and disregarding velocity components in any other direction but the direction of flow (z direction), $v_z = v_\theta = 0$,

Equation 10.1 can be simplified to:

$$-\frac{\partial P}{\partial z} + \rho g_z \frac{1}{r}\frac{\partial}{\partial r}\left(r \cdot \frac{\partial \eta v_z}{\partial r} + \frac{1}{r}\frac{\partial \tau_{\theta z}}{\partial \theta}\right) = 0 \tag{10.2}$$

Disregarding gravitational influence and deformation due to normal stresses, Equation 10.2 can be rewritten:

$$\frac{1}{r}\frac{\partial}{\partial r}\left(r \cdot \frac{\eta \partial v_z}{\partial r}\right) = \frac{\partial P}{\partial z} \tag{10.3}$$

Equation 10.3 can be integrated by assuming the pressure gradient to be independent of r:

$$\frac{2\eta}{r} \cdot \frac{\partial v_z}{\partial r} = \frac{\partial P}{\partial z} \tag{10.4}$$

Solving for the velocity gradient:

$$\frac{\partial v_z}{\partial r} = \frac{r}{2\eta} \cdot \frac{\partial P}{\partial z} \tag{10.5}$$

In reference to Figure 10.3, and given symmetrical flow, it is sufficient to study only one-half of the channel. To facilitate integration, we locate the origin of the coordinate system to coincide with the axial center of the channel, where, we assume, the maximum velocity (v_{max}) is located. Given the condition of adhesion to the wall of the channel, the velocity profile for the surface layer is described by:

$$v_1(-r_1) = 0$$

$$v_1(r) = \frac{1}{4\eta_1}\frac{\partial P}{\partial z}\left(r^2 - r_1^2\right) \tag{10.6}$$

With the condition of adhesion at the boundary between the core and surface layer, the velocity profile for the core layer is described by:

$$v_2(-r_2) = v_1(-r_2)$$

$$v_2(r) = \frac{1}{4\eta_2}\frac{\partial P}{\partial z}\left[r^2 - r_2^2 - \frac{\eta_2}{\eta_1}\left(r_1^2 - r_2^2\right)\right] \tag{10.7}$$

To determine the pressure gradient $\partial P/\partial z$ as a function of volumetric flow rate, $\dot{V}$:

$$\dot{V}_1 = \int_{r_2}^{r_1} v_1(r)\,dr \tag{10.8}$$

$$\dot{V}_2 = \int_{0}^{r_2} v_2(r)\,dr \tag{10.9}$$

Executing the integration based on Equations 10.6 and 10.7,

$$\dot{V}_1 = \frac{\partial P}{\partial z} \cdot \frac{r_1^3}{\eta_1} \cdot \frac{1}{12} \left[-2 - \frac{r_2^3}{r_1^3} + \frac{3r_2}{r_1} \right] \tag{10.10}$$

$$\dot{V}_2 = \frac{\partial P}{\partial z} \cdot \frac{r_2^3}{\eta_2} \cdot \frac{1}{12} \left[-2 - \frac{\eta_2}{\eta_1}\left(\frac{3r_1^2}{r_2^2} - 3 \right) \right] \tag{10.11}$$

The volumetric flow component ratio $\dot{V}_{cr} = \dot{V}_2 / (\dot{V}_1 + \dot{V}_3)$, where $\dot{V}_1 = \dot{V}_3$, can be calculated from Equations 10.10 and 10.11:

$$\dot{V}_{cr} = \frac{\dot{V}_1}{2\dot{V}_2} = \frac{\eta_1 r_2^3 + \eta_2 r_2^3 \dfrac{3}{2}\left(\dfrac{r_1^2}{r_2^2} - 1 \right)}{2\left(\eta_2 r_1^3 + \dfrac{1}{2}\eta_2 r_2^3 - \dfrac{3}{2}\eta_2 r_2 r_1^2 \right)} \tag{10.12}$$

The pressure loss is obtained by solving Equation 10.10 or 10.11 for $\partial P/\partial z$; taking Equation 10.11:

$$\frac{\partial P}{\partial z} = \frac{12\dot{V}_2\eta_2}{r_2^3\left[-2 - \dfrac{\eta_2}{\eta_1}\left(\dfrac{3r_1^2}{r_2^2} - 3 \right) \right]} \tag{10.13}$$

Conceptualization of these parameters in the context of stable multi-layer extrusion flow is very important, especially the ones attendant to flow component ratio, and pressure drop. Both will be of value when we discuss the development of the co-injection die in a later section.

Flow Behavior Instability and Extruded Food System

Starch, protein, and fiber are major components in most product formulations in food extrusion; they constitute the so-called bio-polymeric structures that provide, to an extent, the continuous phase, or product structure in food extrudates. Study of the flow extrusion behavior of food bio-polymers is complicated by the diversity and natural variability of the raw materials used. Study of thermoplastic extrusion and the more predictable behavior of their polymeric structures, relative to food bio-polymers, provide great insight into the flow phenomena inside the extruder and die. Strong analogies can be drawn between thermoplastic and food extrusion to explain flow instabilities such as melt fracture, typical of single layer material flow. In multiple layer flow, thermoplastic extrusion theory identifies two singular types of instability: encapsulation and interfacial instability.

Interfacial instability is explained by the absence of macromolecular entanglements during the merging of two polymeric melts and the forces of flow acting on the interface creating a weak spot on the bond of the adjacent melts. These adjacent melts can slip on each other much easier than the individual layers within both melts, creating flow disturbances, and depending on the severity of the instability, possibly impacting the product quality at the interface. Experimental studies have confirmed this hypothesis, which explain its occurrence even during the multi-layer extrusion processing of identical materials (Michaeli 1984, 2003). Encapsulation, on the other hand, seems more relevant to our study as its occurrence is related to the multi-layer extrusion of materials exhibiting different flow behavior properties.

Encapsulation, as a flow instability phenomenon, has been studied by several authors (Han 1975; Southern and Ballman 1975; Uhland 1977; Schrenk 1978; Wortberg 1978). This type of multi-layer flow instability is found in the flow of polymer melts with different viscosities through a common flow channel (co-extrusion). If the viscosity difference between two polymeric melts is large enough, and the residence time is sufficient, the melt with the lower viscosity will tend to migrate to the region of highest shear stress (periphery of the flow path) and wrap around or encapsulate the melt with the higher viscosity. This tendency is explained by the principle of "minimum dissipated energy." The rearrangements of the polymer melts relative to their viscosity occur in such a way to ensure the pressure loss of the flow is minimal. In stable flow, the low viscosity melt flows along the periphery of the conduction while the high viscous melt flows in the center. In contact with the wall, the low viscous melt flows forming a slip film for the high viscosity melt in the middle of the conduction, therefore, ensuring a low pressure drop for the net flow. In the instable flow, the low viscosity melt flows in the middle of the conduction at higher flow velocity than the surrounded melt. The pressure loss for the net flow in the instable case is greater than for the stable one, and, therefore, energetically less favorable (Michaeli 1984; Michaeli 2003).

Co-injection Flow Instability in Cereal-Based Extruded Foods

From the analysis presented in the previous section, it becomes evident that, in principle, the addition of liquid food colors or flavors into cereal-based extrudates at the die creates conditions of instability given their substantially different flow behavior properties, particularly viscosity. As an example, let us consider extrusion conditions easily reached during direct expanded products; say approximately 3040 kPa die pressure, and 120°C die temperature. Assuming steady state and isothermal flow, one can also assume that injected liquid food color into the die will quickly equilibrate to the extrusion conditions, and therefore, will exhibit dynamic viscosity similar to sub-saturated water at given conditions. At 3040 kPa and 120°C the dynamic viscosity of sub-saturated water is approximately 2.8×10^{-4} Pa·s. So we designate η_{lc} to refer to the viscosity of the liquid food color and enunciate $\eta_{lc} = 2.8 \times 10^{-4}$ Pa·s at 3040 kPa and 120°C.

On the other hand, the viscosity of cereal-based extruding dough can be estimated using the definition of apparent viscosity η and the Ostwald-de Waele or power law model; with apparent viscosity defined as:

$$\eta = \frac{\tau}{\gamma} \tag{10.14}$$

where η = apparent viscosity, Pa·s; τ = shear stress, N/m^2 or Pa; $\dot{\gamma}$ = shear rate, s^{-1}.

And the Ostwald-de Waele or power law model:

$$\tau = m\gamma^n \tag{10.15}$$

where τ = shear stress, N/m^2 or Pa; m = consistency index, N·s^n/m^2; $\dot{\gamma}$ = shear rate, s^{-1}; n = flow behavior index.

Substituting for τ in Equation 10.14 using Equation 10.15:

$$\eta = m\gamma^{n-1} \tag{10.16}$$

From rheological measurements using a rheometer die, and assuming 20% in-barrel moisture at the given pressure and temperature conditions (Harper 1981): $m = 10,800$ (N·s$^{0.42}$)/m^2, $n = 0.42$, $\dot{\gamma} = 192$ s^{-1}.

Applying Equation 10.16:

$$\eta_d = \left(10,800\,\text{N}\cdot\text{s}^{0.42}/\text{m}^2\right) \times \left(192\,\text{s}^{-1}\right)^{0.42-1} = 281.3\,\text{N}\cdot\text{s/m}^2 = 511.8\,\text{Pa}\cdot\text{s}$$

The estimated viscosity ratio between the liquid color (η_{lc}) and the extruding dough (η_d) at 120°C and 3040 kPa can now be estimated: $\eta_{lc}/\eta_d = (2.8 \times 10^{-4}/511.8) = 5.4 \times 10^{-7}$.

A temperature rise due to viscous dissipation cannot commonly be ignored due to the high viscosity of cereal-based plasticized extruding dough. Therefore, the assumption of isothermal flow is not a realistic one, however, the axial development of temperature field due to viscous dissipation and the pressure loss as the flow progresses towards the die exit will affect both extruding dough and liquid color. Hence, even at more realistic nonisothermal flow conditions, the estimated viscosity ratio calculated above is not expected to change considerably and will remain within the same order of magnitude.

Under these conditions, the much lower viscosity liquid color will try to encapsulate the large viscosity extruding dough to minimize the pressure loss of the total flow (Michaeli 2003).

According to Southern and Ballman (1975), conditions exist at which one can reverse encapsulation when extruding two thermoplastic polymer melts with viscosity curves that cross each other (apparent viscosity vs. shear rate). As illustrated in the example above, given the substantially different viscosities between liquid food dyes (similar to water) and extruding dough at direct expanded conditions, one cannot realistically expect to find food co-injection conditions at which this may happen. Therefore, our hypothesis for the development of a co-injection die relied on the basis of mechanical considerations to minimize encapsulation by designing for uniform, low component ratio of the low-viscosity flow, and control of the pressure drop.

DEVELOPMENT OF CO-INJECTION DIE FOR PRODUCTION OF NOVEL CEREAL-BASED EXTRUDATES

A co-injection die technology capable of imprinting and adding color defined patterns onto cereal-based direct expanded extrudates will be presented. To this end, such a technology has to inject and integrate uniform streams of liquid food color into the flow of cereal-based dough undergoing extrusion, at a wide range of extrusion pressures and temperatures, into the die, before expansion. The injected colored patterns have to maintain their relative position across the material flow throughout the rest of the extrusion process which normally includes a converging flow path to compress the extruding material even further, before finally exiting the die. The color patterns remain through the expansion process resulting in a color patterned, direct expanded product. Different injection patterns can be designed to achieve different ornamental effects throughout the extrudate product, allegoric to different motifs or designs. Novelty can be increased by utilizing different design patterns and colors. Critical considerations in the mechanical design of a co-injection die for food extrusion applications are:

1. Flow uniformity of the injected liquid food color.
2. Development of mechanical seals to avoid food liquid color leaks which would impair injection uniformity. This also makes the utilization of o-rings or similar type of external seals unnecessary.
3. Ready supply of the injected liquid color to the immediate location where it incorporates to the cereal-based dough, to avoid interruptions due to normal pressure variations.

This can be accomplished by an extruder co-injection die assembly such as the one shown in Figure 10.4 and patented by Keller et al. (2003) (U.S. Patent 6,620,448 B2) and Keller, Morales-Alvarez, and Ouellette (2005) (U.S. Patent 6,854,970 B2).

This assembly comprises an injection nozzle (600), a co-injection insert set (200 and 300), and a converging front insert (400). The assembly fits into the extruder die head (510). The co-injection insert is a matched set consisting of a forming

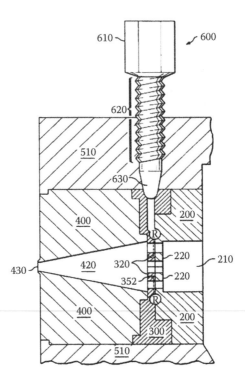

FIGURE 10.4 Co-injection die assembly for development of direct expanded product with color patterns. U.S. Patent 6,854,970 B2.

section (200) and an injection section (300). They form an internal peripheral reservoir manifold (R) through which the fluid color can be readily supplied to capillary channels (352) arranged in distinct designed patterns. The liquid food color flows out of the capillary channels into the cereal-based dough, which flows through the main extruder channel (210), imparting the designed cross-sectional pattern. A more detailed view of the co-injection insert stack (100) is shown in Figure 10.5.

Flow Uniformity
Referring to Figure 10.5, a forming element (220), congruent to the desired color pattern to be applied, is located at the outlet of the main extrudate channel (210). The forming element mates and seals with the center piece of the injection section (300). We refer to this center piece as the injection element (320). The injection element contains, at its downstream exit face, the arrangement of capillary channels in designed alignment with the forming element. Both forming and injection elements form a plurality of passageways congruent to the desired pattern profile. The forming element divides the main extrudate flow into several adjacent secondary flows that momentarily separate through the passageways at the back of the forming element to re-unite at the front part of the injection element where each capillary channel discharges a continuous flow of food liquid color. This results in a continuous band of food liquid color being injected into the temporary gaps formed by the adjacent

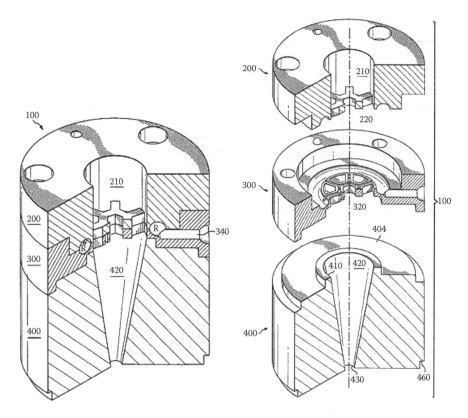

FIGURE 10.5   Cut out and exploded views of co-injection insert stack showing internal details of forming section (200), injection section (300) and converging front insert (400). U.S. Patent 6,620,448 B2.

secondary flows as they exit the passageways and re-unite in front of the capillary channel. Upon exiting from the individual passageways, the adjacent secondary flows unite to enclose the injected bands of food liquid color within a single flow mass, thereby imparting a distinct colored pattern into the main extrudate flow. The color patterned main extruded flow exits the extruder through the converging channel (420) of the front insert (400).

The flow regime of the color patterned dough through the converging channel (420) constitutes a multi-layer flow where a low viscous fluid is surrounded by a large viscous fluid. As described, this constitutes conditions for instability. And it is the precise delivery of the low-viscosity liquid flow, the large component ratio of the two fluids, as well as the development of favorable conditions in pressure drop between stable and unstable cases which prevents instability and encapsulation from occurring within this flow regime at otherwise unstable conditions.

Mechanical Seal

Another important consideration in the design of the co-injection die assembly, and critical to ensuring uniform food liquid color flow, constitutes the corresponding annular seal between the forming section and the matching injection section. The carefully designed stepwise shape of the contact surfaces in both matching sections creates an effective mechanical seal without the use of o-rings or similar. When properly aligned, the forming section and injection section form not only the internal peripheral reservoir manifold (R), but also the stepwise shape of their annular periphery design, which closely match each other according to very small and precise tolerances provided in their mechanical design and manufacture. Once assembled, forming and injection sections are fastened by equally spaced counter-sunk coupling bolts around the periphery of the assembly. This forms the required mechanical seal and ensures that the pressurized liquid food color does not leak through the internal contact surfaces.

Ready Supply of Food Liquid Color

The pressurized food liquid color continuously recharges the internal peripheral reservoir manifold (R) through conduit (340) in fluid communication with the standard positive displacement pumping system. The internal peripheral reservoir manifold makes the liquid food color readily available to the interconnected capillary channels defining the color pattern. This is particularly important to prevent flow interruptions due to normal pressure variations.

PELLET TECHNOLOGY

Pellets or third generation snacks offer a number of distinctive product attributes including unique textures, denser bite than many direct expanded snacks, a variety of formulations and shapes, as well as the ability to generate half products or shelf stable intermediates. This allows manufacturers to produce the intermediate product at one location and rethermalize to complete puffing or expansion at a different time and/or location. A general review of the role of extruders, ingredients, and additional process steps in pellet technology for the development of half products, or third generation snacks, has been accomplished by other authors (Harper 1981; Matz 1993; Moore 1999; Sevatson and Huber 2000; Guy 2001). The objective of this section is

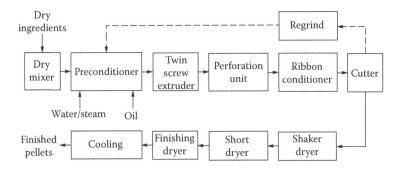

FIGURE 10.6 Schematic representation of process technology for corn masa- and/or potato-based pellets, illustrating key elements of the process. U.S. Patent 6,643,463 B1.

to present a technology for the production of corn masa- and potato-based pellets capable of maintaining product design and desirable product quality by overcoming variations in the characteristics of starting raw materials. Figure 10.6 provides a schematic diagram of the technology illustrating the most important elements of the process. As described by Bhaskar, Rao, and Warren (2001) and Bhaskar et al. (2002), the identification and manipulation of critical process parameters providing precise control over hydration and gelatinization levels of the raw materials is key in this pellet technology.

FORMULATION AND RAW MATERIAL SPECIFICATIONS

Typical formulations for corn masa- and potato-based pellet are summarized in Tables 10.1 and 10.2.

The characteristics of masa and potato flour directly influence the quality of the pellets produced. Therefore, it is important that initial masa-, potato-flour, and other ingredients specifications be established and monitored. As an example, Table 10.3 summarizes key performance attributes and characteristics of the initial masa flour.

Characterization of the pasting properties of the major cereal components of the formulation is important for predicting processing performance and constitutes a significant aspect of the pellet technology hereby described. Several pasting analytical methods can be used to this end. Key criteria include simplicity, practicality, accuracy, and speed of response. Adjustments to the operation of preconditioner and/ or extruder can be made based on the performance of ingredients and processing samples.

BLENDING OF INGREDIENTS

Different formulations require different mixing times (Tables 10.1 and 10.2). A batch blender is typically used. No water is added at this stage, and therefore, no hydration is pursued but the development of a homogenous blend of the dry ingredients. Note the potato-based pellet formulation requires a longer blending time because it is more heterogeneous. The type of blender used will also impact the blending time and quality of blend.

TABLE 10.1
Masa-Based Pellet Formulation and Mixing Time

Ingredient	Proportion (%)
PreCooked Corn Flour	90.0–95.0
Granulated Sugar	1.0–2.5
Salt	0.5–1.0
Sodium Bicarbonate	0.4–0.8
Mono/Diglycerides	0.2–0.3
Blending time	7 min

TABLE 10.2
Potato-Based Pellet Formulation and Mixing Time

Ingredient	Proportion (%)
Potato Starch	30–50
Potato Flakes	25–40
Potato Granules	10–30
Corn Starch	5–15
Salt	1.5
Mono/Diglycerides	0.3
Oil	1–3
Blending time	15 min

PRECONDITIONING

In general, the main objective of this operation is to hydrate and plasticize or soften the ingredient materials (Strahm 2000) without losing their granular nature. To this end, a continuous co-rotating twin-shaft paddle mixer is normally used. Precisely controlling hydration extent and partial gelatinization of the blend during preconditioning constitutes a very important step of the process and is critical in defining the quality of the product. This is accomplished by regulating the direct injection of heated water and/or steam into the blend as it is mixed and conveyed through the preconditioner. Typically

TABLE 10.3
Corn Masa Product Specifications

Key Attributes	Method	Unit of Measurement	Aim	Low Limit	High Limit
Moisture (130°C, 60 min)	AOAC (14.004)	%	11.5	10.0	13.0
RVA[a] Peak (at 80°C)	ASQC	RVU	3100	2800	3400
pH	AOAC		7	6.8	7.2
Fat (Soxhlet Chloroform)	AOAC	%	3.0	2.0	4.5
Particle Size (Screen Size)	Ro–Tap				
Over U.S. #25		%	0	0	—
Over U.S. #40		%	0	0	<0.1
Over U.S. #60		%	25	20	35
Over U.S. #80		%	40	35	45
Over U.S. #100		%	15	10	20
Extraneous Matter	FDA	%	—	0	
Extraneous Matter	Visual	Count	0	0	0

[a] Rapid Visco Analyzer, Newport Scientific Pty. Ltd., Australia.

the dry-ingredient blend enters at 12% moisture content and exits at approximately 24%–30%. The ratio of steam to water added is adjusted depending on the gelatinization level of the entering blend (as measured by Rapid Visco Analyzer, RVA, Newport Scientific Pty. Ltd.). However, the total combined rate of water and steam is maintained the same to ensure consistent moisture content of the blend as it exits the preconditioner.

Steam injection has the most influence, given its higher specific energy and higher diffusivity relative to water. This allows for easier hydration and uniform distribution. Steam addition rate can vary from 1.3% to 7.0% of the rate of blend added into the preconditioner. A typical steam addition rate is 90 lbs/h for every 2200 lbs/h of added blend into the preconditioner. In general, milder preconditioning conditions are required for potato-based formulas relative to corn masa-based formulas.

Typical residence time is 60 to 90 s, and temperature is maintained at approximately 71°C inside the unit to inhibit microbial growth, which allows longer production cycles before sanitation. The inlet water to the preconditioner is preheated to approximately 65°C to 71°C. The temperature of the blend as it exits the unit can also be adjusted by regulating the amount of steam within the constraints explained above for consistent exit moisture content. Additionally, control of the hot water jacket around the unit can assist in regulating the temperature of the blend.

EXTRUSION

A twin-screw extruder with nine heating and cooling barrel zones for an approximate *L/D* ratio of 30 is a preferred configuration. This allows cooking and cooling/forming sections in one machine, and provides flexibility to produce both potato- and corn masa-based pellets. The preconditioned meal and additional water are fed into zone 1. Zones 2 through 5 are heated to maintain barrel temperature from 50°C to 80°C; they are dedicated to achieve additional hydration and thermalization of the blend to develop homogenous viscoelastic dough. Corn masa-based formulas use a moderately high-shear screw configuration to ensure the desired level of heat generation via viscous dissipation, with screw speed ranging from 350 rpm to 380 rpm. Milder cooking conditions are used for potato-based formulas requiring low-shear screw configuration and screw speed ranging from 160 rpm to 220 rpm. The cooking section of the extruder is followed by a vacuum vent in zone 6, and forming section from zones 7 through 9. A low-shear screw configuration is used in the forming section mostly to convey and densify the extruding dough. Zones 6 through 9 are cooled to reduce extrudate temperature and steam flashing at the die. Evaporative cooling provided by the vacuum vent in zone 6 marks the transition between thermalization and forming zones and helps mitigate excess steam flashing at the die. Excess steam flashing at the die exit causes undesirable air bubbles in the resulting extrudate ribbon. A typical vacuum level is achieved between 5.33 kPa and 7.99 kPa (40 mm Hg and 60 mm Hg). The evaporative rate achieved varies between 15 kg to 30 kg of water per hour.

Infeed water to the extruder and screw speed constitute important quality control features in this stage of the process. They both influence extrudate dough viscosity and bed pack (filled section of the barrel) which, in turn, impact residence time and extrudate dough temperature. Optimal cooking in the extruder will be achieved by controlling both residence time and extrudate temperature. Typical extrudate dough temperature in the die ranges from about 88°C to 101°C.

The die is a 60 inch coat-hanger design with adjustable choker bars and die lips. Coat-hanger die designs are typically used for production of extrudates showing large width to thickness ratios, such as sheets or films. The extruding dough flows from the end of the twin-screws to the center of the die where it gets symmetrically distributed on the horizontal plane by a manifold in the shape of coat-hanger, hence the name of the die design. This transforms the shape of the round extruding dough strand into an even flat extrudate. After the manifold, the spread dough flows into a flat area or resistance region, also called land. Choking bars and die lips complete the assembly providing uniform flow distribution and fine flow adjustment. Uniformity across the width of the die is controlled by fine tuning of the aperture between the die lips. Nominal ribbon thickness is about 0.90 mm to 1.35 mm and measured by online thickness gauges before entering a conditioning tunnel. Uniform ribbon thickness is very important to ensure uniform pellet drying and expansion at rethermalization operations such as frying or microwave heating.

RIBBON PRETREATMENT AND CUTTING

Moisture evaporation, as it exits the die, helps the ribbon to quickly transit into a pliable, flexible structure that can be mechanically manipulated without significant permanent deformation. The complexity of the pretreatment and cutting operations will depend upon whether the ribbon will be used for single sheet pellet or a laminated double sheet pellet. More sophisticated cutting effects can be achieved with a laminating process, which, in turn, impacts finished pellet texture and appearance, resulting in complex shapes and denser products. The following description corresponds to a process to produce a laminated perforated double sheet pellet. As it exits the die and becomes more pliable the ribbon is slit by a stationary blade into two sheets of equal width. The two sheets may be transported to perforation units for ornamental purposes. The two parallel sheets are immediately transported by a transfer conveyor into a seven pass belted cooler for conditioning before embossing and cutting. The cooler is maintained at a temperature of approximately 30°C to approximately 60°C. Cold air at approximately 6 m/s and preferably 35°C is applied to both sides of the sheets, additional air temperature in the tunnel is manipulated to achieve product sheet temperature between 35°C and 40°C at the embosser. Cooling of the sheets is necessary to prevent them from sticking and wrapping on the embosser rollers or cutter.

After cooling, the product sheets are delivered by conveying rollers to separate embosser and anvil-cutter roller pairs. Each sheet is embossed and quickly brought together to the nip of anvil and cutter rollers with the embossed surface of the sheets oriented to form the external face of the pellet. The joined sheets are then laminated and cut into pellets by the anvil and cutter rollers; different product shapes are possible depending upon the design on the cutter roll. The small portion of the sheets that are not formed into pellets can be trimmed, chopped, and ground into the so-called regrind. Aim particle size for the regrind pieces is approximately 3.2 mm × 3.2 mm × 2.0 mm. The regrind is recycled back into the process at the inlet of the preconditioner, at a rate of approximately 3% of the total blend feed rate.

Pellet Drying

Careful drying is critical for good pellet quality. A three-step drying operation with shaker dryer, short predryer, and finish dryer steps is preferred. Pellets are pneumatically transported from the cutting operation to the belted shaker dryer. The temperature set-point of the shaker dryer is approximately 55°C and the objective is to reduce the surface moisture of the pellet to prevent them from sticking together, compacting, and deforming in subsequent drying operations. The moisture content of the pellets entering the shaker dryer is about 24%, which is reduced to approximately 20%–22% upon exit. Out of the shaker dryer the pellets are spread into the receiving belt of the short predryer with an oscillating spreader. The short predryer is set to approximately 50°C with approximately 90% relative humidity (RH), and the objective of this stage is to reduce the pellet moisture content from approximately 20%–22% down to 17%–19%. The last drying step is a five pass finish dryer comprising two stages. Stage one is a proper drying zone set at approximately 45°C with approximately 18% RH. The second stage is a tempering zone to equilibrate the moisture gradient within the pellet. It is set at approximately 40°C with 30% RH. Both stages combined reduce and stabilize the moisture content of the pellet from approximately 19% down to approximately 12%. An ambient cooler conveyor can be added to the drying operation, right after stage two to cool the pellets to room temperature.

At this stage, the pellets can be further processed to produce a finished product or packed in bulk producing a stable intermediate product than can be stored or shipped for later processing. Controlled rethermalization of the pellets via frying, baking, or microwave heating would achieve the desired expansion. The expanded pellets can be seasoned and packaged for finished product distribution.

CONCLUSION

Insight on the design of co-injection and pellet technologies for production of direct expanded snacks has been presented. Analysis of multi-layer flow in polymer extrusion has helped understand and overcome instability phenomena observed when co-injecting liquid food substance into cereal-based extruding dough for direct expanded snacks. On pellet technology, characterizing and controlling hydration and gelatinization level of ingredients constitute key control parameters for achieving target product design and quality.

Extrusion is a very versatile process with wide applications in the food industry. The need for innovation provides great stimulus to food extrusion scientists and engineers to leverage fundamental knowledge from polymer extrusion, physical chemistry and other disciplines for their developments, as well as to develop new knowledge for creating new processes and applications.

REFERENCES

Benson, J.O. 1958a. Cereal product with striped effect and method of making same. US Patent 2,858,217.

Benson, J.O. 1958b. Cereal product with honeycomb-like appearance and method of making same. US Patent 2,858,219.

Bhaskar, A.R., Rao, M.V.N. and Warren, D.R. 2001. Process for expanded pellet production. US Patent 6,224,993 B1.

Bhaskar, A.R., Cogan, K.C., Kaafarani, B.M. and Rao, M.V.N. 2002. Process for producing expandable pellets. US Patent 6,432,463 B1.

Fries, E.W. and Kaufman, H.B. 1967. Apparatus for filling baked products. US Patent 3,314,381.

Guy, R. 2001. Snack foods. In *Extrusion cooking, technologies and applications*, ed. R. Guy, 170–173. Boca Raton, FL: CRC Press LLC.

Han, C.D. 1975. A study of co-extrusion in a circular die. *J Appl Pol Sci* 19: 1875–1883.

Harper, J.M. 1981. *Extrusion of foods*, Vol. II. Boca Raton: CRC Press, Inc.

Keller, L.C., Morales-Alvarez, J.C., Ouellette, E.L., Rao, V.N.M. and Terry, V.S. 2003. Extruder die with additive reservoir. US Patent 6,620,448 B2.

Keller, L.C., Morales-Alvarez, J.C. and Ouellette, E.L. 2005. Extruder die injection nozzle. US Patent 6,854,970 B2.

Matz, S.A. 1993. Puffed snacks. In *Snack food technology*, 3rd ed., 159–172. McAllen, Texas: Pan-Tech, International, Inc.

McCulloch, M.G. 1984. Apparatus for incorporating additives in extruded foods. US Patent 4,454,804.

Michaeli, W. 1984. *Extrusion dies: Design and engineering computations*. New York: Macmillan Publishing Co., Inc.

Michaeli, W. 2003. Extrusion dies for plastics and rubber: Design and engineering computations. 3rd Edition. Cincinnati: Hanser Gardner Publications, Inc.

Millauer, C. 1999. Crispy shell-soft filling by co-extrusion technology. In *Advances in extrusion technology, aquaculture/animal feeds and foods*, eds. Y.K. Chang and S.S. Wang, 361–375. Lancaster, Pa: Technomic Publishing Co., Inc.

Moore, G. 1999. Snack food extrusion. In *The technology of extrusion cooking*, ed. N.D. Frame, 110–143. Gaithersburg, MD: Aspen Publishers, Inc.

Parsons, M.H. and Chatel, R.E. 2003. Device system and method for fluid additive injection into a viscous fluid food stream. US Patent 6,509,049 B1.

Schrenk, W.J. 1978. Interfacial flow instability in multilayer coextrusion. *Polymer Eng Sci* 18: 8–9.

Sevatson, E. and Huber, G.R. 2000. Extruders in the food industry. In *Extruders in food applications*, ed. M.N. Riaz, 167–204. Lancaster, PA: Technomic Publishing Co., Inc.

Southern, J.H. and Ballman, R.L. 1975. Additional observations on stratified bicomponent flow of polymer melts in a tube. *J Polym Sci Polym Phys Ed* 13: 825.

Strahm, B.S. 2000. Preconditioning. In *Extruders in food applications*, ed. M.N. Riaz, 115–126. Lancaster, PA: Technomic Publishing Co., Inc.

Uhland, E. 1977. Stratified two-phase flow of molten polymers in circular dies. *Pol Eng Sci* 17(9): 671–681.

Wortberg, J. 1978. Die Design for Single- and Multilayer Extrusion. Dissertation at the Institut für Kunststoffverarbeitung, Rheinisch-Westfälische Technische Hochschule Aachen, Germany.

11 Thermal and Nonthermal Extrusion of Protein Products

Charles I. Onwulata

CONTENTS

EXTRUSION PROCESSING OF PROTEINS

Extrusion cooking is a high temperature short time shear process used to modify food structure, imparting texture. Successful use of single-screw extrusion processing technology to manufacture texturized food products in the food industry dates back to the 1960s, but the introduction of double extrusion and the use of twin-screw extruders made possible the creation of extrusion texturized proteins in the late 1970s (Atkinson 1970; Kinsella 1978; Harper 1979). Better understanding of the extruder configuration parameters necessary for developing a layered and meat-like texture with consistent characteristic properties made soy and other vegetable proteins the first successful and most widely available textured protein food (Cheftel, Kitagawa, and Queguiner 1992). Texturization is the unraveling of the globular structure of native proteins, accompanied by breakage of intramolecular bonds, and re-alignment of disulfide bonds, with the use of mechanical shear, heat, and pressure as requisites (Bhattarcharya and Padmanabhan 1999; Shimada and Cheftel 1988). The use of food protein sources, including dairy products, in texturization remained in the developmental stage up to the recent past due to difficulties associated with maintaining consistent properties over a wide range of temperature and pH conditions in the extruder. For example, milk and egg proteins are temperature and shear sensitive, and are easily destroyed above 120°C, far below the characteristic extrusion texturization temperature range of 160°C to 180°C range for soy and vegetable proteins (Riaz 2004).

Technological difficulties and protein interaction chemistry hurdles are still limiting extrusion texturization of proteins such as casein, whey (a product of cheese manufacturing), egg, and fish proteins; the reasons are that the precise mechanisms of protein texturization through extrusion are still unclear, because the complex protein–protein interaction involved is still poorly understood. The main reason for the sub-par commercial performance of nonplant proteins in texturization has been failure in their textural quality, temperature response, and lack of cost competitiveness. By contrast, texturized soy protein, presently the most commercially successful extruded protein, is finding wide use in a variety of applications, and is available at reasonably low prices (Areas 1992). As an illustration of the effect of thermal and nonthermal extrusion on globular proteins, the characteristic behavior of texturized dairy proteins, in particular whey products, are described further in this chapter.

Proteins from dairy products such as whey protein concentrate (WPC) or whey protein isolate (WPI) present a wide range of extrusion behavior related mostly to the many protein fractions that make them up, each with different association properties, leading to dissimilar behavior at the same temperature and pH conditions. For example, whey products can contain protein amounts ranging from 35% to 95%; these products behave differently at the same extrusion processing conditions depending on the concentration of proteins (Onwulata et al. 2005). The variation in behaviors is largely due to differences in concentration of the major fractions such as α-lactalbumin (α-LA), β-lactoglobulin (β-LG), bovine serum albumin (BSA), the heavy and light chain immunoglobulins (Ig), and minor proteins such as lactoferrin (LF) and lactoperoxidase. In whey products the protease–peptone components glycomacropeptides (GMP), and other low molecular weight fractions are sometimes formed by enzymatic activity on caseins during the cheese-making process (Bonnaillie and Tomasula 2008; De Wit 1989). The complexity of whey protein interactions depends on the concentration and composition of the constituent proteins (Table 11.1).

Extrusion at elevated temperatures and high moistures serves to unfold, denature, and cross-link the proteins into a new molten state (Farrell et al. 2002; Harper 1979), imparting fibrous structure, and improving textural characteristics such as elasticity, chewiness, and toughness (Riaz 2004). During the extrusion cooking of proteins, new peptide bonds are formed by free amino and carboxylic groups of the protein fractions (Table 11.1), and are said to be responsible for the cross-linking that takes place in protein texturization. The primary mechanism of protein texturization is the formation of disulfide bonds, and electrostatic and hydrophobic interactions (Areas 1992).

Through these interactions dense fibrous structures are created from globular proteins such as whey protein or soy protein under high moisture conditions (Shen and Morr 1979). High moisture extrusion processing conditions have been used successfully to transform soy, wheat gluten, or vegetable proteins into meat-like

TABLE 11.1
Composition and Physical Properties of Whey Proteins

Protein	Content (%)	Molecular Weight (kg/mol)	Isoelectric pH
β-lactoglobulin (β-LG)	48–58	18	5.4
α-lactalbumin (α-LA)	13–19	14	4.4
Glycomacropeptide (GMP)	12–20	8.6	<3.8
Bovine serum albumin (BSA)	6	66	5.1
Immunoglobulin (IG)	8–12	150	5–8
Lactoferrin (LF)	2	77	7.9
Lactoperoxidase (LP)	0.5	78	9.6

Source: Table was reproduced from Bonnaillie and Tomasula, 2008, with permission from IFT Press and Wiley-Blackwell Publishing.

products (Kinsella 1982). Other proteins that can be texturized include egg albumin, corn zein, peanut, rapeseed, canola, animal collagen, and others such as casein, which form fibrous structures (Riaz 2004).

A range of texturized protein foods have been developed to mimic or simulate various meats by combining proteins with different ingredients producing different meat-like textures. For instance, wheat gluten may be added to create fibrous structures with improved mechanical strength; oils may be added to stimulate a marbling effect and flavor; starch may be added for improved stiffness, and salts may be added to create water channels, and for improving water absorption and water holding capacities (Linden and Lorient 1999). Textured meat analogs from protein sources such as soy isolates, wheat glutens, egg albumin, or other extrusion cooked vegetable proteins can be shaped into meat-like forms such as patties, or strips (Riaz 2004). The same extrusion forming process can be used to create striated and layered structures using vegetable proteins. The range of structures created from extruded proteins can be widened by changing the physical processing conditions, the chemical reactions in the extruder, or by varying the formulations. For example, in order to obtain a uniform stable meat-like texture, a long cooled die 600 to 900 mm in length is attached to the extruder. A layered fibrous soy meat analog may contain up to 60% to 70% moisture, 10% to 15% protein, and 2% to 5% oil (Riaz 2004; Linden and Lorient 1999). The formulation may contain soy concentrates (~70% protein), soy isolate (~90% protein), corn or wheat starch (10%–20%), and vegetable oil (Riaz 2004).

Recent efforts in protein texturization involve using similar techniques to create whey protein fortified meat analogs (Walsh and Carpenter 2008). Earlier, work on extrusion of whey proteins was limited to the coagulation of whey proteins as the end product (Queguiner et al. 1992), and later as gels (Martinez-Serna and Villota 1992). WPIs were extruded at screw speed of 150 rpm and barrel temperatures ranging from 20°C to 110°C to create firm, spread-like thermocoagulated gels (Queguiner et al. 1992; Szpendowski et al. 1994; Wolkenstein 1988); these structured extruded whey gels can be aerated and used in low-fat low-calorie cheese spreads or ice cream (Cheftel, Kitagawa, and Queguiner 1992; Queguiner et al. 1992). Casein, a nonglobular milk fraction, is generally extruded at moderate temperatures (<80°C) to create caseinate-based products for the dry spinning industry. Fichtali, Van-De-Voort, and Diosady (1995) reported the use of twin-screw extrusion to neutralize acid caseins. A similar extrusion process was adapted for extrusion cooking of cheddar cheeses (Mulvaney et al. 1997). The use of twin-screw extrusion to combine whey proteins and carbohydrate food polymers such as corn or wheat starches to create crunchy snacks has been amply documented (Aguilera and Kosikowski 1978; Harper 1986; Holay and Harper 1982; Matthey and Hanna 1997; Singh et al. 1991; Onwulata et al. 1998).

EXTRUSION-INDUCED CHANGES IN PROTEINS

Whey proteins are modified by heat-induced polymerization during extrusion shear processing, greatly increasing the number of new protein cross-links, the basis for new protein products. Whey proteins may be modified by texturization to enhance their functionality in foods, such as to improve gelation properties, thermal stability, foaming, or emulsification; the result is whey protein ingredients that can be tailored for different applications in a wide range of foods (Schmidt, Packard, and Morris 1984). Temporary or permanent changes can be induced on protein structures readily by relatively minor changes in extrusion processing conditions of moisture, temperature, ionic strength, pH, and others, resulting in sub-structural rearrangements and new steady state intermediate structures (Qi, Brown, and Farrell 2001).

Changes in the native protein structure modify protein functionality, and are the basis for new structure-based functional protein foods. For example, changes in whey protein structure due to enzymatic activity results in the textured product, ricotta cheese. Physical interactions such as noncovalent electrostatic repulsion and attractions, and chemical interactions such as covalently linked disulfide bonds produce the aggregation and gelation (Fitzsimons, Mulvihill, and Morris 2007). Similar changes can be achieved by careful manipulation of the extrusion texturization process conditions to convert the original native structure of proteins into a new state, without altering the amino acid sequence (Li-Chan 2000). A stable gel similar to a ricotta cheese-like product was developed using texturized whey proteins (TWP) (Onwulata et al. 2005). Several of the complex interactions that take place during the extrusion of globular proteins can be illustrated using the interactions of WPI.

Thermal denaturation of the two major protein fractions in WPIs, β-LG (50%) and α-LA (22%), takes place mostly between 50°C and 75°C, and is preceded by unfolding and unmasking of the SH groups (Walstra et al. 1999). Most globular proteins are said to undergo two-stage unfolding (endothermic) and aggregation (exothermic) processes (Fitzsimons, Mulvihill, and Morris 2007). The first is the partially unfolded state or the molten globular state with loss of secondary structure (Farrell et al. 2002), and then the second stage is the aggregation of the protein sub-units mediated by both physical and chemical interactions. The physical and chemical processes include high shear, high temperature or strong acid or alkali pH conditions. For example, Schmidt, Packard, and Morris (1984) reported that extrusion modified whey proteins were denatured by the unfolding and aggregation steps and were stabilized by hydrophobic interactions and disulfide bonds.

That the partially modified protein exists in a metastable unfolded molecular conformation state has been described by Qi et al. (2001), as a molten globular state; an intermediate protein folding state characterized by native-like secondary structures with limited fluctuating tertiary structures (Farrell et al. 2002). This molten globular state is the basis for the new functional directly-puffed extruded whey products, TWP (Onwulata and Tomasula 2004). Changing the protein structure may not change the conformation of proteins or their polypeptide backbone (Li-Chan 2000). When whey proteins are denatured, they become insoluble and aggregate, but are ultimately degraded by prolonged heating above 160°C. There are a number of other nonthermal processes that are used to modify the structure or texture of whey proteins, changing their functionality by attenuating or amplifying particular effects, for example, nonthermal extrusion is used to impart texture and other physical properties such as enhanced solubility (Kilara and Panyam 2003). Further modifications can be achieved by chemical means in tandem with the physical extrusion process, for example adding salts to change pH.

THERMAL GELATION OF GLOBULAR PROTEINS

Thermally induced gels of whey proteins are primarily characterized by the rheology of their aggregated mass. The shape of the aggregates, as fine filaments or particulates, influence rheological behaviors and these changes in behavior can be characterized using small strain and large strain shear properties. Gelation involves the construction of a continuous protein network of macroscopic dimensions with measurable elastic modulus higher than its viscous modulus in the linear strain range, independent of frequency. The values of stress and strain at the fracture point reflect either the strength or deformability of the gels. As Alting et al. (2002) showed, smooth gels are made of small and homogeneous building blocks, while particulate gels appear to be made of large lumps of protein aggregates with large interstitial voids. Particulate aggregate gels contain large strands, usually associated with a significant decrease in water holding capacity.

Whey proteins develop rheological properties upon heating or acidification by the addition of salt. Heat-induced aggregation proceeds relatively slowly, leading to the formation of a fine-stranded gel, comprised mostly of nanometer-thick network strands. If the ionic strength of the gels is low, rubbery textured protein molecules with a transparent or translucent appearance are formed at or near the isoelectric point (near pH 5), or at neutral pH. At low ionic strength, intermolecular electrostatic repulsion is dominant (Ikeda 2003). Such particulate gels easily release liquids and require relatively large stresses to deform. Heat-induced aggregation of β-LG and WPI follows the two-step process at neutral pH; formation of granular primary aggregates is the first step, then subsequent aggregation of the primary aggregates, regardless of the ionic concentration (Ikeda 2003). The growth and aggregation of primary particles are concurrent processes, which causes rheological transitions of heat-induced gelation of whey proteins to be quite different from those of ordinary nonthermal gelling systems, such as gels induced by salts.

Typical thermal gelation of other globular proteins follow the same classical gelation pattern for whey proteins in a two-step mechanism, involving an initial denaturation or unfolding of the protein molecule, then followed by rearrangements and aggregation of functional groups which become available for intermolecular interactions under appropriate ionic conditions, forming three-dimensional gel networks (Krešić et al. 2008). This phenomenon is important in the food industry because of the need to manipulate the rheological and textural properties of food, to create an assortment of products. The properties of thermally induced gels are influenced by the concentration of proteins, heating rate, extent of denaturation, ionic strength, pH, and the presence of specific ions and salts. Processing steps can be used to alter the characteristics of whey protein products modifying their structure and functionality (i.e., foaming, emulsification, and gelation); these functionality changes can be generated by the action of heat, pressure, and by divalent cations (Schmidt, Packard, and Morris 1984).

Heating increases aggregation of proteins, and this tendency is at the highest around pH 7.0 where protein dimers and octamers form larger aggregates than at either pH 3.5 or pH 8.0 (Bonnaillie and Tomasula 2008). Kazimierski and Corredig (2003) showed that the aggregation of WPI depended on heat treatment at neutral pH and low ionic strength, but the β-LG fraction was twice as likely to aggregate than the α-LA fraction regardless of the temperature; there was a higher α-LA reactivity at 65°C that at 75°C or greater. Haque and Sharma (1997) reported that pH changes altered the initial distribution of the different aggregates at low temperatures (25°C–65°C), much below the reported denaturation temperature of approximately 70°C or higher. It has been theorized that intermolecular hydrogen bonding is responsible for the gel formation in globular protein, and is the initiator of the interaction of the secondary protein structures. These secondary structures, β-sheets, when in close proximity are aligned in parallel or anti-parallel configurations, interlinking at nodes, to form gel networks under high temperature or ionic strength conditions (Wang and Damodaran 1991). For heat-induced gelation, the protein must be heated above the denaturation temperature of the globular protein to open the functional groups for optimal gel network formation.

Heating an aqueous solution of sufficiently high concentration of globular protein induces denaturation and aggregation of the molecules. In the case of heat-induced whey protein gel networks, its rheological properties can be controlled by varying gelling conditions such as pH or the ionic strength (Clark 1998). If fibrils are formed in such gels, shear thinning behavior occurred at a pH lower than 3.5 (Akkermans et al. 2008). Puyol, Perez, and Horne (2001) showed that gradual heating of WPIs from 60°C to 90°C induced gelation under uniform ionic strength, but that increasing ionic strength increased gel strength (higher modulus) at lower temperatures.

The extent of conformational changes in proteins and their relation to the physical properties of globular protein gels was investigated by Wang and Damodaran (1991). They used salts to modify the water structure around the proteins, inducing variations in the molecular conformation of proteins; resulting in two gelling dynamics: kosmotropic (structure stabilizing) and chaotropic (structure destabilizing). The salts were shown to affect the denaturation temperature and conformation of several proteins through electrostatic shielding effect at or below the ionic strength of 0.2; the presence of ions affected the stability of hydrophobic interactions of the proteins at higher concentration (Damodaran 1989).

NONTHERMAL (COLD) GELATION OF PROTEINS

The stability of globular proteins is temperature dependent; past the point of maximum stability, the structures are disrupted by cooling, leading to the phenomenon called cold denaturation and gelation. Cold denaturation is a temperature-dependent hydrophobic protein folding process, particularly affected by the ionic strength of the surrounding water (Davidovic et al. 2009). Davidovic et al. (2009) reported a cold denaturation temperature of −20°C for ubiquitin at low temperature and water content. Cold denaturation of proteins in aqueous solution was observed only at temperatures below 0°C under neutral pH (Jonas 1997).

In the food processing industry, cold gelation of globular whey protein is accomplished after an initial thermal treatment to induce protein aggregation, followed by rapid cooling at ambient temperature while gradually lowering the pH. Denaturation of the protein is a prerequisite for cold gelation. Cold gelation induced by lowering pH produces electrostatic interactions of the aggregated primary amino acids, particularly the amino acids on the β-LG fractions. The mechanisms of cold gelation are described as succinylation and methylation. Succinylation results from decreasing the isoelectric point caused by lowering the pH below 2.5, while methylation of the carboxylic acid groups results from increasing the isoelectric point to more alkaline pH values, resulting in gelation (Alting et al. 2002). For example, Alting et al. (2000) reported that at low pH, aggregates of WPI gels showed both syneresis and spontaneous gel fracture because of modified disulfide bonds. For protein aggregation, the net electric charge of the protein and its ability to form disulfide bonds was important for aggregation to occur during pH-induced cold gelation.

The mechanisms of gelation at the molecular level are still being researched, particularly the effects of the physical (electrostatic and hydrophobic) and chemical (disulfide bond formation) interaction processes. Privalov (1990) postulated that the formation of disulfide bonds under these conditions was attributable to a large increase in protein concentration and therefore of the increase in the concentrations of thiol groups. It was shown that formation of disulfide bonds increased the molecular weight of the aggregates formed during gelation, and that the disulfide bonds were involved in stabilizing and strengthening the network. It is thought that noncovalent interactions play a dominant role in the initial formation of acid-induced cold gel aggregates, and that chemical modification of aggregates controls the balance between the role of electrostatic and chemical interactions in cold gelation. This type of gelation is what typically occurs when milk is acidified into cheese curd (Bryant and McClements 1998). Prior to cold gelation, denaturation is required to expose the reactive functional groups of the protein. Interactions between the exposed functional groups are sustained by chemical bonding and physical linkages (Ziegler and Foegeding 1990; Sullivan, Khan, and Eissa 2008). Chemical bonding typically involves disulfide bond formation, which is crucial for protein aggregation (Sawyer 1968; Shimada and Cheftel 1989; Roefs and de Kruif 1994). Physical interactions are maintained by van der Waals forces, hydrogen bonding, electrostatic, and hydrophobic interactions (Alting et al. 2003). Britten and Giroux (2001) showed that when the primary aggregates are close to the isoelectric point or when charges are screened via salt addition, the aggregates are aligned, forming fine gels are obtained. However, if the protein concentration is less than the minimum concentration needed for gelation, a soluble protein polymer aggregate forms.

THERMAL EXTRUSION OF PROTEINS

The most common extrusion food processing design is thermal high temperature short time cooking, with most of the energy in the system coming from screw friction against the heated barrels (Harper 1986). The heat is needed to convert water to superheated steam at high pressure, and to drive the thermochemical reactions in the temperature range of 120°C to 180°C with the typical process lasting <120 s. The formation of fibrous protein networks or texturization, is the result of shearing by extrusion at elevated temperatures (Harper 1979). When the native globular conformations are simply unfolded, or denatured, as a result of some physical process, e.g., shear, there is no change in the primary structure or bond cleavage (Yada et al. 1999). With milk proteins, denaturation occurs at temperatures ranging from 60°C to 90°C with an extended time domain of 30–180 min (Figure 11.1). The extent of denaturation is not directly measurable, but is assessed by measuring changes in properties of proteins such as its heat capacity, but in food science, a functional measure is to determine solubility before and after physical treatment, reporting the amount of insoluble mass as denatured

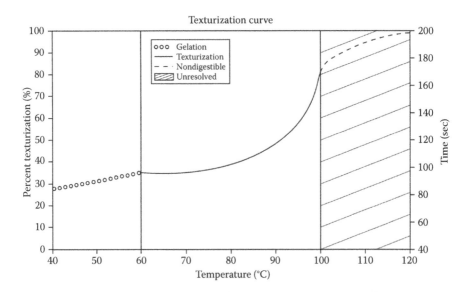

FIGURE 11.1 A typical texturization profile for whey proteins over time and temperature, with unresolved areas.

(Kilara 1984). Texturization can be inferred from the difference in rates of moisture uptake between the native protein and the (extruded) texturized protein, the so-called water solubility index (Figure 11.1). In the process extrusion temperature range of 40°C to 60°C, the WPI is in the gel state. From approximately 60°C to 100°C, rapid texturization occurs. Past 100°C, extremely structured protein networks are formed; these networks are sometimes biologically indigestible. Such structures formed above extrusion melt temperature of 120°C are not easily digestible with urea, dithiothreitol (DTT), or 2-mercaptoethanol, hence they are unresolved.

The protein texturization process involves imparting a structure to protein-rich food ingredients so that visible forms such as fibers or crumbles are created. A restructuring of the protein molecules into a layered cross-linked mass resistant to disruption upon further heating and/or processing occurs during texturization (Harper 1986). Soy, whey, and other proteins are denatured and texturized before they are included in food products, especially those produced through severe cooking, such as by extrusion, to minimize the effects of further heat processing and/or to have a better control on the extent of heat denaturation (Onwulata et al. 2001). Kester and Richardson (1984) and Ryan (1977) suggested that a partial denaturation of native proteins or combining partially denatured with native protein whey proteins was a practical way to produce a unique and desirable blend of functional properties. During texturization, the whey proteins interact forming macroscopic three-dimensional structures and alters their chemical and physical reactivity with other ingredients in a food product, particularly in food systems. Texturized vegetable protein analogs are products made from edible protein sources such as wheat, cotton, and peanuts, and are characterized by having structural integrity and identifiable texture that enables them to withstand hydration during cooking and other preparations (Lockmiller 1972).

Textured soy or textured vegetable products are made from the combination of soy proteins and starches from wheat flour or other proteins such as wheat gluten to create chewy or stringy textured products by using extrusion processing (Shen and Morr 1979). Using the extrusion process to texturize soy proteins for use as meat extenders has been one of the most successful developments in the food industry in the last 60 years (Atkinson 1970). The process involves constraining plant proteins within shear fields and forcing a directional flow within the extruder barrel in the presence of excess moisture to force the protein mass into folded shapes (Bhattacharya, Hanna, and Kaufman 1986). The process of making texturized protein products, and their food applications have been discussed in many reviews and articles (Harper 1986; Kearns, Rokey, and Huber 1989; Kinsella 1978; Noguchi and Isobe 1989).

The extruder is the only device that accomplishes texturization directly; the rotating screws, heated barrel, and die restriction generate shear, heat, and pressure. Extrusion texturization occurs at temperatures of 50°C to 180°C, screw speed of 50 to 450 rpm, and pressure of 1 to 130 atm (Figure 11.1). Texturization of proteins can be accomplished through several different processes (Harper 1986), however, the single extrusion operation has the clear advantage because several food processing unit operations are accomplished simultaneously in the process. Extrusion has been the favored process for texturization starting with the soy proteins (Pordesimo and Onwulata 2008).

Riaz (2004) describes a soy protein texturization process consisting of an initial preconditioning step, followed by the actual extrusion cooking phase and a cooling and forming stage. The preconditioning stage ensures uniform hydration of the solid particles and enhanced heat penetration if steam is injected. In the preconditioner, the preferred moisture content is approximately 20% to 25%, and the material temperature between 70°C and 80°C. Culinary steam can be introduced directly to the product for better mixing and to shorten the retention time to <30 s. After preconditioning, the protein mass is pumped into the extruder for texturization. The preconditioner-plasticized protein/starch mass can consist of defatted soy flour (40% to 50% protein), vegetable oil (5% to 8%), and soy fiber (3% to 5%). Extrusion cooking conditions for a solid mass high moisture meat analog dwell time in the extruder is decreased; the moisture content increases to 30% while pressure is maintained below 10 atm. Extrusion processing conditions such as shorter retention times (10 to 20 s), product temperature (100°C to 180°C), and moisture levels (15% to 30%) all influence the protein dough quality just behind the die and the final product upon exiting the die (Rokey, Huber, and Ben-gera 1992).

Protein–protein cross-linking during extrusion texturization is responsible for the final structure of extruded soy products formed at high temperatures (>180°C). These insoluble structures are held together by hydrophobic electrostatic and disulfide bonds (Mitchell and Areas 1992). Riaz (2004) states that wheat gluten and wheat starch can be used as co-products in meat analogs, for example, wheat flour and wheat gluten are used during wet processing of texturized vegetable protein for meat applications. Wheat products are low cost raw materials, and wheat gluten can be extruded independently forming stringy long fibers with bland taste that are used as meat replacers. When hydrated, wheat glutens form a bland tasting, very extensible, elastic structure that can replace textured soy in most products. Other protein sources for the extrusion process for texturization include cottonseed, canola, rapeseed, and peanuts. These sources of protein have peculiar shortcomings such as the presence of gossypol, an anti-nutrient compound in cotton; allergenicity with peanuts, and irregular supply for rapeseed and canola (Riaz 2004).

Physically modifying whey proteins by extrusion for an intended food application is relatively new. Extrusion processing of whey produces a heat and acid stable physically cross-linked product with extended functional applications in snacks, meats, and sports drinks (Walsh and Carpenter 2008; Faryabi et al. 2008; Onwulata and Tomasula 2004). Texturization changes the folding of whey proteins improving their interaction with other ingredients (Onwulata et al. 2006). In snack applications, expanded products containing up to 25% WPI were demonstrated (Onwulata et al. 1998). Extruded product blend containing 20% to 40% starch and 60% to 80% WPC can be added as ingredients for puffed snacks or serve as meat extenders. Texturized WPC was used to formulate a high-protein product that simulated meat texture and was used to replace meat by up to 50% without noticeable differences in texture (Hale, Carpenter, and Walsh 2002). Walsh and Carpenter (2003) developed a process for directly texturing WPC (WPC80) with an edible biopolymer such as corn starch in the extruder. Further details on this process can be found in later sections for snack and for meat applications using TWP.

NONTHERMAL (COLD) EXTRUSION OF PROTEINS

Elevated extrusion cooking temperatures favor Maillard reactions that may cause discoloration or browning of proteins in the presence of starch, but may be undesirable in some applications. Furthermore, severe food processing conditions such as high temperatures and extreme pH changes may induce transformations and racemization of proteins during cross-linking, destroying their nutritive qualities (Friedman 1999). These are drawbacks with thermal texturization and high temperature cooking extrusion (thermal texturization). Walkenstrom, Windhab, and Hermansson (1998) showed that shear alone was adequate to induce structuring of particulate whey to create gels. Therefore, the basis of cold extrusion is to effect whey protein texturization by shear and minimal or no heat.

Cold or nonthermal extrusion minimizes the loss of protein quality caused by high heat reactions (Camire 1990). In cold extrusion, the molten gel temperatures are not achieved and only shear-induced aggregations, which are similar to cold set acid gels, are encountered (Cho et al. 1995). This makes mechanical energy input a critical factor in nonthermal extrusion. In their review of cold denaturation of proteins under high pressure, Kunugi and Tanaka (2002) pointed out that cold denatured proteins are in a new state, similar to the molten globular state of heat denatured proteins, and that cold denaturation also follows a two-step process: primary disassociation and unfolding, and secondary refolding and re-alignment of the native protein. The most stable gels were produced at 10°C with gels made at temperatures greater than 54°C proving to be unstable. Bryant and McClements (1998) reviewed the molecular basis of protein functionality with special consideration of cold-set gels derived from denatured whey proteins, and showed unique functionalities such as increased gelation, thickening, and water-binding. To create cold-set gels, whey proteins were first denatured by heating to achieve unfolding and aggregation, before rapid cooling to form gels (Resch and Daubert 2002). The particular applications for these cold-set gels were in comminuted meats, fish products, desserts, sauces, and dips.

Beckett et al. (1994) reported the use of extrusion shear to plasticize milk chocolate isothermally below its normal melting point (27°C–32°C), showing that a continuous cold extrusion process can be used to produce a textured milk product. Cho et al. (1995) used cold extrusion to develop natural flavors in a starch and methionine or cysteine matrix. Methionine and cysteine would have been destroyed by high temperature extrusion reducing their nutritional value. Osburn, Mandigo, and Kuber (1995) showed that cold extruded restructured porcine protein products had desirable sensory and textural properties resulting from partial realignment of muscle fibers. We have found that WPI denatured within 90 s in a twin-screw extruder at 50°C (Onwulata et al. 2005). The degree of texturization of WPI may be adjusted through proper selection of such extrusion conditions as moisture, temperature, and shear rates. It was anticipated that because of the low extrusion temperatures, texturized WPI will maintain its nutritive quality when extruded together with corn or wheat flour to make products that retained their structural integrity, without collapsing after die expansion (Onwulata et al. 2001). The technology of cold extrusion cooking is relatively new and offers the food industry the opportunity to modernize, shorten processing times, and ultimately achieve significant cost savings (Reifsteck and Jeon 2000). Extrusion is termed nonthermal or cold extrusion cooking when the process temperature is kept below 50°C. Nonthermal extrusion creates favorable conditions such as diminished thermal effect, increased viscosity, and shear (Cho et al. 1995).

EXTRUDED WHEY PROTEIN PRODUCTS

Extrusion processing optimizes the conditions for whey protein denaturation and development of enhanced specific functional attributes such as sugar tolerance in sugar-rich jelly-type products (Faryabi et al. 2006). Ordinarily, the unmodified spray dried powdery whey powders cannot be used for these applications because they are not suitable when used in amounts greater than 2% in products such as cakes or breads. Whey proteins can be processed into crunchy directly-expanded products such as crispy snacks or as meat analogs using two formulation pathways involving twin-screw extrusion. The first pathway, an essentially finished product, is to blend whey proteins and cereal carbohydrates, such as corn or wheat starch, through twin-screw extrusion under high shear at moderate temperatures (Aguilera and Kosikowski 1978; Harper 1986; Singh et al. 1991). The second approach involves texturizing the whey proteins to produce ingredients that would have better functionality in a fabricated snack product or meat-like structure.

Texturized proteins are defined as food products made from edible protein sources characterized by having structural integrity and identifiable texture that enables them to withstand hydration in cooking and other preparations (Liu 1997). TWP have been shown to work nicely in directly expanded snacks, and depending on the pH, the effect of texturization varies. For instance, alkaline conditions increase insolubility and pasting properties of WPI, while acidic conditions increase solubility. Subtle structural changes occur under acidic conditions but are more pronounced under alkaline conditions. In general, alkaline conditions increase denaturation of extruded WPI resulting in stringy texturized meaty fibrous products, which could be used in snack and meat applications (Hale, Carpenter, and Walsh 2002; Onwulata et al. 2006).

Kester and Richardson (1983) showed that modification of whey proteins to improve functionality can be accomplished by chemical, enzymatic, or physical means, and that the physical means of changing whey protein functional performance can be achieved through thermal treatment, biopolymer complexing, or texturiza- tion. By employing a combination of thermomechanical treatment, extrusion and/ or biopolymer complexing, a physical modification of whey proteins and a conse- quential change in their functionality was achieved. In the extruder, whey proteins can be modified using chemicals, heat, or shear. Chemical treatment alone alters the reactive groups of the amino acids, resulting in changes in the noncovalent forces that influence conformation, such as van der Waals forces, electrostatic interactions, hydrophobic interactions, and hydrogen bonding. Heating and shear alter the confor- mational structure of the whey protein through partial denaturation of the protein, thereby exposing groups that are normally concealed in the native protein. Extrusion texturization of proteins combines the effects of chemical, heat and shear, resulting in the physical and functional modification of the proteins, opening up opportunities for their wider use for different products (Figure 11.2). The typical texturization continuum for WPI in the extrusion temperature range of 0°C to 140°C would show zones of enhanced solubility (<20°C), gelation (20°C–40°C), texture formation (40°C–100°C), and irreversible networks (>120°C). The extrusion shear conditions can be moderate (<150 rpm) or high (>300 rpm).

Physicochemical attributes that make proteins useful in foods are called functional properties, or functionality. The physical functionality of proteins represents those properties that influence their usefulness in foods based on transient structural states, mostly from the properties of the amino acids, their constitution, composition, conformation, and transformation (Figure 11.2). The primary structure, the sequence of amino acids, and the different side chains, produce these functional differences: denaturation, untanglement, alignment, structuring, and texturing. The effect of shear and heat on the primary functional attributes are the water properties such as solubility or insolubility, water holding capacity, and dispersibility which play a role in product thickness and viscosity (Kilara and Vaghela 2000). Protein solubility is influenced by temperature and pH, and as a fundamental property, affects other protein properties. Interactions of whey proteins with primarily water may result in increased viscosity if

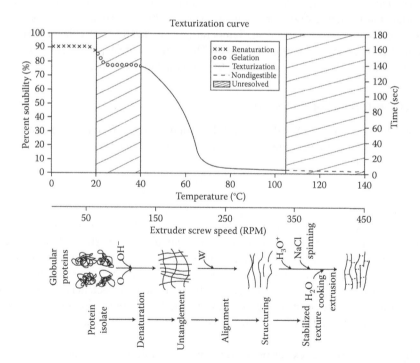

FIGURE 11.2 Extraction and texturization processes. CRC Press, Boca Raton, FL. (Adapted from Linden, G. and Lorient, D. 1999. The exploitation of by-products. In *New ingredients in food processing: Biochemistry and agriculture*, eds. G. Linden and D. Lorient, 184–210. London: Woodhead Publishing.)

the proteins are soluble or partially denatured. Other resultant physicochemical functions such as foaming or gelation are derived from the water properties. Changes in these parameters are reflected in the protein solubility, whereby upon modification by heating, its degree of hydrophobicity (unfolding) increases. Physical processes modify whey proteins faster at elevated temperatures; for example, in food applications, elevated temperatures from 35°C to 85°C change the aggregation and denaturation of proteins, and in turn, affect the physical properties of functioning of whey proteins in foods (Linden and Lorient 1999).

Denaturation and aggregation of whey proteins is likely from disulfide-bonded strands formed by the aggregation of α-LA and β-LG, catalyzed by heat-initiated unfolding of the other serum proteins, forming initial aggregates of different sizes (see Table 11.1). These aggregates are cross-linked by intermolecular disulfide bonds and by noncovalent interactions forming β-LG disulfide-bonded networks. Structural stability of functional whey proteins are maintained by the unfolding of α-LA and β-LG over the hydrophobic residues and by the presence of thiol and disulfide groups. Heat processing results in exchange of disulfide bridges, which produces the changes in functionality, beginning with its solubility, then others like water absorption and foaming. Thermal denaturation unmasks the sulfhydryl groups during unfolding of the protein molecule, and depending on the pH, the mode of intermolecular or intramolecular re-association produces different functionality (Linden and Lorient 1999). For example, foaming ability correlates with solubility and is also temperature dependent. Temperatures greater than 50°C favor foaming, while at 10°C or below, foaming decreases. This temperature dependency of foaming is attributable to the effect of heat on β-LG (Kilara and Vaghela 2000).

WHEY PROTEIN IN PUFFED SNACKS

Research efforts at USDA-ARS, Eastern Regional Research Center, Wyndmoor, PA have focused on using the twin-screw extrusion process to incorporate whey proteins into expanded snack products for the purpose of boosting the protein content in the finished product. For example, increasing the protein content of corn puffs was increased from 5 to 30 g/100 g. The challenge has been to increase the protein content while maintaining the quality of expanded snacks by careful adjustment of extrusion process variables to effect expansion of the complex protein–starch matrices (Onwulata et al. 1998; Onwulata et al. 2005; Onwulata et al. 2006). In a separate work, whey product in the form of WPC and sweet whey solids was used to displace (15 to 35 wt%) comparable quantities of corn meal, wheat starch, and rice and barley flours (Onwulata et al. 2001). The typical effect of such displacement in expansion of corn meal can be seen in Figure 11.3. As the WPC content progressively displaces corn starch by 15, 25, or 35 wt%, the expansion decreases and surface texture changes.

By controlling the extrusion processing temperatures, or operating at high shear, high temperatures (100°C–150°C), and low moisture (10%–15%), whey protein substitution was made possible up to 25 wt% of the carbohydrate with WPC (Figure 11.3). The whey protein supplemented products were not as puffed as the straight corn product because the whey proteins held water and collapsed within

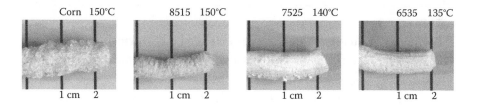

FIGURE 11.3 Expansion characteristics of corn meal substituted with whey proteins weight for weight.

the starch matrices, forming dense and discolored products (Onwulata et al. 2001). Matthey and Hanna (1997) reported similarly reduced expansion and increased hardness in their test products containing WPC and starch blends. Working with WPI, Martinez-Serna and Villota (1992) reported a 30% reduction in expansion ratio due to the addition of 20 wt% WPI. In a more recent effort involving the texturization of WPC (WPC80), whey lactalbumin, and WPI were extruded at cook temperatures below 80°C to achieve varying degrees of melt temperatures and denaturation for the different whey protein products (Table 11.2).

Varying extrusion cook temperature allowed a controlled rate of denaturation, indicating that TWP with a predetermined functionality based on the degree of denaturation can be created. Thermally denatured whey proteins are unique ingredients that have the potential to be used in large amounts.

WHEY PROTEINS IN MEAT ANALOGS

Walsh and Carpenter (2008) described the process of using TWP as meat extenders. Extruded soy products from moderate moisture extrusion are usually the norm for expanded meat-like structures that absorb moisture rapidly (Lin, Huff, and Hsieh 2002). Twin-screw extrusion under high moisture conditions (>50% water) produced TWP with fibrous structures (Noguchi and Isobe 1989; Lin, Huff, and Hsieh 2000, 2002). Extruders used to produce protein meat analogs are routinely fitted with a cooling die, a requisite for proper texture formation in meat replacers and analogs. The cooling die at the end of the extruder helps align the protein mass. Proteins

---

TABLE 11.2

Extrusion Melt Temperatures of Whey Protein Products Below 80°C

Product	Melt Temperature (°C)	Preextrusion (%)	Postextrusion (%)
WPC80	70[b]	40.9[b]	59.9[b]
WLAC	75[a]	68.7[a]	94.4[a]
WPI	74[a]	28.0[a]	94.8[a]

*WPC*80: whey protein concentrate, 80% protein. *WLAC*: whey lactalbumin. *WPI*: whey protein isolate: Number reported is mean of three samples. Means with different letters within a column are significantly ($p < .05$) different. WPC80 was the least denatured after extrusion while lactalbumin and WPI were more significantly denatured (Onwulata et al. 2003).

TABLE 11.3

Consumer Evaluation of Beef Patties Extended with Texturized Whey Proteins or Texturized Soy Proteins at 30 wt% Replacement of Beef

Treatment	Tenderness*	Juiciness	Texture*	Flavor	Acceptability
100% beef	6.17[a]	5.87[ab]	6.32[a]	6.45[a]	6.35[a]
30% TWP	5.64[b]	5.65[ab]	4.53[b]	5.27[b]	5.01[b]
30% TSP	5.23[c]	4.78[b]	5.19[c]	4.00[c]	4.23[b]

*Relates to the tenderness of the first bite, and the texture relates to mouthfeel during chewing. Patties were evaluated using a hedonic scale by panelists. All patties were adjusted to 20% fat. Statistics calculated with analysis of variance using SAS (Cary, NC). Within a column, mean values sharing a superscript letter are not different $(p > .05)$. Adapted from Walsh and Carpenter (2008).

texturized for use as meat analogs using extruders undergo solubilization in the preconditioner, melting of the protein dough under shear and high temperature in the extruder, and shaping and forming fibrous structures by laminar flow in the cooling die (Cheftel, Kitagawa, and Queguiner 1992; Riaz 2004).

Meat extenders can be produced by thermoplastic extrusion of WPI at moderate moisture conditions (40% to 60%). The textured whey protein products used as meat extenders in beef or meatless patties (Hale, Carpenter, and Walsh 2002; Taylor and Walsh 2002; Walsh and Carpenter 2003) were a blend of WPC (containing 80% protein) and corn starch (containing 25% amylose and 75% amylopectin) and water. The mix was extruded at a melt temperature of 162°C, low pressures of 17 to 24 atm, with a final moisture content of 35% to 50% for improved fiber formation (Hale, Carpenter, and Walsh 2002; Taylor and Walsh 2002). A typical protein level of 50% was used, and the water was adjusted with 0.2 M NaOH for best fiber formation (Hale, Carpenter, and Walsh 2002). In a consumer test panel, Hale, Carpenter, and Walsh (2002) found that beef patties extended with 30 wt% TWP were as acceptable as 100% beef in tenderness, juiciness, texture, flavor, and overall acceptability (Table 11.3). In the same taste test, beef patties extended with whey proteins texturized with pH neutral water were not as acceptable as those with 0.2 M NaOH; patties extended with texturized soy protein fared worse (Walsh and Carpenter 2008).

NONTHERMALLY EXTRUDED WHEY PROTEIN PRODUCTS

We have described a novel process that creates stretchable extensible protein-dough containing small amounts of lactose and other carbohydrates. This textured product bakes into fluffy pastry or cake or bread-like products (Figure 11.4).

This product is comprised essentially of mostly whey proteins (60% to 95%) and can contain other animal or plant proteins as substitutes for parts of the cold-textured protein-dough. In the unmodified native state, these proteins are not stretchable and do not form dough-type gels, but when extruded with pH adjusted water under conditions of temperatures lower than 60°C with sufficient stretching shear, the proteins are

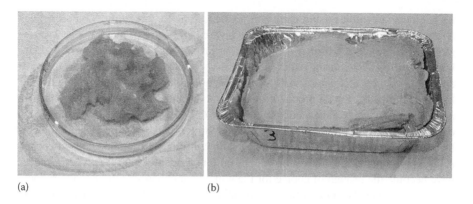

(a)   (b)

FIGURE 11.4   Nonthermally prepared gels (b) and its baking application (a).

stretched and texturized; and, in the presence of water, form dough. This texturized protein dough, when baked at 140°C for 30 min, expands and forms a pastry of any desired shape (Onwulata 2008). The taste is bland, and it can be easily flavored with WPC based sweeteners such as Simplesse® for more compatible dairy flavors. Traditionally, the food industry uses flours (carbohydrates) from different sources such as wheat or potato in formulating baked goods. With the demand for low- and reduced carbohydrate foods increasing in the U.S. and throughout the world, the use of these texturized protein dough products is invaluable. This product replaces traditional baked goods which are made mostly from carbohydrates. Bryant and McClements (1998) reviewed the molecular basis of protein functionality with special consideration of cold-set gels derived from denatured whey and showed unique functionalities such as increased gelation, thickening, and water-binding. The particular application for these cold-set gels is in comminuted meats, fish products, desserts, sauces, and dips. Nonthermally extruded protein gels perform like thickening agents and act like binders, for example, in comminuted meat. Preserving the functionality and nutritive quality of proteins by nonthermal extrusion is a real benefit. Nonthermal extrusion also protects the flavor of whey proteins and permits the creation of specifically denatured or texturized cross-linked protein fibers that can be used as meat extenders, in addition to their application as denatured or texturized ingredients in snack foods.

FUTURE OF THERMAL AND NONTHERMAL EXTRUSION PROCESSING OF WHEY PROTEINS

Extrusion processing is an effective method for denaturing whey proteins, a representative globular protein, for creating a wide range of texturized products. Textured WPI retains its native protein value, functionality, and digestibility even when extruded at 80°C; though permanent changes in functionality occur at approximately 100°C. A better understanding of texturization and the process variables affecting functionality would lead to the development of improved textured protein products tailor-made for specific food applications. Further progress has been made in directly extruding native whey proteins and converting them into forms suitable for use as meat substitutes using extrusion texturization. Texturization, which involves a restructuring of the whey proteins into a layered cross-linked mass, makes it resistant to disruption upon further heating and/or processing. The direct nutritional and health benefits of texturing whey proteins as a means of increasing the quantity added to products characteristically low in protein content, will improve the overall the protein profile of such products and boost their nutritional value. The consumer appeal of such "whey enhanced" products will only grow as more health properties of whey proteins are ascertained by clinical studies. Dairy whey proteins have a healthy appeal to consumers. Increasing health consciousness among consumers has resulted in an increased number of "whey" protein foods; for example, whey proteins have become the protein of choice for the nutritional enhancement of food products such as snack bars because of the accumulating body of evidence supporting impressive health benefits of whey proteins (O'Donnell and O'Donnell 2006). Because of the health benefits of using whey proteins more research in this area will result in more innovative products in the future.

ACKNOWLEDGMENT

The help of several collaborators, scientists, and technicians in accomplishing several related projects in this subject area is appreciated.

REFERENCES

Aguilera, J.M. and Kosikowski, F.V. 1978. Soybean extruded products: A response surface analysis. *J Food Sci* 41: 1200–1212.
Akkermans, C., Van der Goot, A.J., Venema, P., Van der Linden, E. and Boom, R.M. 2008. Properties of protein fibrils in whey protein isolate solutions: Microstructure, flow behavior and gelation. *Int Dairy J* 18: 1034–1042.
Areas, J.A. 1992. Extrusion of Food Proteins. *Crit Rev Food Sci Nutr* 32(4): 365–392.
Atkinson, W.T. 1970. Meat-like protein food products. U.S. Patent 3488770.
Alting, A.C., Hamer, R.J., de Kruif, C.G. and Visschers, R.W. 2000. Formation of disulphide bonds in acid induced gels of preheated whey protein isolate. *J Agric Food Chem* 48: 5001–5007.
Alting, A.C., De Jong, H.H.J., Visschers, R.W. and Simons, J.F.A. 2002. Physical and chemical interactions in cold gelation of food proteins. *J Agric Food Chem* 50: 4682–4689.
Alting, A.C., Hamer, R.J., de Kruif, C.G., Paques, M. and Visschers, R.W. 2003. Number of thiol groups rather than the size of the aggregates determines the hardness of cold set whey protein gels. *Food Hydrocolloids* 17: 469–479.
Bhattarcharya, M. and Padmanabhan, M. 1999. Extrusion processing: Texture and rheology. In *Wiley encyclopedia of food science and technology*. 2nd Edition, ed. F.J. Francis, 706–720. New York: John Wiley & Sons.
Bhattacharya, M., Hanna, M.A. and Kaufman, R.E. 1986. Textural properties of extruded plant protein blends. *J Food Sci* 51(4): 988–993.
Beckett, S.T., Craig, M.A., Gurney, R.J., Ingleby, B.S., Mackley, M.R. and Parsons, T.C.L. 1994. The cold extrusion of chocolate. *Food Bioproducts Process* 72: 47–54.
Bonnaillie, L.M. and Tomasula, P.M. 2008. Whey protein fractionation. In *Whey processing, functionality and health benefits,* eds. C.I. Onwulata and P.J. Huth, 15–38. Danvers, MA: Wiley-Blackwell and Institute of Food Technologists Press.

Britten, M. and Giroux, H.J. 2001. Acid-induced gelation of whey protein polymers: Effects of pH and calcium concentration during polymerization. *Food Hydrocolloids* 15: 609–617.

Bryant, C.M. and McClements, D.J. 1998. Molecular basis of protein functionality with special consideration of cold-set gels derived from heat-denatured whey. *Trends Food Technol* 9: 143–151.

Camire, M.E. 1990. Chemical and nutritional changes in foods during extrusion. *Crit Rev Food Sci Nutr* 29: 35–57.

Cheftel, J.C., Kitagawa, M. and Queguiner, C. 1992. New protein texturization processes by extrusion cooking at high moisture levels. *Food Reviews Int* 8: 235–275.

Cho, M.H., Zheng, X., Wang, S.S., Kim, Y. and Ho, C.T. 1995. Production of natural flavors using a cold extrusion process. ACS Symposium Series 610, p. 120–128. *Am Chem Soc*, Washington, D.C.

Clark, A.H. 1998. Gelation of globular proteins. In *Functional properties of food macromolecules*. 2nd Edition, eds. S.E. Hill, D.A. Ledward, and J.R. Mitchell, 77–142. Gaithersburg, MD: Aspen Publishers, Inc.

Damodaran, S. 1989. Influence of protein conformation on its adaptability under chaotropic conditions. *Int J Biol Macromol* 11: 2–8.

Davidovic, M., Mattea, C., Qvist, J. and Halle, B. 2009. Protein cold denaturation as seen from the solvent. *J AM Chem Soc* 131: 1025–1036.

De Wit, J.N. 1989. Functional properties of whey proteins. In *Developments in dairy chemistry - 4*, ed. P.F. Fox, 285–321. New York: Elsevier Applied Science.

Farrell, H.M., Jr., Qi, P.X., Brown, E.M., et al. 2002. Molten globule structures in milk proteins: implications for potential new structure-function relationships. *J Dairy Sci* 85: 459–471.

Faryabi, B., Mohr, S., Onwulata, C.I. and Mulvaney, S.J. 2008. Functional foods containing whey proteins. In *Whey processing, functionality and health benefits*, eds. C.I. Onwulata and P.J. Huth, 213–226. Danvers, MA: Wiley-Blackwell and Institute of Food Technologists Press.

Fichtali, J., Van-De-Voort, F.R. and Diosady, L.L. 1995. Performance evaluation of acid casein neutralization process by twin-screw extrusion. *J Food Eng* 26: 301–318.

Fitzsimons, S.M., Mulvihill, D.M. and Morris, E.R. 2007. Denaturation and aggregation processes in thermal gelation of whey proteins resolved by differential scanning calorimetry. *Food Hydrocolloids* 21: 638–644.

Friedman, M. 1999. Chemistry, nutrition, and microbiology of D-amino acids. *J Agric Food* 47: 3457–3479.

Hale, A.B., Carpenter, C.E. and Walsh, M.K. 2002. Instrumental and consumer evaluation of beef patties extended with extrusion-textured whey proteins. *J Food Sci* 67: 1267–1270.

Haque, Z.U. and Sharma, M. 1997. Thermal gelation of β-lactoglobulin AB purified from cheddar whey. 1. Effect on association as observed by dynamic light scattering. *J Agric Food Chem* 45: 2958–2963.

Harper, J.M. 1979. Extruder not prerequisite for texture formation. *J Food Sci* 44: ii.

Harper, J.M. 1986. Extrusion texturization of foods. *Food Technol* 40: 70–76.

Holay, S.H. and Harper, J.M. 1982. Influence of the extrusion shear environment on plant protein texturization. *J Food Sci* 47: 1869–1875.

Ikeda, S. 2003. Heat induced gelation of whey proteins observed by rheology, atomic force microscopy, and Raman spectroscopy. *Food Hydrocolloids* 17: 399–406.

Jonas, J. 1997. Cold denaturation of proteins. In *Supercooled liquids: Advances and novel applications,* eds. J.T. Fourkas, D. Kivelson, U. Mohanty, and K.A. Nelson. ACS Symposium Series 676, Orlando, FL. August 25–29, 1996.

Kazimierski, M. and Corredig, M. 2003. Characterization of soluble aggregates from whey protein isolate. *Food Hydrocolloids* 17: 685–692.

Kearns, J.P., Rokey, G.J. and Huber, G.R. 1989. Extrusion of texturized proteins, In text: Proceedings of the World Congress: Vegetable Protein Utilization in Human Foods and Animal Feedstuffs, ed. T.H. Applewhite, 353. Champaign, IL: American Oil Chemists' Society.

Kester, J.J. and Richardson, T. 1984. Modification of whey proteins to improve functionality. *J Dairy Sci* 67: 2757–2774.

Krešić, G., Lelas, V., Jambrak, A.R., Herceg, Z. and Brncic, S.R. 2008. Influence of novel food processing technologies on the rheological and thermophysical properties of whey proteins. *J Food Eng* 87: 64–73.

Kilara, A. and Panyam, D. 2003. Peptides from milk proteins and their properties. *Crit Rev Food Sci Nutr* 43(6): 607–633.

Kilara, A. and Vaghela, M.N. 2000. Whey Proteins. In *Proteins in food processing*, ed. R.Y. Yada, 72–99. Boca Raton, FL: CRC Press.

Kilara, A. 1984. Standardization of methodology for evaluating whey proteins. *J Dairy Sci* 67: 2734–2744.

Kinsella, J.E. 1978. Texturized proteins: fabrication, flavoring and nutrition. *Crit Rev Food Sci Nutr* 10: 147–207.

Kinsella, J.E. 1982. Protein structure and functional properties: Emulsification and flavor binding effects. In *Food protein deterioration*, ACS Symposium Series, Vol. 206, 301–326.

Kunugi, S. and Tanaka, N. 2002. Cold denaturation of proteins under high pressure. *Biochimica et Biophysica Acta - Protein structure and molecular enzymology* 1595: 329–344.

Linden, G. and Lorient, D. 1999. The exploitation of by-products, In *New ingredients in food processing: Biochemistry and agriculture*, eds. G. Linden and D. Lorient, 184–210. London: Woodhead Publishing.

Li-Chan, E.C.Y. 2000. Properties of proteins in food systems: an introduction. In *Proteins in food processing*, ed. R.Y. Yada, 2–25. Boca Raton, FL: CRC Press.

Lin, S., Huff, H.E. and Hsieh, F. 2000. Texture and chemical characteristics of soy protein meat analog extruded at high moisture. *J Food Sci* 65(2): 264–269.

Lin, S., Huff, H.E. and Hsieh, F. 2002. Extrusion process parameters, sensory characteristics, and structural properties of a high moisture soy protein meat analog. *J Food Sci* 67(3): 1066–1072.

Liu, K. 1997. Soybeans: chemistry, technology and utilization. London: Chapman and Hall.

Lockmiller, N.R. 1972. Texture protein products. *Food Technol* 26: 56.

Martinez-Serna, M.D., and Villota, R. 1992. Reactivity, functionality, and extrusion performance of native and chemically modified whey, In *Food extrusion science and technology*, eds. J.L. Kokini, C.T. Ho, and M.V. Karwe, 387–414. New York: Marcel Dekker, Inc.

Matthey, F.P. and Hanna, M.A. 1997. Physical and functional properties of twin-screw extruded whey protein concentrate-corn starch blends. *Lebens Wiss U Technol* 30: 359–366.

Mitchell, J.R. and Areas, J.A.G. 1992. Structural changes in biopolymers during extrusion. In *Food extrusion science and technology*, eds. J.L. Kokini, C.T. Ho and M.V. Karwe, 345–360. New York: Marcel Dekker, Inc.

Mulvaney, S., Rong, S., Barbano, D.M. and Yun, J.J. 1997. Systems analysis of the plastication and extrusion processing of Mozzarella cheese. *J Dairy Sci* 80: 3030–3039.

Noguchi, A. and Isobe, S. 1989. New food proteins, extrusion process and products in Japan. In Proceedings of the World Congress: *Vegetable protein utilization in human foods and animal feedstuffs*, ed. T.H. Applewhite, 375. Champaign, Ill: American Oil Chemists' Society.

O'Donnell, J.A. and O'Donnell, C.D. 2006. Building better foods and supplements. Prepared Foods 175:NS3-NS4, NS6, NS8, NS10-NS11.

Onwulata, C.I., Konstance, R.P., Smith, P.W. and Holsinger, V.H. 1998. Physical properties of extruded products as affected by cheese whey. *J Food Sci* 63: 814–818.

Onwulata, C.I., Konstance, R.P., Smith, P.W. and Holsinger, V.H. 2001. Incorporation of whey products in extruded corn, potato or rice snacks. *Food Res Int* 34: 679–687.

Onwulata, C.I. and Tomasula, P.M. 2004. Whey texturization: A way forward. *Food Technol* 58: 50–54.

Onwulata, C.I., Isobe, S., Tomasula, P.M. and Cooke, P.H. 2005. Properties of whey isolates extruded under acidic and alkaline conditions. *J Dairy Sci* 89: 71–81.

Onwulata, C.I. 2008. Protein containing composition produced by cold extrusion. U.S. Patent #20080280006

Onwulata, C.I., Konstance, R.P., Cooke, P.H. and Farrell, H.M. Jr. 2006. Functionality of extrusion - Texturized whey proteins. *J Dairy Sci* 86: 3775–3782.

Osburn, W.N., Mandigo, R.W. and Kuber, P.S. 1995. Utilization of twin screw cold extrusion to manufacture restructured chops from lower-valued pork. Bulletin #94–219-A. University of Nebraska, College of Agriculture Extension and Home Economics, Lincoln, Neb.

Pordesimo, L.O. and Onwulata, C.I. 2008. Whey texturization for snacks. In *Whey processing, functionality and health benefits,* eds. C.I. Onwulata and P.J. Huth. Danvers, MA: Wiley-Blackwell and Institute of Food Technologists Press.

Privalov, P.L. 1990. Cold denaturation of proteins. *Crit Rev Biochem Molecular Bio* 25(4): 281–305.

Puyol, P., Perez, M.D. and Horne, D.S. 2001. Heat-induced gelation of whey protein isolates (WPI): effect of NaCl and protein concentration. *Food Hydrocolloids* 15: 233–237.

Qi, P.X., Brown, E.M. and Farrell, H.M. Jr. 2001. "New views" on structure-function relationships in milk proteins. *Trends Food Sci Technol* 12: 339–346

Queguiner, C., Dumay, E., Salou-Cavalier, C. and Cheftel, J.C. 1992. Microcoagulation of a whey protein isolate by extrusion cooking at acid pH. *J Food Sci* 57: 610–616.

Reifsteck, B.M., and Jeon, I.J. 2000. Retension of volatile flavors in confections by extrusion processing. *Food Rev Int* 16: 435–452.

Resch, J.J. and Daubert, C.R. 2002. Rheological and physicochemical properties of derivatized whey protein concentrate powders. *Int J Food Prop* 5: 419–434.

Riaz, M.N. 2004. Textured Soy protein as an ingredient. In *Proteins in food processing*, ed. R.Y. Yada, 517–558. New York: Woodhead Publishing Limited.

Roefs, S.P.F.M. and de Kruif, C.G. 1994. A model for the denaturation and aggregation of β-lactoglobulin. *Eur J Biochem* 226: 883–889.

Rokey, G.J., Huber, G.R. and Ben-gera, I. 1992. Extrusion-cooked and textured defatted soybean flours and protein concentrates. In Proceedings of the World Conference on Oilseed Technologies and Utilization, ed. T. H. Applewhite, 290–298. Champaign, IL: AOCS Press.

Ryan, D.S. 1977. Determinants of functional properties of proteins and protein derivatives in foods. In *Food proteins: Improvement through chemical and enzymatic modification,* eds. R.E. Feeney and J.R. Whitaker, 67–91. Washington, D.C.: American Chemical Society.

Sawyer, W.H. 1968. Heat denaturation of bovine β-lactoglobulins and relevance of disulfide aggregation. *J Dairy Sci* 51: 323–329.

Schmidt, R.H., Packard, V.S. and Morris, H.A. 1984. Effect of processing on whey protein functionality. *J Dairy Sci* 67: 2723–2733.

Shen, J.L. and Morr, C.V. 1979. Physicochemical aspects of texturization: fiber formation from globular proteins. *J Am Oil Chem Soc* 56: 638–708.

Shimada, K. and Cheftel, J.C. 1988. Texture characteristics, protein solubility, and sulphydryl group/disulfie bond contents of heat-induced gels of whey protein isolate. *J Agric Food Chem* 36: 1018–1025.

Shimada, K. and Cheftel, J.C. 1989. Sulfhydryl group/disulfide bond interchange reactions during heat-induced gelation of whey protein isolate. *J Agric Food Chem* 37: 161–168.

Singh, R.K., Nielsen, S.S., Chambers, J.V., Martinez-Serna, M. and Villota, R. 1991. Selected characteristic of extruded blends of milk protein raffinate of nonfat dry milk with corn flour. *J Food Proc Pres* 15: 285–302.

Sullivan, S.T., Khan, S.A. and Eissa, A.S. 2008. Whey proteins: functionality and foaming under acidic conditions. In *Whey processing, functionality and health benefits*, eds. C.I. Onwulata and P.J. Huth, 99–132. Danvers, MA: Wiley-Blackwell and Institute of Food Technologists Press.

Szpendowski, J., Smietana, Z., Chojnowski, W. and Swigon. J. 1994. Modification of the structure of casein preparations in the course of extrusion. *Nahrung* 37: 1–4.

Taylor, B.J. and Walsh, M.K. 2002. Development and sensory analysis of a textured whey protein meatless patty. *J Food Sci* 67: 1555–1558.

Walsh, M.K. and Carpenter, C.E. 2003. Textured whey protein product and method. U.S. Patent 6607777.

Walsh, M.K. and Carpenter, C.E. 2008. Whey protein-based meat analogs. In *Whey processing, functionality and health benefits*, eds. C.I. Onwulata and P.J. Huth, 185–200. Danvers, MA: Wiley-Blackwell and Institute of Food Technologists Press.

Walkenstrom, P., Windhab, E. and Hermansson, A.M. 1998. Shear-induced structuring of particulate whey protein gels. *Food Hydrocolloids* 12: 459–68.

Walstra, P., Geurts, T.J., Noomen, A., Jellema, A. and van Boekel, M.A.J.S. 1999. Heat treatment. In *Dairy technology: Principles of milk properties and processes*, eds. P. Walstra, T.J. Geurts, A. Noomen, A. Jellema, and M.A.J.S. van Boekel, 189–209, NY: Marcel Dekker.

Wang, C.H. and Damodaran, S. 1991. Thermal gelation of globular proteins: Influence of protein conformation on gel strength. *J Agric Food Chem* 39(3): 433–438.

Wolkenstein, M. 1988. CALO fats, cholesterol and calories, In *Low-calories products*, eds. G.G. Birch and M.G. Lindley, 43–61. London: Elsevier Applied Science Publishers.

Yada, R.Y., Jackman, R.L., Smith, J.L., Payie, K.G. and Tanaka, T. 1999. Proteins: denaturation and food processing. In *Wiley encyclopedia of food science and technology* 2nd Edition, ed. F. J. Francis, 2000–2014. New York: John Wiley & Sons.

Ziegler, G.R. and Foegeding, E.A. 1990. The gelation of proteins. *Adv Food Nutr Res* 34: 203–298.

12 Quality Control Parameters of Extrudates and Methods for Determination

Sibel Yağcı and Fahrettin Göğüş

CONTENTS

INTRODUCTION

This chapter is written to summarize fundamentals of the latest methods commonly used for quality control of extruded food products. Quality is difficult to define precisely, but it refers to the degree of excellence of a food and it includes all the characteristics which have significance in determining the degree of acceptability of that food. The overall quality of extruded products can be described by physical, chemical, microbiological, nutritional, and sensory parameters. These quality factors depend on the composition of the product, process variables, expected deteriorative reactions, packaging used, and shelf life of product.

Diverse grains, formulated in countless ways and subjected to varying process conditions, can be used to make a wide spectrum of extruded food and feed products. Quality of extruded products must be monitored on a regular basis during production and also storage to ensure that a uniform product is produced and that it meets the required quality control standards. There exists much literature about routine quality

297

control of these products, but it has not been compiled in one place. The authors have made an attempt to compile all the available information and present it in a comprehensive manner. This chapter will cover basic principles and methodology for the physical, chemical, and sensory quality attributes of extruded food and feed products. The effect of process parameters and feed formulations on the extrudate quality attributes and nutritional quality will not be discussed here because most of the details regarding this subject are covered in other chapters of this book.

CLASSIFICATION OF THE METHODS USED FOR QUALITY CONTROL OF EXTRUDATES

Raw materials for extruded products are similar in their general nature to the ingredients used in all other types of food products. They contain materials with different functional roles in the formation and stabilization of the extruded products, and provide color, flavor, and nutritional qualities. The transformation of raw materials during particular extrusion processing is one of the most important factors that distinguish one food type from another (Guy 2001). Extrusion of raw food ingredients, containing mainly starches, proteins, and lipids leads to many physical and chemical transformations in granules which in turn cause changes in properties of the extruded products. In most extrusion processing, feed mixes are processed at high temperatures for relatively short residence times, at high pressure under shear forces, and in most cases, at relatively low water content. The dry and expanded nature of products, such as puffed snacks and breakfast cereals, are often produced by this kind of process depending on the type and severity of the extrusion process. On the other hand low temperature and high residence times are necessary for the production of cold formed products, such as pasta products. Some extruded products need medium temperatures and/or shear forces. For example, most confectionery products are produced using the liquorice extruder, equipped with a heating jacket and one or more screws that push the liquorice paste through a cooling tunnel and towards a cutting system, which gives the desired shapes and sizes to the final product. Companies use many methods to evaluate ingredient functionality, process control, and extrudate attributes. These methods are classified as subjective and objective methods. Subjective methods are conducted by one or more persons using sensory (sight, taste, smell, or touch) procedures to assess product attributes. Objective methods provide numerical information about physical, chemical, microbiological, and toxicological attributes of the extruded products.

SUBJECTIVE METHODS

Subjective methods require the individual evaluators or investigators to give their opinions regarding the qualitative and quantitative values of the characterisitics of extruded products. These methods consist of a physiological reaction resulting from prior training experiences of the individual, the influence of personal preference, and powers of perception. Subjective methods usually involve sensed perceptions of texture, flavor, odor, color, and touch. Sensory panels are a traditional way to evaluate quality of extruded products in the course of product development.

Shape and Size

Most of the extruded foods are made to a size and shape determined by the manufacturer's process. Shapes and sizes are specified to fulfill marketing requirements and very often the shape and size are the only changes of many of the extruded snack products (Chessari and Sellahewa 2001). The shape and size of the extruded products are mainly dependent on the die geometry and cutting system and the degree of expansion. Die shape and design influences finished piece geometry and texture. The speed of the rotating knife assembly regulates the length of the final extrudate. Die shear rates can be changed from single die to a triple and quadruple die with multiple openings and flow channels. Dies with high shear rates cause starch-bearing products to have increased stickiness, water absorption, and solubility properties. Tapered die holes create extrudates with a smoother surface; a die insert having an abrupt cross-sectional change and short land length causes greater mechanical damage to food ingredients and leads to a finer cell structure and softer, pithier texture (Harper 1989). Most of the direct expanded cereals are simple in shape characteristics, but they can be made in more complex forms with a degree of three dimensional shaping from flow deformations and puffing. This is achieved by the application of new cutting technology in such a way that two knives are used to cut the edge of the extrudate at the die exit (Guy 2001). For the half products such as nonexpanded corn pellets, commonly processed under moderate or low shear input, the final shape is provided by using a cold forming die, which contains a sufficient open area to prevent expansion of the precooked viscoelastic dough. After a proper drying process (70°C–80°C, 1–3 h), a smooth, brittle, and glassy structure is formed, then this can be expanded into light foam in hot oil or air (Huber 2001). Dry expanded pet food can be made in a variety of shapes, and colors; more commonly found in today's marketplace are a mixture of shapes and colors sold together in a package. The multicolor, multishape, multitexture extrusion die assembly is recently used as an extruder accessory; this allows the continuous and simultaneous production of a blend of colored products at the extruder die. Visual observation of a shape of the extruded products during processing can permit a rapid response by adjusting the processing variables to maintain quality criteria. Uniformity of structure, quantity of broken piece, homogeneity of air size and shape are generally controlled parameters during processing. People may compare their evaluations of extrudate properties to reference standards or photos to determine whether attributes meet established quality criteria.

Sensory Characteristics

The consumer requirements of extruded foods essentially consist of sensory attributes. The extruded products can be characterized by their sensory properties such as appearance, color, flavor, texture, and mouthfeel. Flavor and color are generated by a number of reactions (especially the Maillard reaction) which are controlled by the composition, temperature, and residence time in extrusion cooking as in traditional cooking process. However, these reactions are accelerated in extrusion cooking because of shear forces (Chessari and Sellahewa 2001). Many extruded foods are bland since there is little time for flavor development. Volatile flavors may flash off with water vapor when the food exits the extruder at the die (Camire 2000). Extrusion cooking is essentially a short-time high-temperature process and a number of traditional cooked flavors are not produced. Therefore, flavors are usually added after extrusion to make the product more palatable. Breakfast cereals are usually coated with sugar or sugar syrups and flavors, especially cocoa and fruit flavors, which improve sensory acceptance. Coating also extends the crunchy texture of breakfast cereals as the product is consumed after adding milk. Puffed cereals usually have a mild taste, more than 200 compounds that contribute to the flavor have been reported (Jacoby and King 2001). Extruded foods are generally toasted to improve the color of the final product. Some caramelization and Maillard reactions take place, which introduces flavors as well as a dark color (Chessari and Sellahewa 2001). However, excess reactions are detrimental to product quality. If heat treatment is too extensive, undesirable colors and flavors may appear (Ilo and Berghofer 1999).

The hardness and crispness of expanded extrudate is a perception of the human being and is associated with the expansion and porosity of the product. Crispy products, which are often porous and brittle, constitute a large category of breakfast cereals and snack foods. Texture in these products arises from an incremental and progressive fracturing of cell wall components in response to deformation. During mastication such fracturing gives rise to specific sensory perceptions such as crispness (Barrett, Rosenberg, and Ross 1994). It is the one of the most important sensory characteristics of snack products. If a crisp food does not produce the characteristic "snap" and the vibrations associated with it, then it is considered to be of poor quality (Duizer and Campanella 1998). It is generally accepted that crispness, which is perceived through a combination of tactile, kinesthetic, visual, and auditory sensations, represents the key texture attributes of dry snack products such as breakfast cereals or extruded rice crisps (Szczesniak 1990).

Sensory Evaluation Methods

Sensory evaluation is made by assessing the eating quality of extrudate under controlled conditions. It is a quantitative method in which numerical data are collected to establish a specific relationship between perceptions resulting from human senses and product characteristics. Sensory evaluation methods are used for new product development, ingredient and process modifications, cost reduction, quality maintenance, and product optimization. Sensory evaluation of extruded products may be carried out by several hundred consumers or by panels of a smaller number of highly trained people depending on the type of information required. Different sensory methods are used, depending on the questions asked and resources available. According to a primary exponent of the method, there are three different testing methods used for sensory evaluation: *discrimination, descriptive,* and *hedonic testing methods* (Lawless and Heymann 1999).

Discrimination Tests

Simple discrimination testing methods attempt to answer whether any differences at all exist between two types of products. These tests may be used when a company considers substituting lower cost ingredients in the formulas, or when reformulation is needed for a new and improved statement. If a product is being improved, a discrimination test is used to determine if a difference exists, then alternative methods are used to determine if the changed product is better than the original (Jacoby and King 2001). There are three common discrimination tests used: *triangle test, duo–trio test,* and *paired comparison test.* The *triangle test* consists of three samples presented simultaneously to the panelist. The panelist is asked to identify the different sample. In the *duo–trio test,* a reference sample is given with two test samples; then the panelist match the test sample identical to the reference sample. In the *paired comparison test,* panelists are asked to choose which of two products is stronger in a given attribute. So, panelist's attention is directed to a specific attribute (Lawless and Heymann 1999). However, it is not widely used during product development studies because it is difficult to change only one parameter of extruded products which commonly consist of complex formulations. These tests are generally performed using 10–25 semitrained or 5–10 trained panelists. Examples of discrimination tests are illustrated in Figure 12.1.

Descriptive Analyses

Descriptive sensory analysis techniques provide complete sensory descriptions of products, determine how ingredient or process changes affect product characteristics, and identify key sensory attributes that promote product acceptance. These techniques are often used to monitor a competitor's offerings such as how a competitor's product is different from yours. Descriptive analysis techniques are also used for shelf-life testing, product development and for quality assurance. These tests cannot be used with consumers, because in all descriptive methods, the panelists should be trained at the very least to be consistent and reproducible (Lawless and Heymann 1999). There are several kinds of testing methods, based on descriptive analyzing principles. Texture profile analyses and quantitative descriptive analyses are generally used for sensory evaluation of extruded food products.

Texture Profile Analyses   The texture profile method was developed at General Foods Corporation in the early 1960s. This technique aims to allow the description of

Triangle test
Place an "X" under the sample which is different than the others.

762   181   543

Duo–trio test
Place an "X" under the sample that matches the reference. First inspect the reference sample, then test other samples.

Reference   543   181

Paired Comparison test
Which sample is better for you? Place an "X" under the sample.

543   181

FIGURE 12.1   Examples of discrimination testing methods.

mechanical, geometric, and other textural sensations associated with a product from first-bite through complete mastication and also accounts for the temporal aspect of attributes (Murray, Delahunty, and Baxter 2001). In the texture profile method, an analysis is done at three stages of ingestion. In the initial stage (the first one to five chews depending on the product type), mechanical properties of hardness, brittleness, and viscosity are measured. The masticatory stage involves gumminess, chewiness, and adhesiveness, as well as hardness, and viscosity. At the residual stage, the changes made during mastication and the condition of the residue are evaluated. These changes are—rate of breakdown, type of breakdown, moisture absorption, and mouth coating. At all three stages, the geometrical properties, such as grainy, gritty, cellular, crystalline, powdery, fibrous, etc., and other properties such as greasy, oily, moist, watery, etc., are noted (Bourne 1982). Definitions of some terms used in texture profile analyses are given in Table 12.1 (tabulated from Bourne 1982).

Texture profile analyses are made quantitative by means of standard rating scales which are anchored with specific food products. The major texture parameters were determined for a wide variety of food products. Anchored rating scales were set up with uniform intervals at which the texture of one food product served as the reference standard. The original texture profile analyses had scales of varying length; for example, the scale for chewiness had seven points, gumminess had five points, and viscosity had eight points. The hardness scale has nine references ranging from cream cheese (point 1) to peanuts (6) to rock candy (9) (Bourne 1982). Texture profile tests are conducted by professionals with experience in sensory evaluation techniques. A minimum of 10 panelists are trained, with the number of training hours for a texture profile panel being as many as 130 h over a 6–7 months period. The panelists practice using the rating scales on a series of reference food products (Lawless and Heymann 1999). The extent of panel training and absence of food products used to anchor the scales are major limitations of this technique.

Quantitative Descriptive Analysis Quantitative descriptive analysis (QDA) provides a comprehensive description of sensory properties of the product. This

---

TABLE 12.1

Definitions of Some Parameters Measured in Texture Profile Analyze

Terms	Definition
Hardness	Force required to compress a substance between molar teeth (solid food), or between the tongue and palate (semisolid food)
Brittleness or fracturability	Force with which a sample crumbles, cracks, or shatters
Viscosity	Force required to draw a liquid from a spoon over the tongue
Chewiness	Length of time required to masticate a sample at a constant rate of force application to reduce it to a consistency suitable for swallowing
Adhesiveness	Force required to remove material that adheres to the mouth during the normal eating process
Gumminess	Denseness that persists throughout mastication or the energy required to disintegrate a semisolid food to a state ready for swallowing

technique has many uses including product reformulation, determination of consumer reaction, quality maintenance and evaluation, establishing product definitions or standards. The vocabulary used to describe the product and the product evaluation itself is achieved by reaching agreement among the panel members. The panelists (10–12 judges) are trained in developing a descriptive language to fit the products they evaluate (Lawless and Heymann 1999). The first step in training descriptive panelist on a product is to standardize terminology describing the attributes of the product (Jacoby and King 2001). Panelists generate a set of terms that describes differences among the products, and decide on the reference standards and/or verbal definitions that should be used to anchor descriptive terms. Then the next step is the profiling product which is commonly represented by a spider diagram. Once definitions are defined, the product is fully characterized by determining the amount of each attribute present. Different attributes such as flavor, appearance, and texture can be shown on separate plots. Profiling fully characterizes a product, and is considered the fingerprint defining that product (Jacoby and King 2001). Dansby and Bovell-Benjamin (2003) measured the sensory properties of ready-to-eat breakfast cereal samples produced by incorporation of various amounts of wheat bran into sweet potato flour using the QDA technique. In that study, 12 different sensory characteristics of a breakfast cereal sample including appearance, flavor, and texture attributes are generated by the panelists. Sensory attributes of breakfast cereal and control samples were quantified with a numerical intensity scale from 0 (not detected) to 10 (very intense) and a spider diagram clearly indicates relative differences in sensory attributes among the samples. The authors concluded that this data could be used to optimize ready-to-eat breakfast cereal formulation and for designing studies to test its consumer acceptance. QDA is relatively simple over the other descriptive techniques. The design for descriptive analysis based on repeated measures and the statistical analysis is generally conducted using analysis of variance (Murray, Delahunty, and Baxter 2001).

Hedonic Testing Methods

Hedonic scales are extensively used in measurement of acceptance or liking of a consumer, the consumer panelist rates their liking for the product on a scale. This is also known as a degree-of-liking scale. Frequently the most efficient procedure is to determine consumers' acceptance scores in a multiproduct test and then to determine their preferences indirectly from the scores (Lawless and Heymann 1999). Hedonic testing focuses on pleasant and unpleasant features of the food tested, measures like/dislike, and is used for overall, texture, and flavor acceptability (Jacoby and King 2001). There are different types of hedonic scales used in sensory testing of foods (Table 12.2). These scales are easily used with nontrained panelists, facial hedonic scales are extensively used for children where use of verbal scales is limited. The results of hedonic scales can be easily processed; a mean score and standard deviation can be calculated for each product.

OBJECTIVE METHODS

These methods consist of determinations from which personal influences of the investigators are entirely excluded. The objective methods used for extrudate quality

TABLE 12.2

Examples of Hedonic Scale Used in Sensory Analysis of Food

Nine Point Hedonic Scale	*Facial Hedonic Scale*
Check the box that describes your overall opinion about this product	**Check the box under the figure which best describes your opinion about this product**

☐ Like extremely

☐ Like very much

☐ Like moderately

☐ Like slightly

☐ Neither like nor dislike

☐ Dislike slightly

☐ Dislike moderately

☐ Dislike very much

☐ Dislike extremely

Dislike very much	Dislike a little	Not sure	Like a little	Like very much
☐	☐	☐	☐	☐

evaluation are based on recognized standard scientific tests including physical, chemical, and instrumental methods.

Expansion Indices

The open cell structure of many extruded products such as breakfast cereals, second generation snacks, pet food etc., are formed by direct expansion at the die of the extruder. Expansion indices describe the degree of puffing undergone by the sample as it exits the extruder (Asare et al. 2004). The amount of expansion in extrudate food depends on the pressure differential between the die and the atmosphere (Harper 1989) as well as the ability of the exiting product to sustain expansion. Both of these are related in part to dough viscosity. High material moisture content usually provides less viscosity than lower moisture content. Therefore, the pressure difference is larger for low moisture content materials and consequently gives high expansion product (Suknark, Philips, and Chinnan 1997).

In the extrusion process; moistened, starchy, and/or proteinaceous feed materials are compressed, sheared, and heated to form a melt that is forced through a die opening to the atmosphere. By the time the feed material reaches the die opening, it has been heated and pressurized, resulting in a semiliquid mass of gelatinized starch containing superheated water. The pressure instantly drops when the product exits the die opening and the superheated water drop flashes to steam, causing the product to expand into a network of cells (Burtea 2001). In the extrusion of pellets or third generation snacks, expansion occurs by frying, by microwave heating, or by oven heating at 160°C–190°C for 10–20 s. Physical transformation in these products is based on water vaporization inside the amorphous starch matrix (Colonna, Tayeb, and Mercier 1989). The degree of expansion is quite variable and depends on both the process parameters employed and the properties of the feed material. Processing conditions and major ingredients have the ability to influence the degree of expansion significantly, since they dictate the type and extent of physical and chemical modifications that take place during extrusion which, in turn, affect expansion (Moraru and Kokini

2003). It is important to understand that extruded products need to be characterized in terms of bulk properties such as expansion ratios, bulk density, and porosity.

Sectional and Longitudinal Expansion

The extrudate expansion followed two different directions that are sectional (radial) and longitudinal (axial) expansion depending on the viscoelastic properties of the melt. The measurement of sectional and longitudinal expansion indices allowed a better description of the porosity of expanded extrudates (Alvarez-Martinez, Kondury, and Harper 1988). Expansion can be expressed either as the ratio between the cross-sectional area of the rod-shaped product and the area of die (Mercier and Feillet 1975) or as the ratio between the diameters of the extruded product and the die, which is called the expansion ratio (Faubion and Hoseney 1982). Alvarez-Martinez, Kondury, and Harper (1988) defined that the extrudate expands in two preferential directions: the direction parallel to the flow which can be expressed by the longitudinal expansion index (LEI), and the direction perpendicular to the flow which can be expressed by the sectional expansion index (SEI). Longitudinal expansion has been defined as the ratio of the exiting velocity of the extrudate after expansion to the velocity of mass inside the die and can be expressed as:

$$\text{LEI} = \left[\pi D^2 / 4 L_{se} \rho_d \right] \times \left[(1 - MC_d) / (1 - MC_e) \right] \tag{12.1}$$

where D is the diameter of the die, L_{se} is the specific length of extrudate (length of extrudate per unit mass in mm $\cdot$ g^{-1}), ρ_d is the density of the dough behind the die, MC_e is the extruder moisture content, and MC_d is the dough moisture content. SEI is defined as the ratio between the cross-sectional area of the extrudate and the aperture area of the die. The expansion phenomenon is not isotropic for most studies in the literature, as SEI is much more favored than LEI. More SEI is generally associated with lower LEI (Colonna, Tayeb, and Mercier 1989). Anisotropy of melt expansion at the die causes structural anisotropy of the resulting extrudates, with an impact on the mechanical properties and sensory attributes of directly expanded cereals. High SEI values, or conversely low LEI values, lead to relatively large cell sizes in the cell structure with high sensory hardness and crispness of the extrudates (Bouvier 2001). The volumetric expansion index (VEI) is inversely proportional to the extrudate density and is defined as:

$$\text{VEI} = \text{SEI} \times \text{LEI} \tag{12.2}$$

During calculating expansion parameters, a caliper is used to measure the dimension of the extrudates. Replicated dimensional measurements are essential, because dimensions can vary depending on the size and location of individual, locally protruding cells near the extrudate surface.

Bulk Density and Porosity

The bulk density is a very important product quality attribute from the view of commercial production of extruded products. Bulk density, expressed as g $\cdot$ cm^{-3}, is a measure of expansion. The sectional expansion ratio considers only the direction perpendicular to extrudate flow, while bulk density considers expansion in all directions. Bulk density is expressed as:

$$\rho_b = \frac{M}{V_e} \tag{12.3}$$

where M is the sample mass and V_e is the expanded sample volume. Accurate determination of sample volume requires appropriate estimation of geometric structure. The shape of this structure largely depends on the design characteristics of the die and can be assumed as oval, cylindrical or ellipse. For example, for cylindrical shaped extrudate the following formula can be used to calculate the bulk density.

$$\rho_b = 4/\pi D_e^2 l \tag{12.4}$$

where ρ_b is the bulk density (g/cm^3), D_e is the diameter of the extrudate (cm), and l is the length per gram of the extrudate (cm/g). When it is difficult to calculate the expanded volume of the extrudate, volumetric displacement procedure can be used to measure bulk density. In this method, glass beads with a diameter of approximately 1–2 mm or sand are used to fill small air gaps present in the expanded structure of the extrudate sample. The difference in weight between the glass jar filled with beads plus extrudate and with only beads is used to determine the volume displaced by the extrudate samples. A high bulk density is generally associated with a low expansion index (Suknark, Philips, and Chinnan 1997; Rayas-Duarte, Majewska, and Doetkott 1998) because more compact material is obtained after milling a less expanded product.

Air cells were created during extrusion giving expanded products with variable pore size and number. Porosity created during extrusion can be used to describe the expansion properties of the extruded product (Thymi et al. 2005; Yağcı and Göğüş 2008). Porosity can be calculated from bulk and solid volumes using the equation:

$$\text{Porosity} = \frac{\text{Bulk Volume} - \text{Solid Volume}}{\text{Bulk Volume}} \tag{12.5}$$

where Bulk Volume $= 1/\rho_b$, Solid Volume $= 1/\rho_s$ and ρ_s is the solid density. Solid volume for porosity and true density measurements can be determined by stereopycnometry. This technique is based upon displacement of, and increased pressure in, ideal gas within a chamber of known volume due to the presence of the sample. Stereopycnometry is best performed on ground or crushed samples in order to eliminate errors due to the presence of completely closed cells, which exclude the interpenetrating vapor and thus falsely increase volume determinations (Barrett 2003).

Color

Color is also an important characteristic of extruded foods. The processing conditions used in extrusion cooking such as high temperature are known to favor non-enzymatic browning reactions (Maillard reaction and sugar caramelization) in extruded foods.

The most characteristic consequence of the nonenzymatic browning reactions is the formation of colored compounds which influence the appearance of extruded products. In many instances, it is highly desirable; it gives a suitable appearance and a good flavor. However, nonenzymatic browning is detrimental to product quality if it is too extensive, undesirable colors, and flavors may appear. The changes of color during extrusion may be used as a measurable indication of the nonenzymatic browning reactions and the extent of browning provides a quantitative indicator of the extent of the chemical reaction (Ilo and Berghofer 1999). The color of extruded products is normally monitored during processing; that is, samples are periodically collected from the extruder exit and analyzed in a laboratory. Although there are different color spaces, the most commonly used of these in the measuring of color in food is the international commission on illumination (CIE) L^*, a^*, b^* color space due to the uniform distribution of colors, and because it is very close to human perception of color. L^* is the luminance or lightness factor which indicates how light or dark the sample is; a^* and b^*, the chromaticity coordinates, represent the red(+)/green(−) and yellow(+)/blue(−) color attributes, respectively.

Instrumental Texture Measurement

Texture is one of the major criteria which consumers use to judge the quality and freshness of extruded foods. The rapid flashing of water forms the characteristic texture of extruded products when the starch melt comes out of the extruder die. As the pressure is suddenly reduced from a high pressure in the extruder to atmospheric pressure, water changes from liquid to vapor. As bubbles of water vapor come out of the starch melt, the product stretches and the matrix sets because of evaporative cooling. Air bubbles get trapped in this matrix and the characteristic puffed structure is formed. The texture of the extruded food product is affected by the shear environment in the extruder screw and die, the type of ingredients and the time/temperature history necessary for chemical cross-linking of the molecules (Harper 1986).

Numerous mechanical instruments have been developed over the past century for measuring textural attributes. Despite the large variations in design, these mechanical instruments either measure or control functions of force, deformation, and time. The types of loading by these instruments include: puncture, compression, shearing, twisting, tension, and bending. During mechanical texture measurement, a test specimen is subjected to a controlled and known extent of deformation from which a deformation curve of its response is generated. Then the response force deformation curve is recorded and then analyzed to quantify specific textural attributes (Bourne 1982).

Mechanical properties (shear stress, breaking strength, resistance to rupture, or fracture.) of extrudates have been studied by several researchers. In the study by Grenus, Hsieh, and Huff (1993), extrusion and extrudate properties of rice flour were investigated with the addition of rice bran at different levels. Shear strength of the extrudates was determined by Warner–Bratzler shear equipment. It was found that shear strength of the extrudate showed an inverse relationship to radial expansion. Extrusion of blends of rice and chick pea flours through a single-screw extruder was studied by Bhattacharya and Prakash (1994). The shear strength of the dry extrudates was determined as the maximum force offered by the sample during shearing in an Instron Universal testing machine with a Warner–Bretzler device. Many authors measured the breaking strength to evaluate the textural properties of the extrudate (Rayas-Duarte, Majewska, and Doetkott 1998; Onwulata et al. 2001; Yağcı and Göğüş 2009a). High breaking strength values are usually related to large cells with thicker cell walls, creating a crunchy texture. Reduced breaking strength values are usually related to a large number of small cells per unit area with thinner cell walls, resulting crispy texture and reduced hardness (Rayas-Duarte, Majewska, and Doetkott 1998; Onwulata et al. 2001). Low breaking strength values are usually related to a large number of small pores per unit area with thinner cell walls, resulting crispy texture and reduced hardness (Rayas-Duarte, Majewska, and Doetkott 1998; Onwulata et al. 2001). Moore et al. (1990) measured the breaking stress and deformability modulus of wheat flour extrudates using an Instron compression cell. The breaking strength of extrudates made from durum clear flour, hazelnut flour, and rice grit was measured using texture analyzer equipped with a Warner–Bretzler shear apparatus. Breaking or shear strength was calculated as the maximal peak force divided by the extrudate cross-sectional area (Yağcı and Göğüş 2009a).

Nowadays a texture analyzer is widely used for measurement of textural properties of extruded food products. Duizer and Campanella (1998) used a texture analyzer and acoustic recording and measured mechanical properties of extruded snack food and its sensory crispness. They concluded that as moisture content and water activity of extrudates decreased, they are crisper and require less force to break down. Mendonça, Grossmann, and Verhé (2000) investigated possibilities of corn bran utilization through the production of expanded snacks with high fiber using the extrusion process. They measured the texture of snacks such as the hardness (peak force during first compression) and fracturability (force at first yield) using a texture analyzer. For snacks it was desirable to have low values for fracturability and hardness. The maximum peak force from the texture analyzer represents the resistance of extrudate to initial penetration and is believed to be the hardness of extrudate (Baik, Powers, and Nguyen 2004; Ding et al. 2005; Altan, McCarthy, and Maskan 2008; Yağcı and Göğüş 2009b). In another study the texture parameters such as fracturability, cohesiveness, adhesiveness, hardness and chewiness of oat–corn extrudates were analyzed using texture profile analysis. Increasing the percentage of oat flour resulted in extrudates with a higher hardness, and lower springiness, gumminess, and chewiness (Liu et al. 2000).

Various texture measurement methods are used to describe textural properties of extruded food products. Selection of a suitable testing method and type of data analysis can help better characterization of textural properties of extruded food products. Veillard, Moraru, and Kokini (2003) tested different methods (uniaxial compression, puncture with conical and cylindrical probes, cutting with blades and fracture wedge fixtures) at various testing conditions to assess the texture of expanded extruded foods of different composition and geometry. They concluded that among the various methods uniaxial compression gave the most consistent results with the lowest coefficient of variation. Recently, Doğan, Samuel, and Kokini (2005) used a texture analyzer to investigate the effect of uniaxial compression and puncture testing methods and different testing parameters such as probe speed, deformation amount, and the number of replicates on textural properties of several extruded products. They concluded that force deformation curves conducted using uniaxial compression comprehensively describes the deformation behavior of the sample to quantify hardness, toughness, elasticity, and the crispness of the extrudates (Figure 12.2). Selected test conditions for the extruded food samples were 1 mm/s probe speed and 80% deformation strain. They observed that higher test speeds and a deformation level excessively diminishes the resolution in the jagged part of the force deformation curve. The method they suggested is applicable to extruded breakfast cereals, extruded snacks and extruded pet food products.

Co-extruded products are composed of two completely different phases which make these products harder to analyze in terms of textural attributes. Samuel, Doğan, and Kokini (2005) evaluated various methods for describing textural properties of some co-extruded products. Among the methods, the penetration testing method with punch probe was able to identify differences in the thickness of the crust as well as the extent of the filler between different samples. They also concluded that shape, size and composition of co-extruded products are important criteria for selecting a suitable texture measurement method. There are many methods of data analysis of the force deformation curve obtained from texture analyzer to quantify textural attributes. Table 12.3 summarizes the main approaches used for data analyses of the force deformation curve obtained for different extruded food products.

Acoustic Measurement Techniques

Acoustic emissions are an important aspect of food texture perception. Quality and acceptability of food products are often assessed based on the sounds produced

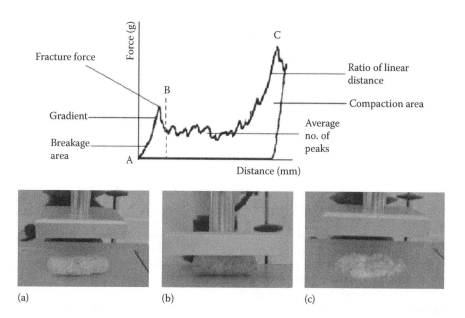

(a)   (b)   (c)

FIGURE 12.2   Sample behavior and force-deformation profile during uniaxial compression. (Modified from Doğan, H., Samuel, L. and Kokini, J.L. 2005. Development of instrumental methods and data analysis tools for objective textural characterization of extruded food products. *Annual Meeting of the Institute of Food Technologists*, July 15–20, New Orleans, Louisiana.)

TABLE 12.3

Basic Methods Used in Data Analysis of Force Deformation Curve

Textural Parameter	Data Analysis Method	References
Hardness	• Maximum peak force in the force deformation curve	Vincent (1998), Mendonça, Grossmann, and Verhé (2000), Baik, Powers, and Nguyen (2004), Ding et al. (2005), Li et al. (2005), Doğan, Samuel, and Kokini (2005), Altan, McCarthy, and Maskan (2008), Yağcı and Göğüş (2009b)
Fracturability	• First significant peak during the compression of the product	Li et al. (2005), Mendonça, Grossmann, and Verhé (2000)
	• Fracturability Parameters: Number of fractures occurring per compression, Mean fracture intensity, Cumulative fracture stress	Barrett and Kaletunç (1998), Barrett, Rosenberg, and Ross (2006)
Crispness	• The slope (N/mm) at which a product breaks were measured from force–distance curve	Jackson, Bourne, and Barnard (1996), Altan, McCarthy, and Maskan (2008)
	• Analyzing the jagged part of the force deformation curve give Ave. Puncturing Force= $\dfrac{\text{Area under the Force Deformation Curve}}{\text{Distance of Puncturing}}$ Number of Spatial Ruptures (NSR) = $\dfrac{\text{Total Number of Peaks}}{\text{Distance of Puncturing}}$ Ave. Specific Force of Structural Ruptures = $\dfrac{\text{Sum of Force Drops per Peak}}{\text{Number of Peak}}$ Crispness Work = $\dfrac{\text{Average Puncturing Force}}{\text{NSR}}$	Barrett, Rosenberg, and Ross (1994), Vincent (1998), Valles-Pamies et al. (2000), Roudaut et al. (2002)
	• Determination of power spectrum using Fast Fourier Transform Analysis	Barrett and Peleg. (1992), Barrett, Rosenberg, and Ross (1994), Roudaut et al. (2002)
	• Determination of Apparent Fractal Dimension using fractal techniques	Barrett et al. (1992), Nuebel and Peleg (1993), Valles-Pamies et al. (2000)
	• Smoothening of jaggedness of force deformation curve and calculation of Ratio of Linear Distance (RLD)	Doğan, Samuel, and Kokini (2005)
Toughness or Stiffness	• Energy required to bite or chew the product and calculated as area under force deformation curve	Vincent (1998), Samuel, Doğan, and Kokini (2005), Ding et al. (2005)
Elasticity	• Gradient of the linear region of the force deformation curve	Doğan, Samuel, and Kokini (2005)

during crushing or biting of the food (Duizer 2004). The loudness and pitch of the audible sound and the high frequency nonaudible sound emitted during fracturing depends among others on the local material that fractures. Therefore an analysis of the sound emitted can give information about the start of the fracturing process, the amount of fracture, the pitch of the emitted sound, which is related to the type of fracturing process and the behavior of the material involved and the sizes of broken pieces (Luyten, Plijter, and Van Vliet 2004).

The sound produced when a food is crushed contains a large amount of information. The cell walls of the product snap and energy is released. It is this energy moving through the air (or other sound medium) which is detected and recorded. Two approaches have been taken to study noisy textures using acoustic techniques. The first one is recording sounds produced by the application of a force on the product during biting and chewing by the teeth, and the second one is to record the sounds produced during application of a force by instrumental shear or compression probes. (Duizer 2001; Duizer 2004). The sound wave recorded during acoustic testing is presented as a plot of amplitude vs time. The amplitude of the sound wave refers to the amount of energy the sound source produces. A high amount of energy is reflected in a high-amplitude wave and a low amount of energy is reflected in a low amplitude wave (Duizer 2004).

Sound emission technique has mainly been used to understand and measure crispness and establish relationships between emitted sound and structural characteristics of extruded snack products. The methods, developed to study crispness, have focused on the sounds generated at fracture, the sound being recorded during instrumental crushing or during mastication. The instrumental approach (using a texture analyzer, for instance) is favored because all aspects of fracture are then controlled. However, the sounds recorded during mastication are more representative of the auditory stimuli related to crispness, especially when bone- and air-conducted vibrations are recorded and analyzed together (Roudaut et al. 2002). The generated sound is analyzed using some transformation methods such as Fourier transform, fractal analysis, blanket algorithm, and Kolmogorow dimension of fractal analysis (Duizer and Campanella 1998). In literature, there are many studies on the acoustic research applied to crispy extruded products. Liu and Tan (1999) concluded that sound signal features can predict sensory crispness of snack products to a good degree of accuracy. Duizer and Campanella (1998) concluded that fractal analysis of acoustic data provides a good method for quantifying the sounds produced as extruded crisp products are bitten. In their study, sensory crispness was well correlated with the acoustic signatures emitted from extruded samples. In a similar study, the mechanism relating sensory perception of expanded snack to their sensory mechanical and acoustic properties during crushing was investigated. The authors reported that the acoustic signatures of the brittle foams were closely related to their microstructural, mechanical, and sensory properties (Cheng et al. 2007). Being of a nondestructive character, as well as rapid and simple, the acoustic method can be recommended for both laboratory and industrial applications.

Measurement of Physicochemical Properties

Starch is the predominant ingredient in extruded snack and ready-to-eat cereals. Starch plays a critical role during extrusion, so the physicochemical properties of the many extruded products are related to structural changes of the starch within the raw material, including gelatinization, melting, and fragmentation (Lai and Kokini 1991). During the extrusion process, the starch is partially hydrated and subjected to increasing shear while it is mechanically conveyed and heated. A major difference between extrusion and other forms of food processing is that gelatinization occurs at much lower moisture levels (12%–22%). Excess water is not available in extrusion and the starch granules do not swell and rupture as in classical gelatinization, but are instead mechanically disrupted by high shear forces and drastic pressure changes. The hot melt is then allowed to expand directly, or is cooled and forced through the die to shape it. Therefore, under extrusion at low moisture content, a mixture of small amounts of gelatinized and melted states of starch as well as fragmentation exist simultaneously (Lai and Kokini 1991). There are many methods to determine if the starchy portions of any extruded products are properly cooked or gelatinized. These methods are based on the determination of solubility properties, pasting behavior, enzyme susceptibility, morphological characteristics, and degree of gelatinization using sophisticated methods such as differential scanning calorimetry (DSC) and x-ray diffraction.

Solubility Indices

When extruded starches are dispersed in an excess of water, their main functional properties are water absorption and water solubility. After grinding, extruded products present some solubility, leading to a thickening behavior. Water solubility and absorption parameters characterize the extruded product and are often important in predicting how the extruded material may behave if further processed. Changes in water absorption and solubility of extruded material may be interpreted on the basis of starch–water interactions that govern the solid phase structure of the processed starch (Colonna, Tayeb, and Mercier 1989).

The water absorption index (WAI) measures the volume occupied by the starch after swelling in excess water, which maintains the integrity of starch in aqueous dispersion (Mason and Hoseney 1986). The WAI can be used as an index of starch gelatinization (Ding et al. 2005). It is the weight of gel obtained per gram of dry sample (Colonna, Tayeb, and Mercier 1989). Water absorption has been generally attributed to the dispersion of starch in excess water, and the dispersion is increased by the degree of starch damage due to gelatinization and extrusion induced fragmentation, that is, molecular weight reduction of amylose and amylopectin molecules (Rayas-Duarte, Majewska, and Doetkott 1998). The amount of water absorbed by the ground extrudate has been used as an indirect estimation of the porosity of the material. As the porosity of the extrudate material increases, the water absorption would also increase (Colonna, Tayeb, and Mercier 1989; Yağcı and Göğüş 2008). The WAI achieved a maximum value at extrusion temperatures of 180°C–200°C. The WAI generally increases along with the increase in temperature, after which it decreases, probably due to increased dextrinization. Product containing dextrinized (degraded) starch tends to stick to the teeth and pack in the molars (Mercier and Feillet 1975). This procedure has a limitation such that it cannot be used for comparison of products having different size and shape properties because physical dimensions of extruded sample affects the rate of water absorption.

The water solubility index (WSI) expresses the percentage of dry matter recovered after the supernatant is evaporated from the water adsorption determination

(Anderson et al. 1969). The WSI is related to the quantity of soluble molecules, which is related to dextrinization. The WSI, often used as an indicator of degradation of molecular components (Kirby et al. 1988), measures the degree of starch conversion during extrusion which is the amount of soluble polysaccharide released from the starch component after extrusion. The WSI is raised due to disintegration of starch granules and low molecular compounds from extrudate melt during the extrusion process, this may cause an increase in soluble material. The reducing effect of increasing feed moisture content on the WSI has been reported in literature for rice-based extrudates (Ding et al. 2005) and maize-finger extrudates (Onyango et al. 2004) and extrudates produced from food by products (Yağcı and Göğüş 2008). It was suggested that the increasing WSI is caused by greater shear degradation of starch during extrusion under low moisture conditions. Increasing severity of thermal treatment in the extruder will progressively increase the WSI, indicating that more starch polymers have been degraded into smaller molecules. The WAI and WSI can also be useful in determining suitability of extruded starchy products for use in suspensions or solutions (Colonna, Tayeb, and Mercier 1989).

Pasting Behavior

Pasting behavior of starch-based extruded products is best evaluated by ViscoAmylograph which involves continuous measurement of shear viscosity during controlled heating and subsequent cooling of an aqueous sample. This instrument records the pressure or torque against a stationary sensor from a rotating sample and container mechanically when sample slurry is subjected to a heating and cooling cycle (Xie et al. 2006). Changes in viscosity profiles often explain how new processes affect the starch properties. These instruments are extremely useful in measuring variation in starch-based ingredients. Final pasting curve of apparent viscosity versus time can be analyzed based on measurements of various parameters which are given in Table 12.4. Colonna, Tayeb, and Mercier (1989) summarized the main observations of the studies which investigated the behavior of extruded starch during heating and cooling cycles in the Brabender ViscoAmylograph. They reported that extruded starches lack a gelatinization peak during the heating process in excess water, whereas native starch exhibits a rapidly rising viscosity peak with the onset of gelatinization. Extruded starches exhibit higher initial uncooked paste viscosity. It was also reported that increasing severity of extrusion treatment (low moisture and high temperature) results in the decrease in initial cold viscosity, and increasing extrusion temperature causes a reduction in the pasting consistency. Extruded starch absorbs water rapidly to form a paste at room temperature. Gels of extruded starch have lower retrogradation values than nonextruded starch.

The Rapid Visco Analyzer (RVA) is another widely used instrument used for the characterization of a wide variety of starch-based products for cooking properties and related information. Similar to the ViscoAmlograph, RVA measures the viscosity developed during cooking and stirring a starchy sample during a preprogrammed temperature change. In the RVA, the resistance to the movement of a rotating sensor is electronically detected. It is relatively fast and has nearly unlimited variables of concentration, heating, and stirring rates. When starch granules are heated in water beyond a certain temperature, the granules absorb a large amount of water and swell

TABLE 12.4
Some Parameters Measured from Pasting Curve Obtained from Viscoamylograph

Parameter	Analysis of Pasting Curve	References
Pasting temperature	• Temperature at which pasting is initiated and varies with type of additives, starch type and processing. This temperature indicates beginning of gelatinization	Limpisut and Jindal (2002), Mariotti et al. (2006)
Peak viscosity	• The maximum viscosity obtained during pasting process	Lai and Kokini (1991), Limpisut and Jindal (2002), Mariotti et al. (2006)
Paste viscosity	• Instantaneous viscosity of 15% aqueous sample at 80°C obtained during pasting process from 25°C to 95°C	Gomez and Aguilera (1984)
Amylographic value	• Rough indication of the degree of gelatinization and calculated as difference between final viscosity of starch suspension at 95°C and initial viscosity of starch suspension at 30°C	Lawton and Handerson (1972)
Breakdown viscosity	• Difference between the maximum viscosity and the viscosity at the end of the first holding period	Limpisut and Jindal (2002), Mariotti et al. (2006)
Setback viscosity	• Difference between the viscosity at the end of the cooling period and the viscosity at the end of the first holding period	
Cold paste viscosity	• Equilibrium viscosity of sample which is suspended in water at room temperature	Colonna, Tayeb, and Mercier (1989),
Hot paste viscosity	• Viscosity of sample after heating and cooling cycles are completed	Iwe et al. (1998)
Time of cooking	• Pasting time	Lai and Kokini (1991), Mariotti et al. (2006)

to many times their original size. When the majority of granules become swollen, viscosity increases rapidly up to a peak viscosity and then due to the rupture of the swollen starch granules viscosity decreases relatively. As the mixture is subsequently cooled, re-association of starch molecules can occur. A final cooking curve, related to the relative rate of hydration and starch gelatinization of the sample, is useful for adjustment of conditions to produce consistent extruded products. Viscosity is generally measured as RVA units (RVU), which are then converted by a multiplying factor (1 RVU ≈ 12 centipoise) (Bryant et al. 2001). The RVA has the advantages of using a small sample size, short testing time, and ability to modify testing conditions.

X-ray Scattering

The physical and chemical change of starch granules can be detected by X-ray diffractometry. X-rays are electromagnetic radiation with a wave length from 0.1 to 1.0 nm which is in the order of molecular spacing in a crystal. The recorded diffracted beams give the information of crystal and molecular structure within the crystal. X-ray measurement detects the semicrystalline character, which reflects the presence of both the ordered and amorphous regions within the starch granule. These will be lost in starch gelatinization or high levels of starch conversion (Sriburi and Hill 2000). With the aid of X-ray diffraction diagrams, native starches can be divided into A-type (cereal), B-type (tuber and roots) and C-type (legume). An additional form, called V-type, occurs in swollen granules. Extrusion cooking destroys the organized crystalline structure either partially, or completely, depending on the amylose–amylopectin ratio and on extrusion variables such as moisture and shear. At higher temperatures, the structure is completely destroyed, leading to an X-ray pattern typical of an amorphous state (Colonna, Tayeb, and Mercier 1989). This amorphous pattern was due to high disorganization of starch molecules caused by heat treatment during the extrusion process (Gonzalez-Soto et al. 2007). Wide-angle X-ray scattering (WAXS) is one of the X-ray diffraction techniques at which the distance from the sample to the detector is shorter and diffraction maxima at larger angles can be observed. This method is useful for determination of the crystalline structure of polymers like extruded products. The X-ray scattering technique is used to analyze starch crystallinity in various starch sources such as cassava starch (Sriburi and Hill 2000); potato starch (Cheyne et al. 2005); banana starch (Gonzalez-Soto et al. 2007); wheat, corn, and rice starches (Bindzus et al. 2002); potato, tapioca, and corn starch (Butrim, Litvyak, and Moskva 2009) during the extrusion process. Crystallinitiy also determines flexibility of the molecular chain of extruded polymers. Polymers with higher crystallinity have better mechanical strength (Guan and Hanna 2006). The crystalline pattern of acetylated starch cellulose foams showed losses in crystallinity of both acetylated starches and cellulose during the extrusion process (Guan and Hanna 2006).

Measurement of Enzyme Susceptibility

Extrusion cooking probably increased the enzymatic availability of starch by way of gelatinization, inactivation of endogenous α-amylase inhibitor, disruption of cellular structure, size reduction, and increased starch surface (Cheftel 1986). Certain enzymes act much more rapidly on cooked starch than native starch, thus many methods are developed based on the utilization of damaged starch granules on enzymatic hydrolysis by specific enzymes. Figure 12.3 shows the basic steps of experimental procedure used for determination of the degree of gelatinization of starch in any sample. These tests are time-consuming, costly, and need skilled personnel to perform them; but they are more sensitive with respect to many other measurement techniques.

In vitro enzymatic digestibility tests, which mimic digestion in the intestinal tract of humans, are useful to monitor starch and protein digestibility in extruded products. Extrusion has been shown to thoroughly gelatinize starch even at very low food moisture levels, thus enhancing starch digestibility (Gomez and Aguilera 1983). It also improves protein digestibility by thermal denaturation and re-orientation of proteins (Onyango et al. 2004). Digestibility tests are especially important in the pet food industry, where the majority of products are extruded and made from cellulosic waste materials. If the product is not extruded properly, the product that contains starch which will not be fully digested causing the potential for "leaks" of starch into the hindgut where microbes digest it very rapidly, can contribute to gastrointestinal

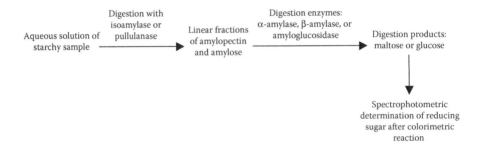

FIGURE 12.3   Flow diagram for determination of the degree of starch gelatinization.

issues for animals ingesting the feed. In these tests, a sample solution is subjected to particular enzyme solutions under conditions similar to the human digestion system, this is provided by the suitable buffer solutions. In starch digestibility tests, reducing sugar content of the sample is commonly determined using a spectrophotometer after a colorimetric reaction with a suitable reducing reagent (Onyango et al. 2004). Most protein digestibility tests need determination of the initial and residual nitrogen content, so analyses take several hours to complete. Preparation of enzyme solutions is a critical step in starch and protein digestibility tests, since activity of enzyme solution can be rapidly decreased during analysis time, conditions should be carefully controlled. Digestibility tests can also be performed using *in vivo* methods which involve a feeding study.

Methods Based on the Formation of Amylose–Iodine Blue Complex

The linear fraction or amylose of starch gives a deep blue complex with iodine and this characteristic has been used as an analytical tool to measure the amylose content. Amylose content can be determined by the iodine-binding procedure through potentiometric titration of dissolved starch with standard iodine, similar amperometric titration or spectrophotometric detection of intensity of blue coloration with iodine. However, iodine also binds with the long β-chain of amylopectin molecules, which causes overestimation of the amylose content. Furthermore, phospholipids and free fatty acids compete with iodine in forming complexes with amylose, which tends to cause underestimation of the amylose content. Subsequently, the amylose content as estimated by classical iodine reaction was designated as "apparent amylase" or "amylose-equivalent." Apparent amylose content has been used to describe the amount of starch damage in different extruded products (Bhatnagar and Hanna 1994; Matthey and Hanna 1997). On the other hand, it was reported that it is difficult to differentiate samples by the dye binding method, because extensive transformation of starch during the extrusion process caused formation of many starch derivatives such as dextrin (Colonna, Tayeb, and Mercier 1989).

The degree of starch gelatinization has also been estimated based on the colorimetric reaction of amylose with iodine. The differential solubility of gelatinized starch in dilute alkali has been utilized in the determination of the degree of starch gelatinization. In this method a starch sample is dispersed in alkali solution commonly potassium hydroxide solution (KOH) for dissolving amylose, subsequently neutralized with acid then treated with iodine and the absorbance of polyiodide–amylose complex is determined (Birch and Priestley 1973; Wootton and Chaudhry 1980). A critical concentration of alkali was found 0.2 N KOH for gelatinized starch 0.5 N KOH for raw starch (Birch and Priestley 1973). This method has been used to determine the degree of gelatinization of starch in various extruded products such as barley extrudates (Köksel et al. 2004); extruded tarhana (Ibanoglu, Ainsworth, and Hayes 1996), and extruded corn starch (Owusu-Ansah, Voort, and Stanley 1983).

Differential Scanning Calorimetry

DSC has proven to be an extremely valuable tool to quantify the gelatinization of starch. DSC uses two pan systems, where the sample is placed in one pan and the other pan is left empty as a reference. It basically measures the amount of energy absorbed or released from the sample during controlled and uniform heating and cooling cycles. In the case of starch gelatinization, the DSC thermograms give the possibility of transition enthalpies occurring during melting of the semicrystallization (or double-helical structure) in the starch as well as precise measurement of the transition temperatures such as initial gelatinization temperature (onset, T_o), endotherm peak temperature (peak, T_p) and the temperature at which gelatinization ceased (end, T_e) (Xie et al. 2006). DSC has been widely used to characterize the thermal behavior of starch in extruded food products and the raw materials (Figure 12.4).

Bhatnagar and Hanna (1996) used the DSC for measuring the degree of starch gelatinization in single and twin-screw extruded corn starch samples which were taken from different locations along the extruder barrel. They reported that the gelatinization measured by DSC indicated 90%–95% starch gelatinization in the samples

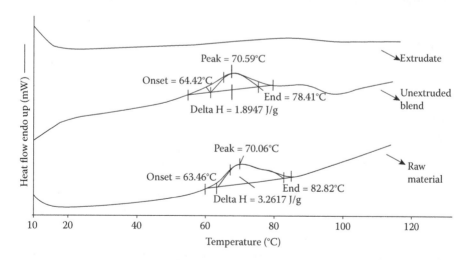

FIGURE 12.4 Example of DSC curves of raw materials, unextruded blend, and extruded sample. (Data was taken from Yağcı, S. 2008. The use of durum clear flour in combination with hazelnut cake and different pomaces in the production of extruded food. PhD thesis in Food Engineering Department, University of Gaziantep, Gaziantep, Turkey.)

which were taken from the die exit in both single-screw and twin-screw extruders. In the study by Gomez and Aguilera (1984), extruded corn starches showed no gelatinization endotherm whereas raw starch exhibited an endotherm, this indicates the gelatinization of starch in extruded products. Recently, DSC studies of native starch and its extrudates indicated that both central and outer regions of the extrudate showed reduced gelatinization endotherms (about 10% and 75%, respectively) compared to unextruded native starch. In addition to this, the DSC samples from the outer regions of extrudate displayed higher temperature endotherms indicating the characteristics of the formation of amylose–lipid complex (Cheyne et al. 2005). Various factors including the sample preparation method, type of pan and measurement conditions, such as heating rate and amount of water added, must be carefully considered during analyses of specimens for starch gelatinization in DSC (Yu and Christie 2001).

Extruded foods are well known examples of amorphous or partially amorphous structures (Table 12.5). These amorphous substances undergo both glass transition and melting at characteristic temperatures T_g and T_m, respectively. Physical changes such as textural attributes of extruded food products are associated with the phenomenon of glass transition and melt transition in starch-based extruded products. Measurement of T_g is a powerful tool to analyze the effect of recipes as well as process and storage conditions on the textural properties of extruded products (Bindzus et al. 2002). During storage, all extruded products may present a slow re-association of macromolecules in the semidry state (i.e., staling), decrease the T_g (Colonna, Tayeb, and Mercier 1989). DSC is a sensitive technique for measurement of T_g and T_m, for instance a large change in specific heat is observed as the material is transformed from immobile glassy state to the more mobile rubbery state (Strahm 1998). Guan and Hanna (2006) investigated the thermal properties

TABLE 12.5

Glass and Melt Transition during Common Extrusion Process (Tabulated from Strahm [1998])

State	Temperature of Material	Example in Extrusion Process	Properties of Product
Rubbery	Between T_g and T_m	Extruder melt at the die exit (before cooling and drying)	Crystalline starch structure has been destroyed. The mass is amorphous, the product can easily be deformed without fracturing the structure, but it is sufficiently coherent, it will not flow through one's finger
Glassy	Below T_g	Dried and cooled extruded products	The structure is amorphous, it fractures easily
Melted	Above T_m	Heated and plasticized melt in the extruder barrel	The structure flows through the barrel easily

of extruded acetylated starch–cellulose foams by measuring the glass transition temperature (T_g) and melting temperature (T_m), which were found to be strongly dependent on the blend composition and extrusion processing conditions such as barrel temperature and screw speed. In another study, Bindzus et al. (2002) obtained the master curve showing the relationship between the water content and T_g, they concluded that this curve can be helpful in modeling of the extrusion process of starch or starch containing blends and in predicting ability to store extruded products.

Measurement of Structural Characteristics

Monitoring morphological and ultra structural view of extruded products under the microscope is widely used for characterization of the cell structure of food components, this may be helpful in controlling the extrusion process so as to produce a desired and uniform extrudate product (Barrett and Peleg 1992). One of the most important microscopic techniques is the scanning electron microscopy (SEM), this is often employed to analyze the internal geometrical structure of a food product such that cell size, the number of cells per unit cross-sectional area and the surface appearance of a sample can be analyzed in detail (Tan, Zhang, and Gao 1997; Choi and Phillips 2004). It was reported that SEM images have a distinct advantage in contrast and clarity to show the details of a product structure. SEM has a large depth of field allowing many areas of a sample to be in focus at the same time. SEM also produces images of high resolution, which means that closely spaced features can be examined at a high magnification (Choi and Phillips 2004). There have been many reported studies on the use of SEM for extruded food structure analysis. Gomez and Aguilera (1984) used SEM to drive a model for structure events occurring during low-moisture extrusion of starch by comparing the physicochemical properties of raw, gelatinized, and dextrinized corn starch with those present after extrusion at varying moisture content. In this manner, they could be able to show the differences in microstructure of extrudates through the extrusion process; in case the structure varied gradually from a dense and smooth one to a highly expanded one characterized by very thin walls and rough, flaky internal surfaces in the most dextrinized extrudate. Tan, Zhang, and Gao (1997) characterized the cellular structure of puffed corn meal extrudates using SEM images. They developed two major algorithms on the basis of run length and void region segmentation in the SEM images and correlate these image features with the directly measured cell size. They reported that image processing techniques are useful in cellular structure characterization and provide a rapid and consistent way for cellular structure quantification and analysis. Choi and Phillips (2004) use SEM to characterize the cellular structure of extruded snack products made from peanut based flour. They measured average cell size and the number of cells per unit area from SEM images and determined the effects of ingredients and extrusion process conditions on cellular structure. They concluded that different formulations and extrusion processing conditions produced snack products with different internal cell structure, which can be clearly revealed with SEM. SEM is also useful to monitor modification of starch structure in extruded products. Under the SEM, starch granules of extruded corn semolina are covered with a proteinaceous film and separated from each other by thin cellulosic walls (Colonna, Tayeb, and Mercier 1989). Disappearance of the intact structure of a starch globule in SEM image show gelatinization of starch in an extruded sample (Ghorpade, Bhatnagar, and Hanna 1997; Butrim, Litvyak, and Moskva 2009). Butrim, Litvyak, and Moskva (2009) compared SEM images of various native and extruded starches having different biological origins. They observed that native grains are completely disintegrated in starch extrusion and large conglomerates are formed, having various shapes and a comparatively smooth and profiled surface. SEM is a highly expensive piece of equipment, the specimens also require an extensive preparation procedure and skilled personnel to perform it.

Extrusion cooking, depending on process conditions and food mix composition, causes swelling and rupture of starch granules and then results in the disappearance of native starch crystallinity and gelatinization of starch granules. Gelatinization of starch granules in extruded product can be detected under a light microscope with polarized filters. Polarized light microscopy is typically used to investigate the physical state of starch granules, which in the crystalline state are characterized by the presence of a Maltese cross pattern. When native starch granules are viewed under the microscope using polarized light, they exhibit a phenomenon known as birefringence. Because of the regular orientation of D-glucosyl units in amorphous and crystalline regions, most of the starch granules have a typical "Maltese cross" (optical anisotropy or double refraction) (Xie et al. 2006). The native starch granule depicting such characteristics under a light microscope during gelatinization begins to fade out its distinct granular boundary as the granule is hydrated and heated. The disappearance of the Maltese crosses (loss of birefringence) in the starch granules is an indication of the irreversible swelling of starch granules that occurs above the gelatinization temperature (Zimeri and Kokini 2003). In this method a ground and sieved extruded sample is hydrated with a water: glycerol solution (about 10%) and then viewed under the microscope using 10× to 100× magnification. This method needs many replications because representative samples for examination are difficult to obtain, and it is also a laborious and tedious procedure to estimate accurately the amount of ungelatinized starch granules in the sample.

In some cases specific staining procedures are used in combination with bright-field light microscopy to distinguish the starch from proteins, cellulose and other components. The familiar blue coloration produced by staining starch with iodine is widely used, allowing the identification and localization of starch even after processing has destroyed the characteristic granule structure. Staining with iodine vapor, Lugols iodine, and toluidine blue are commonly employed to clarify amylose, amylopectin, and cell wall polysaccharides, respectively. Starch in extruded products can be detected as granular bodies in the protein continuum by staining with iodine under light microscopy (Guy 2001). As well as granule rupture, hydration and unfolding can be clearly seen from these images. Imaging techniques are especially helpful for comparison of structural characteristics of extrudates produced by different processing conditions. The effect of extrusion conditions on microstructures of starchy potato pastes were investigated using SEM and light microscopy equipped with bright-field and cross-polarized filters (Cheyne et al. 2005). In this study the extrudate samples were stained using the iodine vapor (for starch) and toluidine blue (for cell wall material) to clarify images obtained from the bright-field light micros-copy. Figure 12.5 shows several cross-section images of extrudate samples obtained using SEM and results of different staining techniques. Berrios et al. (2004) used SEM and light microscopy with a fluorescence filter to examine the effect of sodium bicarbonate addition on the microstructure of extruded bean. In their study, extru-date sections were stained with Safranin O and acid fuchsin stains to display starch and protein portions of extruded bean under fluorescence filter set at a 450–490 nm excitation wavelength.

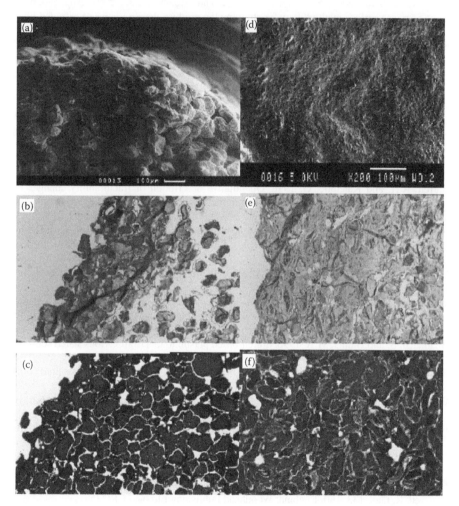

FIGURE 12.5 Cross-section images of extrudate samples: Potato flake extrudate (a) SEM, (b) toluidene blue stained, (c) iodine vapor stained; Potato granule extrudate (d) SEM, (e) toluidene blue stained, (f) iodine vapor stained. (Reprinted from Cheyne, A., Barnes, J., Gedney, S. and Wilson, D.I. 2005. Extrusion behaviour of cohesive potato starch pastes: microstructure-process interactions. *J Food Eng* 66:13–24. With permission.)

REFERENCES

Alvarez-Martinez, L., Kondury, K.P. and Harper, J.M. 1988. A general model for expansion of extruded products. *J Food Sci* 53: 609–615.

Altan, A., McCarthy, K.L. and Maskan, M. 2008. Twin-screw extrusion of barley-grape pomace blends: Extrudate characteristics and determination of optimum processing conditions. *J Food Eng* 89: 24–32.

Anderson, R.A., Conway, H.F., Pfeifer, V.F. and Griffin, E.L. 1969. Gelatinization of corn grits by roll and extrusion cooking. *Cereal Sci Today* 14: 4–12.

Asare, E.K., Sefa-Dedeh, S., Sakyi-Dawson, E. and Afoakwa, E.O. 2004. Application of response surface methodology for studying the product characteristics of extruded rice-cowpea-groundnut blends. *Int J Food Sci Nutr* 55: 431–439.

Bhattacharya, S. and Prakash, M. 1994. Extrusion of blends of rice and chick pea flours: A response surface analysis. *J Food Eng* 21: 315–330.

Baik, B.K., Powers, J. and Nguyen, L.T. 2004. Extrusion of regular and waxy barley flours for production of expanded cereals. *Cereal Chem* 81: 94–99.

Barrett, A.H., Rosenberg, S. and Ross, E.W. 2006. Fracture intensity distributions during compression of puffed corn meal extrudates: Method for quantifying fracturability. *J Food Sci* 59: 617–620.

Barrett, A.H. 2003. Characterization of macrostructures in extruded products. In *Characterization of cereals and flours*, eds. G. Kaletunç and K.J. Breslauer, 369–386. New York: Mercel Dekker Inc.

Barrett, A.H. and Kaletunç, G. 1998. Quantitative description of fracturability changes in puffed corn extrudates affected by sorption of low levels of moisture. *Cereal Chem* 75: 695–698.

Barrett, A.H., Rosenberg, S. and Ross, E.W. 1994. Fracture intensity distribution during compression of puffed corn meal extrudates: Method for quantifying fracturability. *J Food Sci* 59: 617–620.

Barrett, A. and Peleg, M. 1992. Relationship between extrudate cell structure and texture. *J Food Sci* 57: 1253–1257.

Berrios, J.J., Wood, D.F., Whitehand, L.A. and Pan, J. 2004. Sodium bicarbonate and the microstructure, expansion and color of extruded black beans. *J Food Process Pres* 28: 321–335.

Bhatnagar, S. and Hanna, M.A. 1996. Starch-based plastic foams from various starch sources. *Cereal Chem* 73: 601–604.

Bhatnagar, S. and Hanna, M.A. 1994. Extrusion processing conditions for amylose-lipid complexing. *Cereal Chem* 71: 587–593.

Bindzus, W., Livings, S.J., Gloria-Hernandez, H., Fayard, G., Lengerich, B. and Meuser, F. 2002. Glass transition of extruded wheat, corn and rice starch. *Starch* 54: 393–400.

Birch, G.G. and Priestley, R.J. 1973. Degree of gelatinisation of cooked starch. *Starch* 25: 98–100.

Bourne, M.C. 1982. *Food Texture and Viscosity: Concept and Measurement*. London: Academic Press.

Bouvier, J.M. 2001. Breakfast cereals. In *Extrusion cooking: Technologies and applications*, ed. R. Guy, 133–160. England: Woodhead Publishing Ltd. and CRC Press.

Bryant, R.J., Kadan, R.S., Champagne, E.T., Vinyard, B.T. and Boykin, D. 2001. Functional and digestive characteristics of extruded rice flour. *Cereal Chem* 78: 131–137.

Burtea, O. 2001. Snack foods from formers and high-shear extruders. In *Snack foods processing*, eds. E.W. Lusas and L.W. Rooney, 281–314. Boca Raton: CRC Press.

Butrim, S.M., Litvyak, V.V. and Moskva, V.V. 2009. A study of physicochemical properties of extruded starches of varied biological origin. *Russian J Applied Chem* 82: 1195–1199.

Camire, M.E. 2000. Chemical and nutritional changes in food during extrusion. In *Extruders in food applications*, ed. M.N. Riaz, 127–142. Boca Raton: CRC Press.

Cheftel, J.C. 1986. Nutritional effects of extrusion-cooking. *Food Chem* 20: 263–283.

Cheng, E.M., Alavi, S., Pearson, T. and Agbisit, R. 2007. Mechanical–acoustic and sensory evaluations of cornstarch–whey protein isolate extrudates. *J Texture Studies* 38: 473–498.

Chessari, C.J. and Sellahewa, J.N. 2001. Effective process control. In *Extrusion cooking: technologies and applications*, ed. R. Guy, 83–105. England: Woodhead Publishing Ltd. and CRC Press.

Cheyne, A., Barnes, J., Gedney, S. and Wilson, D.I. 2005. Extrusion behaviour of cohesive potato starch pastes: Microstructure-process interactions. *J Food Eng* 66: 13–24.

Choi, I.D. and Phillips, R.D. 2004. Cellular structure of peanut-based extruded snack products using scanning electron microscopy. *J Texture Studies* 35: 353–370.

Colonna, P., Tayeb, J. and Mercier, C. 1989. Extrusion cooking of starch and starchy products. In *Extrusion cooking*, eds. C. Mercier, P. Linko and J.M. Harper, 247–319. St Paul, MN: American Association of Cereal Chemists Inc.

Dansby, M.A. and Bovell-Benjamin, A.C. 2003. Sensory characterization of a ready-to-eat sweet potato breakfast cereal by descriptive analysis. *J Food Sci* 68: 706–709.

Ding, Q.B., Ainsworth, P., Tucker, G. and Marson, H. 2005. The effect of extrusion conditions on the physicochemical properties and sensory characteristics of rice-expanded snacks. *J Food Eng* 66: 283–289.

Doğan, H., Samuel, L. and Kokini, J.L. 2005. Development of instrumental methods and data analysis tools for objective textural characterization of extruded food products. *Annual Meeting of the Institute of Food Technologists*, July 15–20, New Orleans, Louisiana.

Duizer, L. 2001. A review of acoustic research for studying the sensory perception of crisp, crunchy and crackly textures. *Trends Food Sci Technol* 12: 17–24.

Duizer, L.M. 2004. Sound input techniques for measuring texture. In *Texture in food: solid foods*, ed. D. Kilcast, 146–163. New York & Washington, DC: CRC Press.

Duizer, L.M. and Campanella, O.H. 1998. Sensory, instrumental and acoustic characteristics of extruded snack products. *J Texture Studies* 29: 397–411.

Faubion, J.M. and Hoseney, R.C. 1982. High-temperature short-time extrusion cooking of wheat starch and flour. II. Effect of protein and lipid on extrudate properties. *Cereal Chem* 59: 533–537.

Ghorpade, V.M., Bhatnagar, S. and Hanna, M.A. 1997. Structural characteristics of corn starches extruded with soy protein isolate or wheat gluten. *Plant Foods Human Nutr* 51: 109–123.

Gomez, M.H. and Aguilera, J.M. 1984. A physicochemical model for extrusion of corn starch. *J Food Sci* 49: 40–44.

Gomez, M.H. and Aguilera, J.M. 1983. Changes in starch fraction during extrusion-cooking of corn. *J Food Sci* 48: 378–381.

Gonzalez-Soto, R.A., Mora-Escobedo, R., Hernandez-Sanchez, H., Sanchez-Rivera, M. and Bello-Perez, L. 2007. Extrusion of banana starch: characterization of the extrudates. *J Sci Food Agric* 87: 348–356.

Grenus, K.M., Hsieh, F. and Huff, H.E. 1993. Extrusion and extrudate properties of rice flour. *J Food Eng* 18: 229–245.

Guan, J. and Hanna, M. 2006. Selected morphological and functional properties of extruded acetylated starch-cellulose foams. *Biosource Technol* 97: 1716–1726.

Guy, R. 2001. Raw materials for extrusion cooking. In *Extrusion cooking: technologies and applications*, ed. R. Guy, 5–27. England: Woodhead Publishing Ltd. and CRC Press.

Harper, J.M. 1986. Extrusion texturization of foods. *Food Technol* 40: 70–76.

Harper, J.M. 1989. Food extruders and their applications. In *Extrusion cooking*, eds. C. Mercier, P. Linko and J.M. Harper, 1–15. St Paul, MN: American Association of Cereal Chemists Inc.

Huber, G. 2001. Snack foods from cooking extruders. In *Snack foods processing*, eds. E.W. Lusas and L.W. Rooney, 315–367. Boca Raton: CRC Press.

Ibanoglu, S., Ainsworth, P. and Hayes, G.D. 1996. Extrusion of tarhana: Effect of operating variables on starch gelatinization. *Food Chem* 51: 541–544.

Ilo, S. and Berghofer, E. 1999. Kinetics of colour changes during extrusion cooking of maize gritz. *J Food Eng* 39: 73–80.

Iwe, M.O., Wolters, I., Gort, G., Stolp, W. and Zuilichem, D.J. 1998. Behaviour of gelatinisation and viscosity in soy-sweet potato mixtures by single screw extrusion: A response surface analysis. *J Food Eng* 38: 369–379.

Jacoby, D. and King, C. 2001. Sensory evaluation in snack foods development and production. In *Snack foods processing*, eds. E.W. Lusas and L.W. Rooney, 529–546. Boca Raton: CRC Press.

Jackson, J.C., Bourne, M.C. and Barnard, J. 1996. Optimization of blanching for crispness of banana chips using response surface methodology. *J Food Sci* 61: 165–166.

Kirby, A.R., Ollett, A.L., Parker, R. and Smith, A.C. 1988. An experimental study of screw configuration effects in the twin-screw extrusion cooking of maize grits. *J Food Eng* 8: 247–272.

Köksel, H., Ryu, G.H., Baman, A., Demiralp, H. and Ng, P.K.W. 2004. Effects of extrusion variables on the properties of waxy hulless barley extrudates. *Nahrung* 48: 19–24.

Lai, L.S. and Kokini, J.L. 1991. Physicochemical changes and rheological properties of starch during extrusion. *Biotechnol Progress* 7: 251–266.

Lawless, H.T. and Heymann, H. 1999. *Sensory Evaluaion of Food*. New York: Springer.

Lawton, B.T. and Handerson, B.A. 1972. The effects of extruder variables on the gelatinization of corn starch. *The Canadian J Chem Eng* 50: 168–172.

Li, S., Zhang, H.Q., Jin, Z.T. and Hsieh, F. 2005. Textural modification of soya bean/corn extrudates as affected by moisture content, screw speed and soya bean concentration. *Int J Food Sci Technol* 40: 731–741.

Liu, X. and Tan, J. 1999. Acoustic wave analysis for food crispness evaluation. *J Texture Studies* 30: 397–408.

Liu, Y., Hsieh, F., Heymann, H. and Huff, H.E. 2000. Effect of processing conditions on the physical and sensory properties of extruded oat-corn puff. *J Food Sci* 65: 1253–1259.

Limpisut, P. and Jindal, V.K. 2002. Comparison of rice flour pasting propertiesusing brabender viscoamylograph and rapid visco analyser for evaluating cooked rice texture. *Starch* 54: 350–357.

Luyten, H., Plijter, J.J. and Van Vliet, T. 2004. Crispy/crunchy crusts of cellular solid foods: a literature review with discussion. *J Texture Studies* 35: 445–492.

Mariotti, M., Alamprese, C., Pagani, M.A. and Lucisano, M. 2006. Effect of puffing on ultrastructure and physical characteristics of cereal grains and flours. *J Cereal Sci* 43: 47–56.

Mason, W.R. and Hoseney, R.C. 1986. Factors affecting the viscosity of extrusion-cooked wheat starch. *Cereal Chem* 63: 436–441.

Matthey, F.P. and Hanna, M.A. 1997. Physical and functional properties of twin-screw extruded whey protein concentrate-corn starch blends. *LWT - Food Sci Technol* 30: 359–366.

Mendonça, S., Grossmann, M.V.E. and Verhé, R. 2000. Corn bran as a fibre source in expanded snacks. *LWT - Food Sci Technol* 33: 2–8.

Mercier, C. and Feillet, P. 1975. Modification of carbohydrate componenet by extrusion cooking of cereal product. *Cereal Chem* 52: 283–297.

Moore, D., Sanei, A., Van Hecke, E. and Bouvier, J.M. 1990. Effect of ingredients on physical/structural properties of extrudates. *J Food Sci* 55: 1383–1402.

Moraru, C.I. and Kokini, J.L. 2003. Nucleation and expansion during extrusion and microwave heating of cereal foods. *Comp Reviews Food Sci Food Safety* 2: 147–162.

Murray, J.M., Delahunty, C.M. and Baxter, I.A. 2001. Descriptive sensory analysis: Past, present and future. *Food Res Int* 34: 461–467.

Nuebel, C. and Peleg, M. 1993. Compressive stres-strain relationships of two puffed cereals in bulk. *J Food Sci* 58: 1356–1360,1374.

Onwulata, C.I., Smith, P.W., Konstance, R.P. and Holsinger, V.H. 2001. Incorporation of whey products in extruded corn, potato or rice snacks. *Food Res Int* 34: 679–687.

Onyango, C., Noetzold, H., Bley, T. and Henle, T. 2004. Proximate composition and digestibility of fermentedand extruded uji from maize–finger millet blend. *LWT - Food Sci Technol* 37: 827–832.

Owusu-Ansah, J., Voort, F.R. and Stanley, D.W. 1983. Physicochemical changes in cornstarch as a function of extrusion variables. *Cereal Chem* 60: 319–324.

Rayas-Duarte, P., Majewska, K. and Doetkott, C. 1998. Effect of extrusion process parameters on the quality of buckwheat flour mixes. *Cereal Chem* 75: 338–345.

Roudaut, G., Dacremont, C., Vallés Pámies, B., Colas B. and Le Meste, M. 2002. Crispness: A critical review on sensory and material scienece approaches. *Trends Food Sci Technol* 13: 217–227.

Samuel, L., Doğan, H. and Kokini, J.L. 2005. Textural analysis method development for two-phase food products. *Annual Meeting of the Institute of Food Technologists*, July 15–20, New Orleans, Louisiana.

Sriburi, P. and Hill, S.E. 2000. Extrusion of cassava starch with either variations in ascorbic acid concentartion or pH. *Int J Food Sci Technol* 35: 141–154.

Strahm, B. 1998. Fundementals of polymer science as an applied extrusion tool. *Cereal Foods World* 43: 621–625.

Suknark, K., Philips, R.D. and Chinnan, M.S. 1997. Physical properties of directly expanded extrudates formulated from partially defatted peanut flour and different types of starch. *Food Res Int* 30: 575–583.

Szczesniak, A.S. 1990. Texture: Is it still an overlooked food attribute. *Food Technol* 44: 86–95.

Tan, J., Zhang, H. and Gao, X. 1997. SEM image processing for food structure analysis. *J Texture Studies* 28: 657–672.

Thymi, S., Krokida, M.K., Papa, A. and Maroulis, Z.B. 2005. Structural properties of extruded corn starch. *J Food Eng* 68: 519–526.

Valles-Pamies, B., Roudaut, G., Dacremont, C., Leste, M.L. and Mitchell, J.R. 2000. Undertsanding the texture of low moisture cereal products: mechanical and sensory measurements of crispness. *J Sci Food Agric* 80: 1679–1685.

Veillard, P.V., Moraru, C.I. and Kokini, J.L. 2003. Development of instrumental methods for the textural characterization of low moisture extrudates. *Annual Meeting of the Institute of Food Technologists*, July 15–20, Chicago, IL.

Vincent, J.F.V. 1998. The quantification of crispness. *J Sci Food Agric* 78: 162–168.

Wootton, M. and Chaudhry, M.A. 1980. Gelatinization and *in vitro* digestibility of starch in baked foods. *J Food Sci* 45: 1783–1784.

Xie, F., Liu, H., Chen, P., Xue, T., Chen, L., Yu, L. and Corrigan, P. 2006. Starch gelatinization under shearless and shear conditions. *Int J Food Eng* 2: 1–29.

Yağcı, S. and Göğüş, F. 2009a. Selected physical properties of expanded extrudates from the blends of hazelnut flour-durum clear flour-rice. *Int J Food Prop* 12: 405–413.

Yağcı, S. and Göğüş, F. 2009b. Development of extruded snack from food by-products: A response surface analysis. *J Food Proc Eng* 32: 565–586.

Yağcı, S. and Göğüş, F. 2008. Response surface methodology for evaluation of physical and functional properties of extruded snack foods developed from food-by-products. *J Food Eng* 86: 122–132.

Yağcı, S. 2008. The use of durum clear flour in combination with hazelnut cake and different pomaces in the production of extruded food. PhD Thesis in Food Engineering Department, University of Gaziantep, Gaziantep, Turkey.

Yu, L. and Christie, G. 2001. Measurment of starch thermal transitions using differential scanning calorimetry. *Carbohydrate Polymers* 46: 179–184.

Zimeri, J.E. and Kokini, J.L. 2003. Morphological characterization of the phase behavior of inulin–waxy maize starch systems in high moisture environments. *Carbohydrate Polymers* 52: 225–236.

13 Modeling of Twin-Screw Extrusion Process for Food Products Design and Process Optimization

Guy Della Valle, Françoise Berzin, and Bruno Vergnes

CONTENTS

INTRODUCTION

DEFINITION, INTEREST, AND OBJECTIVES OF PROCESS MODELING

The development of new processes or new materials is more and more assisted, besides a classical experimental approach, by the use of theoretical models, able to provide information on the process conditions and their effects on product characteristics. Instead of long and tedious trial and error procedures, process modeling may rapidly help to avoid inappropriate solutions and to concentrate on essential developments. This tendency has largely emerged during the last decades, for example, in the field of injection molding of synthetic polymers, where the use of numerical software for calculating mold filling, residual stresses, or part cooling are now common practices. It has been less common in the food domain, although scientific approaches of numerical modeling have been developed in food processing especially by applying computational fluid dynamics (Norton and Sun 2006).

Process modeling may present different aspects and require different techniques. In the domain of food extrusion, Meuser, Van Lengerich, and Reimers (1984) have been the first to apply response surface modeling (RSM) and experimental design to correlate product temperature T_p and specific mechanical energy (SME) to the functional properties of extruded starch. This approach has shown T_p and SME as important variables of the extrusion process, but the number of experiments required grows exponentially with the number of studied parameters and the validity of polynomial models is restricted to the experimental domain, so it can hardly be used for industrial product design. Despite these drawbacks, the RSM approach has been very popular, mostly because of its simplicity, and results are still found in today's literature; these works generally lead to expected results, for products of various compositions (Pansawat et al. 2008), but the approach hardly copes with the basic mechanisms that govern product change during processing and is restricted to a first experimental approach. Meanwhile, models based on a chemical engineering approach, by fitting the residence time distribution (RTD) curves, described the extruder as an assembly of various chemical reactors (Eerikaïnen and Linko 1989). This technique usually provides a limited quantity of information since it has principally shown that RTD was affected mainly by screw speed, screw geometry and feed rate (Ganjyal and Hanna 2002), but not by temperature or water content, which are well known for their importance on biopolymers structural changes. Moreover, it necessitates a number of adjustable parameters (type and number of reactors, for example), which limits their potentiality in terms of prediction or extrapolation. Using both RTD and SME measurements, the changes of minor components may be predicted by simple models which correlate the thermomechanical history to the retention of thiamine in extruded foods (Cha et al. 2003). Afterwards, process modeling based on expert systems and neural networks was also proposed (Linko et al. 1992). These approaches, where the models improve themselves progressively by experience, are mainly useful for process control, for example, in order to control torque and SME of a twin-screw extruder via screw speed and water input, as illustrated by Wang, Smith, and Chessari (2008).

In fact, the most appropriate approach for developing efficient and deterministic models is to derive the equations based on continuum mechanics. These equations are typically mass and heat transfer balances, completed by a constitutive equation describing the rheological behavior of the studied material. The main advantage of models based on continuum mechanics (knowledge models) is that they do not require any adjustable parameter. Of course, all the parameters involved in the various equations have to be known, but it is a deterministic approach and the models are thus predictive.

An extruder appears generally as a black box, where it is sometimes difficult to establish relationships between entry and exit parameters. The interest of a knowledge model is to make this black box transparent: the model will allow us to calculate data which cannot be measured (for example, shear rates and strains), to clearly understand the relationships between the various parameters (for example, what are the consequences of a change in screw speed?), or to predict the limits of the process (when will the product temperature exceed limit values? When will maximum torque be reached?). Such a model can thus be used for a better understanding of the process, but also for optimization (best processing conditions for desired properties, best screw design for increasing output, and reducing power consumption). It may also be used to solve the complex problems of scale-up: when a process/product has been developed at the laboratory scale (at a few kilograms per hour), how to design the screw profile and how to define the processing conditions to obtain the same product quality at industrial scale (a few tons per hour)? In the case of single-screw extrusion, the geometry is rather simple, and analytical solutions of the flow equations can be provided (Alves, Barbosa, and Prata 2009) and simplified to address problems like oil extraction (Willems, Kuipers, and de Haan 2009).

An efficient model of the twin-screw extrusion process should allow us to obtain a correct estimation of the process parameters (pressure, temperature, residence time, and viscosity), but also the right tendencies when modifying the control values (screw speed, feed rate, and barrel temperature). It is the objective of this chapter to show the potentialities of process modeling based on continuum mechanics in the field of twin-screw extrusion for food applications through a common approach developed over several years.

In fact, two main aspects have to be considered in twin-screw extrusion modeling. We can consider the whole process, from feeding of the raw materials in the hopper until the exit of the transformed product at the die, or just focus on a limited portion of the extruder. The first aspect, called global modeling, necessitates the use of a simplified approach, generally based on one dimensional (1D) flow assumption. The second one can be highly sophisticated and usually uses three dimensional (3D) finite element method (FEM). It allows a very accurate description of the flow field, but is expensive in terms of computational resources (Valette et al. 2007, 2009). In the field of food extrusion, numerical models have been used, for example, to develop filled snacks by co-extrusion, and in particular, to predict the available lysine (de Cindio, Gabriele, and Pollini 2002) and using FEM for cereal food production, without any concern on product quality (Ficarella, Milanese, and Laforgia 2006). In this chapter, we will focus on a global approach of the twin-screw extrusion process, which is much easier to use and more appropriate for considering realistic industrial applications.

MODELING OF TWIN-SCREW EXTRUSION USING A CONTINUUM MECHANICS APPROACH

Raw materials are fed into extruders under a divided solid form (powder and flour) and liquids (water, syrup, and fat) may be added when not previously mixed. The progress of the product along the extruder follows a general sequence of solid conveying, followed by a transition from solid into a macroscopically homogeneous viscous phase, which may be called dough, or a melt, which is then conveyed and submitted to a thermal and mechanical treatment (mainly shear). Finally, this melt undergoes a liquid/solid transition, at the die exit, due to sudden changes in pressure and temperature. These state changes, as well as heat and mass transfers, result from various phenomena which are rapidly described thereafter; they are taken into account by variations of thermal and physical properties, which are rapidly recalled for the overall comprehension of the process, and also because they are important inputs of the model.

Solid/Melt Transition

This transition is mainly due to friction between sheared particles which may be fragmented, in the case of starch granules (Barron et al. 2001), or become rubbery, for proteins (Cuq, Abecassis, and Guilbert 2003). Considering the screw profile shown in Figure 13.1, this transition is generally considered to occur at the first restrictive element (kneading disc and left-handed screw), essentially due to the compression and the friction of the flour particles in this section. These changes are very different from those occurring under an excess of water, where shear stresses are negligible, i.e., gelatinization for instance. Using a specific rheometer with preshearing, Barron et al. (2002) showed that, in first approach, they may be accounted for by the peak temperature measured by differential scanning calorimetry (DSC) as defined for potato starch melting by Donovan (1979). Extensive reviews on starch melting phenomenon as a function of water content may be easily found in the literature for different starch botanical origins and composition of the starchy product, especially sugar content (Colonna, Tayeb, and Mercier 1989; Núñez et al. 2009). The overall energy balance of the melting phenomenon within the extruder may thus be written as:

$$\Delta H + C_p \Delta T = E_{cd} + E_f \qquad (13.1)$$

where ΔH is the specific enthalpy for the state change (melting in the case of starchy product, with an order of magnitude of approximately $10 \text{ J} \cdot \text{g}^{-1}$), ΔT is the overall temperature increase up to peak temperature, E_{cd} and E_f are the terms of specific energy supplied by conduction from the barrel and screw(s), and dissipation by friction, respectively. Taking usual values for heat conductivity and specific heat, as well as the interparticle coefficient measured by Della Valle and Vergnes (1994), Barron et al. (2002) calculated orders of magnitude for these specific energy terms and concluded that energy inputs for solid/melt transition was approximately 500 $\text{J} \cdot \text{g}^{-1}$, a value in agreement with those found for starch conversion during extrusion (Guy and Horne 1988; Zheng and Wang 1994). It is also close to the values found for

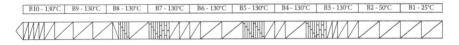

| B10 - 130°C | B9 - 130°C | B8 - 130°C | B7 - 130°C | B6 - 130°C | B5 - 130°C | B4 - 130°C | B3 - 130°C | B2 - 50°C | B1 - 25°C |

FIGURE 13.1 Example of screw profile of a twin-screw extruder, including three restrictive sections (two blocks of kneading discs and one left-handed element, from right to left).

the destructurization by twin-screw extrusion of cell wall rich materials (wheat bran, sugar beet and citrus pulps, oat and pea hulls) in order to solubilize polysaccharides (pectins, arabinoxylans) or improve the functional properties of some byproducts from cereal and fruit processing industries (Ralet, Della Valle, and Thibault 1993).

Conveying Viscous Fluids

Above these values of temperature and energy, the product may be considered as a homogeneous fluid and its flow described by continuum mechanics equations. It is conveyed by the rotation of the screws and the main force applied to its flow, in unit area, is the shear stress (Pa), defined by:

$$\tau = \eta(\dot{\gamma})\dot{\gamma} \tag{13.2}$$

where the shear viscosity η (Pa · s) is generally a decreasing function of shear rate $\dot{\gamma}$ (s^{-1}). Because of intense shear stresses, macromolecular degradation of starch occurs during extrusion (Colonna and Mercier 1983). This phenomenon may be evaluated by measurements of intrinsic viscosity, which reflects product average molecular weight, or more accurately by size exclusion chromatography, which shows that amylopectin is more affected by degradation than amylose (Baud et al. 1999). Starch macromolecular degradation is controlled by the SME received during the thermo-mechanical treatment (Meuser, Van Lengerich, and Reimers 1984; Davidson et al. 1984; Vergnes et al. 1987). Even for flours and grits, the starch original structure is completely destroyed when SME is higher than 140–170 kWh/t (Guy and Horne 1988; Della Valle et al. 1989). In twin-screw extrusion, SME increases with screw speed N and decreases with feed rate Q, and it is, usually, a linear function of the ratio N/Q. Figure 13.2 shows that starch intrinsic viscosity decreases regularly when

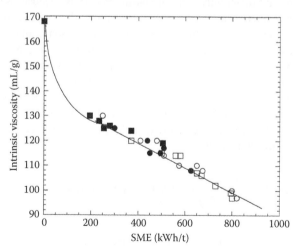

FIGURE 13.2 Change of starch intrinsic viscosity with SME. (From Berzin, F., Tara, A., Tighzert, L. and Vergnes, B. 2010. Importance of coupling between specific energy and viscosity in the modeling of twin-screw extrusion of starchy products. *Polym Eng Sci* 50: 1758–1766. With permission.)

SME increases, independent of the flow conditions (feed rate and screw speed) and of the screw profile. From this figure, starch transformation, and thus starch properties, can be evaluated if SME is correctly known. Intrinsic viscosity [η] and SME are usually related by exponential or power law relationships (Willett, Millard, and Jasberg 1997; Yeh and Jaw 1999; Brümmer et al. 2002). However, according to Figure 13.2, intrinsic viscosity can be correctly estimated in the range 200–800 kWh/t by a simple linear relationship (Berzin at al. 2010).

$$[\eta] = A(\text{SME}) + B \tag{13.3}$$

where $A = -0.05$ and $B = 140$, for SME expressed in kilowatt-hour per ton and [η] in milliliter per gram.

Since SME modifies starch structure, it also affects its rheological behavior, as first shown by Vergnes and Villemaire (1987); it implies an accurate definition of the relationships between SME and the shear viscosity of molten starch. This chapter is not devoted to the rheological behavior of products during extrusion, which is the aim of Chapter 5 by McCarthy, Rauch, and Krochta in this book. However, as it is an important property for the model input, we recall thereafter some significant results. Highly concentrated, and especially starch-based biopolymer products have a viscous behavior which can be defined by a power law:

$$\eta = K\dot{\gamma}^{n-1} \tag{13.4}$$

where n is the flow index ($0 \leq n \leq 1$) and K (Pa · s^n) the consistency. Biopolymer based products often behave differently to thermoplastic materials because, as stated before, a mechanical treatment is necessary prior to flow, and also because of the reduced amount of water. Therefore, the use of conventional rotational and capillary commercial rheometers is not adapted. This is why the accurate determination of their viscous behavior has required the set-up of specific methods based on capillary rheometry, either in-line using a slit die with balanced feed rate (Rheopac) (Vergnes, Della Valle, and Tayeb 1993), or using a preshearing rheometer (Rheoplast) (Vergnes and Villemaire 1987). Overall, the results lead to the conclusion that, in the range 5–500 s^{-1}, the behavior of extruded starchy melts can be described by the following relations (Della Valle et al. 1996):

$$\eta = K \exp\left[\frac{E}{R}\left(\frac{1}{T} - \frac{1}{T_0}\right) + \beta(\text{SME} - \text{SME}_0) + \gamma(WC - WC_0)\right] \cdot \dot{\gamma}^{n-1} \tag{13.5}$$

$$n = \alpha_1 T + \alpha_2 WC + \alpha_3 WC \cdot T \tag{13.6}$$

T is the absolute temperature (K), WC the total water content and SME the volume mechanical energy (J · m^{-3}). T_0, SME$_0$ and WC_0 are reference values, β and γ are parameters whose numerical values are determined experimentally by fitting flow curves; they are available from literature. Coefficient β (>0) reflects the role of ther-momechanical history in decreasing viscosity as shown by Figure 13.3a. Here the shear viscosity curves of molten starch at 85°C and 40% water content are shown for three different values of SME, ranging from 185 to 466 kWh/t. As expected, the viscosity decreases when SME increases, because of the macromolecular degrada-tion. From the experiments shown in this figure, the following results were obtained:

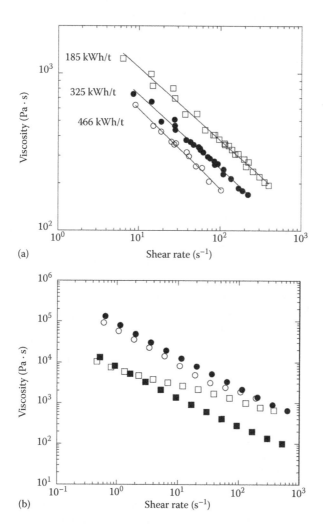

FIGURE 13.3 Viscosity function of shear rate: (a) for different values of SME. (wheat starch, 85°C, 40% water) (From Berzin, F., Tara, A., Tighzert, L. and Vergnes, B. 2010. Importance of coupling between specific energy and viscosity in the modelling of twin-screw extrusion of starchy products. *Polym Eng Sci* 50: 1758–1766. With permission), (b) for plasticized starches (125°C, 27% glycerol) from various botanical origin (□: waxy maize, ■: wheat, ○: pea, ●: amylomaize) (Data from Della Valle, G., Vergnes, B. and Lourdin, D. 2007. Viscous properties of thermoplastic starches from different botanical origin. *Int Polym Proc* 22: 471–480.)

$K = 1920$ Pa · s, $n = 0.53$, $E/R = 2730$ K, $T_0 = 363$ K, $\beta = -0.00158$ t · kWh^{-1}, SME$_0 = 325$ kWh · t^{-1}, $\gamma = -5.78$ and $WC_0 = 0.40$.

The dependence of n with temperature and water content (Equation 13.6) is a crude approximation, deriving from the fact that a power law has been chosen to describe the rheological behavior. In fact, a more accurate description of the viscous behavior could be obtained by applying time-temperature and time-water content superposition (Vergnes, Della Valle, and Colonna 2003), and by selecting the Carreau-Yasuda law (Della Valle and Vergnes 1994):

$$\eta = \phi K \left[1 + (\lambda \phi \dot{\gamma})^a \right]^{\frac{n-1}{a}} \tag{13.7}$$

$$\phi = \exp \left[\frac{E}{R} \left(\frac{1}{T} - \frac{1}{T_0} \right) + \beta (\text{SME} - \text{SME}_0) + \gamma (WC - WC_0) \right] \tag{13.8}$$

The values of the parameters in Equations 13.5 through 13.8 also depend on the amylose content of starch, and the decrease of viscosity when amylose content decreases may be interpreted by a lower ability of amylopectin to create entanglements, as illustrated by the results obtained on starches plasticized by glycerol (Figure 13.3b) (Della Valle, Vergnes, and Lourdin 2007). For more complex compositions, including sugars, salts, various plasticizers (glycerol and fatty acid), viscosity variations are taken into consideration by a factor similar to α, while E/R varies little, and β decreases by the protector effect of the small molecular masses on starch polymers (Willett, Jasberg, and Swanson 1995). However, the influence of proteins cannot be described so directly, given the possible formation of interchain links at high temperatures. Indeed, Redl, Guilbert, and Vergnes (1999) have observed a behavior similar to that of molten starches for gluten in the presence of glycerol, but with a negative value of β ($\approx 3.10^{-8}$ (Jm^{-3})$^{-1}$), which illustrates the structuring role of mechanical energy by creating a network. For corn flour and starch/ zein mixtures, no significant change in shear viscosity could be detected depending on the protein content (0%–15%) (Chaunier et al. 2008), although zein is known to have lower viscosity values (Madeka and Kokini 1992). Moreover, practice teaches that the extrusion of flour is far easier than the corresponding starch. Hence, this behavior cannot be interpreted by simply applying a law of mixture, and some effort has to be made to measure accurately the shear viscosity of these materials and then relate it to the structural changes occurring during the process.

The 1D flow of the viscous fluid in the channels of a twin-screw extruder may be described by solving Stokes equation in cylindrical coordinates (r, θ, z), in different slices where temperature and viscosity are supposed to be locally constant (Figure 13.4). Each slice represents an elementary screw section, so that summing up these successive slices allows us to describe the whole screw channel, from the first restrictive element where the material can be considered as a melt until the die exit, taking into account the variations of temperature and viscosity. Under such conditions, axial and transverse velocity fields (radial velocity is null) can be determined in each slice and the volume flow rate Q_v along a screw channel may be written:

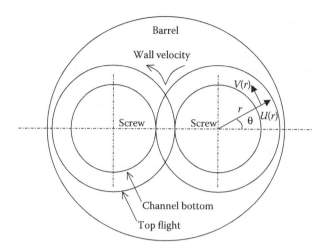

FIGURE 13.4 Schematic view of the transverse section of a twin-screw extruder and illustration of kinematical assumptions made for flow modeling.

$$Q_v = A\Omega + B\frac{1}{\eta}\frac{\Delta P}{\Delta\theta} \qquad (13.9)$$

where A and B are coefficients depending on screw geometry, Ω the screw rotation speed (rd/s^{-1}), ΔP the pressure variation along an angular screw section $\Delta\theta$. η is deduced (Equations 13.4 through 13.8) from the shear rate $\dot{\gamma}$ averaged on the screw section from the velocity field. Summingup relations (Equation 13.9) along the screws, and knowing viscosity variations, it is possible to set the pressure balance of the extruder and to compute the specific energy due to viscous dissipation E_v:

$$\dot{W} = \int_V \eta\dot{\gamma}^2 dV \text{ and } E_v = \frac{\dot{W}}{\rho_m Q_v} \qquad (13.10)$$

where $\dot{W}$ is the power dissipated in the considered volume V of the screw and ρ_m the melt density.

Die Outlet and Liquid/Solid Transition

When the biopolymer based melt leaves the die at high temperature, it expands because of the sudden pressure drop and water vaporization, which cools it, leading to a solid porous material, i.e., a solid foam. This phenomenon, called expansion, deserves attention because it gives the product its attractive texture, mostly crispy. Its amplitude is assessed by the ratio of densities of the melt ρ_m to that of the final extruded product ρ_p:

$$\text{VEI} = \rho_m/\rho_p = \text{SEI} \cdot \text{LEI} \qquad (13.11)$$

where SEI and LEI are the sectional and longitudinal expansion indices, respectively; SEI is estimated by the ratio of extrudate to the die sectional area and LEI by the ratio of velocities of the melt to the extrudate, according to the definitions first given by Launay and Lisch (1983). Alvarez-Martinez, Kondury, and Harper (1988) have used RSM to link these indices to thermorheological variables measured on the extrusion process and underlined the role of die shear rate and moisture content, which was ascertained by Della Valle et al. (1997). The ratio SEI/LEI^2 can be defined as a ratio of anisotropy; large values ($>>1$) reflect the radial expansion of large bubbles, whereas low values ($<<1$) rather characterize expansion favored in the axial direction of small bubbles (Desrumaux, Bouvier, and Burri 1998; Robin et al. 2010). Expansion follows a sequence of phenomena: nucleation, bubble growth, and coalescence before shrinkage or collapse. Scarcely described in literature, nucleation is related either to the initial presence of air in the material or to the incorporation or injection of gas during melting. These nuclei will give rise to bubbles which grow as a result of thermodynamic instability or heterogeneity of the environment. Park, Behravesh, and Venter (1998) show that a rapid drop in die pressure increases the nucleation rate during the extrusion of molten polystyrene and CO_2 in solution. Nucleation determines the number of bubbles per unit volume N_c, and their radius, and provides initial conditions for bubble growth. For this phenomenon, the rate of increase of the bubble radius R is governed by the ratio of driving forces (internal pressure ΔP in the vapor bubble) to resisting forces (matrix bi-extensional viscosity η_e):

$$\frac{1}{R}\frac{dR}{dt} \approx \frac{\Delta P}{\eta_e} \tag{13.12}$$

which is a simplification of the Amon and Denson model (1984). Both terms are functions of temperature and water content. This relation explains the negative correlation between expansion and shear viscosity measured at the die (Della Valle et al. 1997), as seen from Figure 13.5, which illustrates Equation 13.12. Using this relationship with Equation 13.11, the overall expansion may be approached by:

$$VEI = \frac{4}{3}\pi R^3 N_c \rho_m + 1 \tag{13.13}$$

However, during expansion, the wall materials surrounding bubbles often rupture; if this phenomenon occurs before the extrudate becomes solid, i.e., at a temperature where the material is still rubbery ($T > T_g + 30°C$, T_g being the glass transition temperature), then it results in an overall collapse of the extrudate and a lower value of VEI, according to Fan, Mitchell, and Blanshard (1994). In this temperature range, elongational viscosity limits coalescence by cell wall rupture as shown by the finer cell structures found for high amylose starches (Babin et al. 2007); this interpretation may also refer to the importance of the strain hardening index in bread proofing (van Vliet 2008). In line with the study of Fan, Mitchell, and Blanshard (1994) and the Amon and Denson model, the comparison of expansion by extrusion with

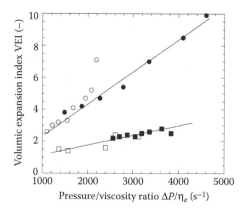

FIGURE 13.5 Variations of the volumetric expansion index of extruded starches with pressure/viscosity ratio at the die for various amylose contents. (□: 0; ■: 0.23; ○: 0.47; ●: 0.7, Data from Della Valle, G., Vergnes, B., Colonna, P. and Patria, A. 1997. Relations between rheological properties of molten starches and their expansion by extrusion. *J Food Eng* 31: 277–296.)

bubble growth during bread making has led to the development of numerical models using 3D FEM (Hailemariam, Okos, and Campanella 2007; Bikard et al. 2008) for isothermal conditions, which, unfortunately, do not cope with the highly transient character of expansion.

Consequently, up to now, the design of extruded products with defined density and cellular structures still relies on the above phenomenological models and on the knowledge of the variables which define the state of the material and its thermomechanical history at the die exit. The model developed before has led to computer software that may be used to calculate these variables.

EXAMPLE OF A GLOBAL MODEL: LUDOVIC© SOFTWARE

GENERAL PRESENTATION: DATA AND ORGANIZATION

We present briefly the Ludovic software.[*] For more details, it is suggested to refer to Vergnes, Della Valle, and Delamare (1998). The bases of Ludovic are the elementary models developed for the various types of elements encountered along a twin-screw extruder, as explained in the previous paragraph. These elements are principally screw conveying elements, either with a positive (right-handed) or negative (left-handed) pitch and blocks of kneading discs.

These elementary models are linked together to obtain a global description of the flow field along the extruder. Solid conveying in the first part of the extruder is not calculated, as no important phenomenon occurs in this region. First, the changes in material begin in the melting section, where the initial powdery product is transformed into a homogeneous molten phase. Even though a melting model has been developed for synthetic polymers (Vergnes et al. 2001),

---

[*] http://sciencecomputerconsultants.com
   http://www.inra.fr/cepia_eng/you_are_looking_for/platforms_and_tools/softwares/ludovic

it is not really adapted to food products such as starch or cereals, for which the melting mechanisms remain only partially understood (Barron et al. 2002). Consequently, we will assume that the melting is instantaneous and takes place in the first restrictive element of the screw profile, which is experimentally observed (Barrès et al. 1990). Afterwards, the material is supposed to be fully molten and it can fill or not fill the screw channel, according to the local geometry and the flow conditions.

Once the screw profile has been defined, its constituting elements are automatically divided into a series of sub-elements (portions of screw channel or kneading discs), in which the calculations will be made. But, as the screws are starve fed, the filling ratio of the system is not known. Consequently, it is not possible to start the calculation in the feeding zone and to proceed downstream. Calculation is bound to start from the die exit and to proceed backwards. As the final product temperature is unknown, an iterative procedure must be used:

- An arbitrary final temperature is selected.
- For the corresponding viscosity, the pressure drop through the die is calculated.
- Sub-element by sub-element and proceeding in an upstream direction, the pressure and temperature evolutions are computed, using the equations presented previously.
- When the pressure falls to zero (starved section), the presence of a restrictive element upstream (left-handed element or kneading discs) is checked. If it is the case, the computation goes on, with a pressure equal to zero, until a new section under pressure is encountered.
- When the first restrictive element is reached, the upstream position where the pressure falls to zero is calculated: it is the place where melting is assumed to take place. At this position, we compare the calculated temperature to the melting temperature. The computation is achieved when both temperatures are equal. Otherwise, the final temperature is modified and the computation restarted. It has been checked that the final result after convergence is independent of the choice of the first arbitrary temperature. The convergence is usually obtained in less than 50 iterations, which takes a few seconds on a PC.

Different data files have to be defined prior to starting a computation. The first one concerns the screw and barrel configurations. For each element, we must indicate geometrical data such as length, pitch, number of flights, disc thickness, staggering angle, etc., as well as the succession of elements along the screw axis. Barrel elements and die characteristics must also be described. The second set of data concerns the material to be processed. The rheological behavior must be defined, according to the various preselected laws—Newtonian, power law (Equations 13.2 and 13.4), Carreau–Yasuda law (Equations 13.7 and 13.8), as well as some physical characteristics such as density, heat capacity, thermal conductivity, melting enthalpy, and temperature. Finally, the processing conditions have to be selected: screw rotation speed, feed rate, and barrel temperatures.

Examples of Global Results

To show and comment on the results obtained with Ludovic, we consider the case of a laboratory scale Clextral BC 21 extruder (centerline distance: 21 mm, screw diameter: 25 mm, ratio L/D: 36), with the screw profile shown in Figure 13.1. It has a first block of kneading discs for melting (five discs at 45°, followed by five discs at 90°). Then, three mixing zones are situated along the screw profile: the first two are made with kneading discs (five discs at 90°, followed by five discs at −45°, and five discs at 45°, followed by five discs at 90°); the last one, more severe, associates a left-handed element and five discs at 90°. The extruded material is a wheat starch with 10% water added, whose behavior is described by a power law (Equation 13.4, with $n = 0.3$, $K = 5000$ Pa · s). The barrel temperature is set at 130°C.

In a first step, we present results obtained for a screw speed of 300 rpm and a feed rate of 5 kg/h. Figure 13.6 shows the changes along the screws of pressure and temperature. We first observe that the screws are only very partially filled: the only parts under pressure are those containing severe restrictive elements, i.e., a left-handed screw element and a block of kneading discs at −45°. As expected, the die is also filled and the head pressure in this case is approximately 0.7 MPa. The temperature increases rapidly in the melting zone, from 130°C (melting temperature of the selected wheat starch for this moisture content: $WC = 0.1$) to 144°C. Then, it changes along the screw profile according to the type of elements: in restrictive elements (left-handed screw element and blocks of kneading discs) temperature increases by viscous dissipation, whereas in screw conveying elements it decreases by heat transfer towards the barrel. Finally, in this case the exit temperature is 144°C.

Figure 13.7 presents the changes along the screws of cumulative SME and residence time. It can be seen that the zones where SME and residence time increase are, once again, the restrictive parts of the screw profile. Shear rate is indeed more important in kneading discs, producing higher energy. As these sections are filled,

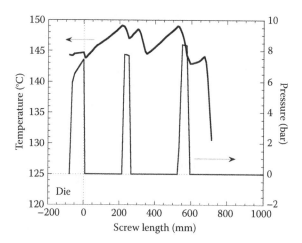

FIGURE 13.6 Changes of pressure and temperature of wheat starch along the screws (5 kg/h, 300 rpm).

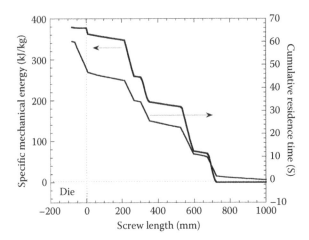

FIGURE 13.7 Changes of cumulative SME and average residence time of wheat starch along the screws (5 kg/h, 300 rpm).

residence time is also longer than along the screw elements, where the material is conveyed at high speed. In the conditions tested, an average residence time of 60 s is expected.

Other parameters such as local shear rate, filling ratio, viscosity, cumulative strain could also be plotted, but it is more interesting to focus on the influence of parameters like screw speed and feed rate. For example, at 5 kg/h, we have calculated the flow conditions for three different screw speeds, 150, 300, and 450 rpm. It can be seen in Figure 13.8 that the temperature is very sensitive to the screw speed. Even if the shape of the temperature profile remains similar, temperature increases a lot

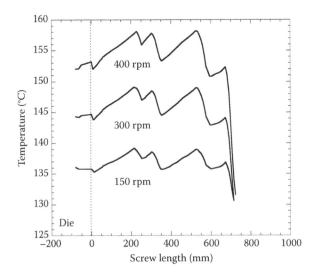

FIGURE 13.8 Changes of temperature along the screws for three different screw speeds (5 kg/h).

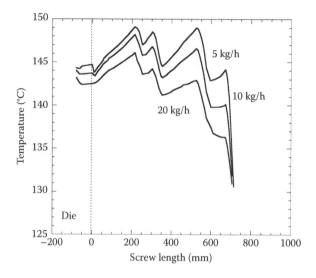

FIGURE 13.9   Changes of temperature along the screws for three different feed rates (300 rpm).

with screw speed: final temperature changes, respectively, from 136°C to 144°C and 152°C. Modeling can thus be helpful for defining processing conditions above which degradation problems could occur.

Figure 13.9 shows the influence of the feed rate, 5, 10, and 20 kg/h, respectively, on the temperature, at a constant screw speed of 300 rpm. The effect is less important than that of the screw speed. Temperature tends to decrease when feed rate is increased, but at the die exit the difference due to feed rate is approximately one or two degrees.

Another important parameter depending on operating conditions is the RTD. It can be theoretically calculated from the values of local average residence times (Poulesquen and Vergnes 2003). It can be seen in Figure 13.10a that an increase in screw speed leads to a shift of the RTD towards the shortest times, but without changing its shape. On the contrary, an increase in feed rate induces both a shortening of the times, but also a narrowing of the distribution (Figure 13.10b). If we characterize the RTD by the average residence time and the variance, which gives an idea of the width of the distribution, we obtain the following values: 80 s and 366 s^2 at 150 rpm, 69 s, and 357 s^2 at 300 rpm, 66 s, and 350 s^2 at 450 rpm, 69 s and 357 s^2 at 5 kg/h, 40 s and 90 s^2 at 10 kg/h, 26 s and 31 s^2 at 5 kg/h. All the data issuing from Ludovic modeling have been validated to numerous experimentations carried out on various types of materials (Carneiro, Covas, and Vergnes 2000, Poulesquen et al. 2003). Concerning food products, validation has been made on maize starch, wheat, and pea flours, first by comparing the computed pressure and temperature profiles to the ones measured on an instrumented twin-screw extruder, secondly, by showing that the computed energy E_v, defined in Equation 13.10, is correlated to starch degradation, protein insolubilization, and can be used to scale-up from pilot to industry processing (Della Valle et al. 1993). In spite of an overall

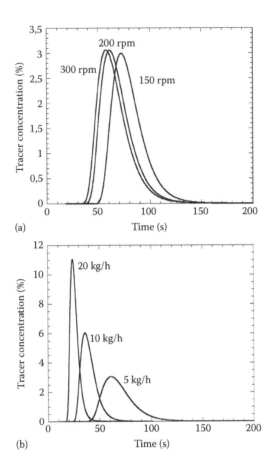

(a)

(b)

FIGURE 13.10 Influence of screw speed (a, 5 kg/h) and feed rate (b, 300 rpm) on the RTD.

satisfactory agreement between computed and experimental results on SME, product temperature and, to a lesser extent on die pressure, the application of Ludovic to the extrusion of a more complex formulation for breakfast cereals, including sugar and fats, has underlined the need for an accurate knowledge of the viscous behavior (Robin et al. 2010).

APPLICATION TO PRODUCT DESIGN

PREDICTION OF STARCH TRANSFORMATION

As recalled in preceding sections, during twin-screw extrusion of starchy products, starch is submitted to high temperatures and mechanical stresses which modify its structure: granular structure is disrupted and macromolecular degradation occurs. As the main properties of the extruded product are controlled by its level of transformation, it is important to be able to predict this parameter. Using Ludovic, we can calculate the various variables of the extrusion process, among which is the

SME, which directly influences the product viscosity (see Equations 13.5 and 13.8). SME has to be accurately calculated if we want to obtain a correct estimation of starch transformation. For this purpose, two procedures can be used, without or with viscosity coupling:

- Without coupling, a global value of SME for the whole process is first estimated and the viscosity parameters corresponding to this value are thus defined. In this case, the viscosity varies along the screw profile only with shear rate and temperature.
- With coupling, the changes in viscosity induced by the local SME are computed step by step, and calculations of viscosity and SME are coupled along the screw profile. In this case, viscosity varies according to local values of shear rate, temperature, and SME.

For the calculation with coupling, the following procedure is adopted: a first simulation is made (from die to hopper) at a fixed value of SME (here, 325 kWh/t) to obtain a first guess of the pressure and temperature distribution along the extruder. Then, a second simulation is carried out (from hopper to die), by coupling step by step, in each sub-element, calculations of flow, SME, and viscosity: temperature T_i, shear rate $\dot{\gamma}_i$, cumulative energy SME_i, and viscosity η_i are supposed to be known in sub-element i; then, from viscosity in sub-element i and processing conditions, temperature and SME changes, ΔT and ΔSME, as well as average shear rate $\dot{\gamma}_{i+1}$ are calculated in sub-element $i + 1$; new values of temperature and cumulative SME are then calculated: $T_{i+1} = T_i + \Delta T$ and $SME_{i+1} = SME_i + \Delta SME$; Equations 13.5 and 13.6 or Equations 13.7 and 13.8 are then used to calculate the new viscosity η_{i+1} in sub-element $i + 1$, and the procedure is repeated until reaching the extruder end.

The comparison of the variations of temperature, SME, and viscosity along the screws, with and without coupling, is presented in Figure 13.11. For each parameter, we observe a crossover between the curves: before this point, the values obtained without coupling are lower than the ones obtained with coupling; after this point, it is the opposite. It can be seen that the crossover corresponds to the SME value that has been chosen for defining the viscosity without coupling. In the present case, the coupling induces a reduction of viscosity by a factor of approximately two, in the final part of the extruder.

As a consequence of coupling, it can be seen in Figures 13.12 and 13.13 that the theoretical results are largely improved. SME follows the same trends, but much closer to the experimental values, even at high screw speed or low feed rate. The variation of temperature with screw speed is well predicted (Figure 13.13a). Moreover, the variation with the feed rate is in agreement with the experiments (Figure 13.13b): calculated temperature increases with the feed rate, with an error of maximum 2%, when it slightly decreases without coupling.

In order to predict starch transformation, we have calculated the intrinsic viscosity from computed SME, according to Equation 13.3, for different screw profiles and various processing conditions, with and without coupling between viscosity and SME. Results are presented in Figure 13.14 which clearly shows the advantage of

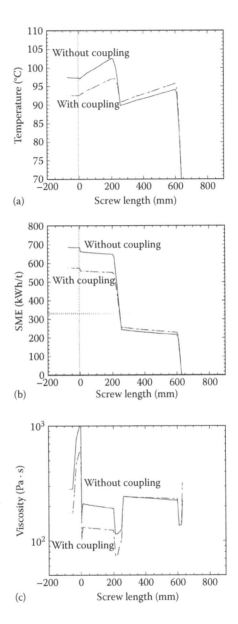

FIGURE 13.11 Changes in temperature (a), SME (b) and viscosity (c) along the screws, calculated with and without coupling. (From Berzin, F., Tara, A., Tighzert, L., and Vergnes, B. 2010. Importance of coupling between specific energy and viscosity in the modelling of twin-screw extrusion of starchy products. *Polym Eng Sci* 50: 1758–1766. With permission.)

taking into account the specific effect of SME on starch viscosity. Whatever the processing conditions, in a range between 90 and 130 mL/g, starch transformation is estimated with an error inferior to 10%, when it could reach more than 30% without coupling. In conclusion, we can say that process modeling allows us to estimate

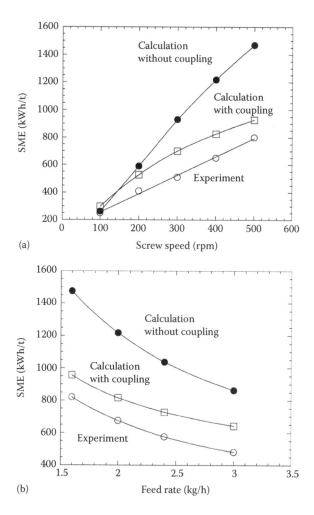

FIGURE 13.12 Change in SME with screw speed (a) and feed rate (b), calculated with and without coupling and comparison with experiments. (From Berzin, F., Tara, A., Tighzert, L., and Vergnes, B. 2010. Importance of coupling between specific energy and viscosity in the modelling of twin-screw extrusion of starchy products. *Polym Eng Sci* 50: 1758–1766. With permission.)

starch transformation in realistic extrusion conditions, as far as accurate data concerning viscosity and relationships viscosity/SME and SME/transformation are available.

PREDICTION OF CELLULAR STRUCTURE AND MECHANICAL PROPERTIES

Although not related directly to extrusion modeling, the texture of extruded products is such a part of their quality that it is important to be able to predict it. In this purpose, mechanical properties of extruded products may be addressed by considering

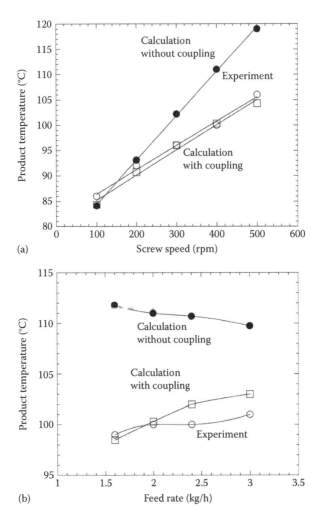

FIGURE 13.13 Change in temperature with screw speed (a) and feed rate (b), calculated with and without coupling and comparison with experiments. (From Berzin, F., Tara, A., Tighzert, L. and Vergnes, B. 2010. Importance of coupling between specific energy and viscosity in the modelling of twin-screw extrusion of starchy products. *Polym Eng Sci* 50: 1758–1766. With permission.)

them like solid foams or cellular solids. In this case, it has been shown by Gibson and Ashby (1997) that any mechanical property of the foam E^* can be defined by:

$$\frac{E^*}{E_s} = a \left(\frac{\rho^*}{\rho_s} \right)^n \tag{13.14}$$

in which E_s is the corresponding property of the constitutive solid material, ρ_s its density and ρ^* the density of the foam. a and n are coefficients which respectively

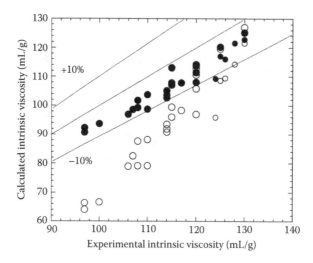

FIGURE 13.14 Comparison between experimental and calculated (with and without coupling) values of intrinsic viscosity. (From Berzin, F., Tara, A., Tighzert, L. and Vergnes, B. 2010. Importance of coupling between specific energy and viscosity in the modelling of twin-screw extrusion of starchy products. *Polym Eng Sci* 50: 1758–1766. With permission.)

depend on the cellular structure and the type of property and solicitation considered (compression, traction, or bending). Due to its simplicity, the model has become very popular and it leads to fair correlations for a wide range of extruded products (Hayter, Smith and Richmond 1986; Agbisit et al. 2007), although different values of a and n are found along the literature. The discrepancy may arise from a lack of knowledge of the property of the dense material E_s and also of the cellular structure. Using images determined by X-ray microtomography, Babin et al. (2007) have found that extruded foams with more heterogeneous cellular structure, in terms of void cells, have higher mechanical properties. This effect was confirmed by finite element simulations on cellular materials which have allowed the computation of E^*/E_s for various cellular structures of an elastic material (Guessasma et al. 2008). All these results finally suggest that the crispiness of an extruded starchy product may be favored by a uniform cellular structure with numerous cells and walls of low size. In line with the preceding section on expansion, such a structure is favored by extrusion conditions which favor longitudinal expansion, promoted by large values of elongational viscosity and high temperatures, a purpose for which it is useful to first compute the extrusion variables.

APPLICATION TO PROCESS OPTIMIZATION

OPTIMIZATION OF STARCH TRANSFORMATION

We have seen in the previous section that it is possible to accurately predict starch transformation during twin-screw extrusion. Now, we will show that the use of process modeling also allows us to optimize a process in order to obtain relevant product transformation. Let us assume for instance, that a target product has been obtained on

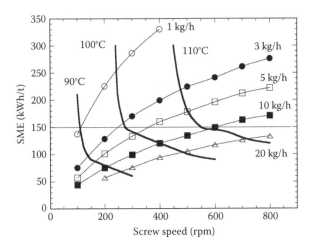

FIGURE 13.15 Variations of SME with screw speed, for different feed rates. Full lines are isovalues of the exit temperature (laboratory scale extruder).

a laboratory scale extruder Clextral BC 21 (diameter 25 mm) at a SME of 150 kWh/t and a maximum temperature of 100°C. The question is now to find the best processing conditions, i.e., the maximum possible output, for a selected screw profile. In this purpose, we will calculate extrusion conditions by varying feed rate, between 1 and 20 kg/h, and screw speed, between 100 and 800 rpm, i.e., a wide range of values. Results are shown in Figure 13.15. We observe first that SME increases with screw speed and decreases with feed rate, as expected. The maximum product temperature follows the same trends. Depending on the chosen processing conditions, this maximum value can be reached either at the die exit or in a restrictive element, located along the screw profile. If we just consider the SME, the target of 150 kWh/t can be reached for different conditions: for example, 1 kg/h–110 rpm, or 3 kg/h–250 rpm, or 5 kg/h–360 rpm, or 10 kg/h–600 rpm. But the increase in screw speed also induces important viscous dissipation and thus an increase in temperature, the last condition leading to a temperature of 110°C. Consequently, the maximum value of 100°C is rapidly overtaken and only the two first conditions of the preceding series can be considered as valid. Finally, the best choice in this case would be 3 kg/h at 250 rpm. Figure 13.15 shows also that higher outputs could only be reached to the detriment of SME: for 20 kg/h and 500 rpm, for example, a SME of only 100 kWh/t would be attained. If we admit now that a maximum temperature of 110°C can be supported by the extruded product, the optimum output could be increased up to 10 kg/h, for a screw speed of 550 rpm.

A diagram like the one shown in Figure 13.15 is a useful tool for optimizing process conditions according to quality criteria. Obviously, more sophisticated optimization techniques could be used; such methods, based on a coupling between Ludovic and genetic or evolutionary algorithms, have recently been developed in the case of optimization of screw profile and processing conditions, in the case, for example, of polycaprolactone polymerization and starch cationization (Gaspar-Cunha, Covas, and Vergnes 2005, Teixeira et al. 2010).

FROM LABORATORY TO INDUSTRIAL PRODUCTION: EXAMPLE OF SCALE-UP

Another possible use of process modeling concerns scale-up. If we consider the previous example of starch transformation, we have shown that a product corresponding to the selected criteria (SME = 150 kWh/t, maximum temperature 100°C) could be produced on a laboratory scale extruder at 3 kg/h and 250 rpm. Let us assume that the industrial target is now to elaborate the same product on a bigger extruder, for example, a Clextral Evolum 53 of 50 mm diameter. Although extrusion is a powerful process, its unit cost may still appear high when compared to other treatments (Rosentrater et al. 2003). Consequently, before any investment, it is essential to estimate the maximum possible production output. After having extrapolated the screw profile, for the same types of restrictive elements, located at equivalent positions along the screw length, we will use Ludovic in the same way as for optimization. For the industrial scale, we calculate the flow conditions for various values of screw speed (between 100 and 600 rpm) and feed rate (between 10 and 200 kg/h). The results are presented in Figure 13.16. We observe that the general trends are similar to the ones shown in Figure 13.15. The range of temperatures and SME are also, for the tested conditions, close to the previous ones. From Figure 13.16, we can deduce that the maximum output for satisfying the quality conditions are 24 kg/h at 210 rpm, i.e., a factor eight compared to the laboratory scale extruder (3 kg/h at 250 rpm). As previously, if the limit temperature could be increased up to 110°C, the maximum output would be 50 kg/h at 380 rpm.

In Figure 13.17, we have compared the results obtained for the two machines. We can see that the curves of maximum temperature as a function of SME follow exactly the same trends, and that some are perfectly superimposed. It means that, when varying screw speed, some feed rates are equivalent and lead to the same results: for example, 10 kg/h on the BC21 are identical to 50 kg/h on the EV53, or 20 kg/h equivalent to 100 kg/h. Finally, by looking at the results of Figure 13.17, we

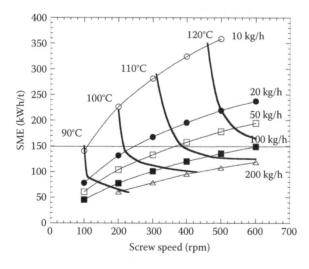

FIGURE 13.16 Change in SME with screw speed, for different feed rates. Full lines are isovalues of the exit temperature (industrial scale extruder).

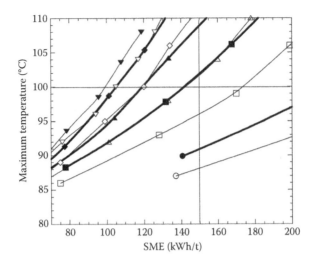

FIGURE 13.17 Change in maximum temperature with SME. Comparison between labora-tory (open symbols, small lines) and industrial (filled symbols, bold lines) scale extruders (○: 1 kg/h; ●: 10 kg/h; □: 3 kg/h; ■: 30 kg/h; △: 5 kg/h; ▲ 50 kg/h; ◇: 10 kg/h; ◆: 100 kg/h,; ▽: 20 kg/h; ▼: 200 kg/h).

can conclude that, for the criteria we have selected, the scale-up rule between the two machines is a factor 5 on the flow rate. This is far away from the classical extrapola-tion rule saying that the flow rates scale like the ratio of diameters to the third power. In the present case, this rule of thumb would have predicted a factor 9.5, leading to a flow rate of 95 kg/h on the industrial machine.

CONCLUSION AND PROSPECTS

Processing biopolymers by extrusion can take advantage of methods implemented in different areas of industrial materials processing, often synthetic polymers. This is particularly the case for flow modeling inside twin-screw extruders, where rheol-ogy and continuum mechanics can be applied as illustrated by the example of the software described before. For clarity reasons, we have not restricted the illustration only to examples directly linked to the food industry. Indeed, one of the main chal-lenges in applying such tools for food design remains the definition of the quality of the product, because this concept is not always easy to express in terms of structural features. For instance, from a nutritional point of view, one of the drawbacks of extruded starchy products, like breakfast cereals for instance, is their high glyce-mic index. This is due to the macromolecular degradation, evaluated by the intrinsic viscosity in the examples above. However, it is known that under certain conditions of temperature, compatible with those promoted by extrusion, starch may create amylose–lipid complexes, which are less easily digestible. It may then be suggested that the extrusion conditions suitable to promote this relevant range of temperature could be computed through the approach envisioned in this chapter. The next issue is the knowledge of the thermal and rheological properties of the real recipe for more accurate predictions, but as mentioned in this chapter, the methodological tools are available for this purpose.

REFERENCES

Agbisit, R., Alavi, S., Cheng, E., Herald, T. and Trater, A. 2007. Relationships between microstructure and mechanical properties of cellular cornstarch extrudates. *J Text Studies* 38: 199–219.

Alvarez-Martinez, L., Kondury, K.P. and Harper, J.M. 1988. A general model for expansion of extruded products. *J Food Sci* 53: 609–615.

Alves, M.V.C., Barbosa, J.R. Jr. and Prata, A.T. 2009. Analytical solution of single screw extrusion applicable to intermediate. *J Food Eng* 92: 152–156.

Amon, M. and Denson, C.D. 1984. A study of the dynamics of the growth of closely spaced spherical bubbles. *Polym Eng Sci* 24: 1026–1034.

Babin, P., Della Valle, G., Dendievel, R., Lourdin, D. and Salvo, L. 2007. X-ray tomography study of the cellular structure of extruded starches and its relations with expansion phenomenon and foam mechanical properties. *Carb Polym* 68: 329–340.

Barrès, C., Vergnes, B., Tayeb, J. and Della Valle, G. 1990. Transformation of wheat flour by extrusion-cooking: Influence of screw configuration and operating conditions. *Cereal Chem* 67: 427–433.

Barron, C., Bouchet, B., Della Valle, G., Gallant, D.J. and Planchot, V. 2001. Microscopical study of the destructurating of waxy maize and smooth pea starches by shear and heat at low hydration. *J Cereal Sci* 33: 289–300.

Barron, C., Della Valle, G., Colonna, P. and Vergnes, B. 2002. Energy balance of low hydrated starches transition under shear. *J Food Sci* 67: 1–12.

Baud, B., Colonna, P., Della Valle, G. and Roger, P. 1999. Macromolecular degradation of extruded starches measured by HPSEC-MALLS. In *Biopolymer science: Food and non food applications*, eds. P. Colonna and S. Guilbert, 217–221. INRA Editions, Paris.

Berzin, F., Tara, A., Tighzert, L. and Vergnes, B. 2010. Importance of coupling between specific energy and viscosity in the modelling of twin screw extrusion of starchy products. *Polym Eng Sci* 50: 1758–1766.

Bikard, J., Coupez, T., Della Valle, G. and Vergnes, B. 2008. Simulation of bread making process using a direct 3D numerical method at microscale. Analysis of foaming phase during proofing. *J Food Eng* 85: 259–267.

Brümmer, T., Meuser, F., Van Lengerich, B. and Niemann, C. 2002. Effect of extrusion-cooking on molecular parameters of corn starch. *Stärke* 54: 1–8.

Carneiro, O.S., Covas, J.A. and Vergnes, B. 2000. Experimental and theoretical study of twin-screw extrusion of polypropylene. *J Appl Polym Sci* 78: 1419–1430.

Cha, J.Y., Suparno, M., Dolan, K.D. and Ng, P.K.W. 2003. Modeling thermal and mechanical effects on retention of thiamin in extruded foods. *J Food Sci* 68: 2488–2496.

Chaunier, L., Chanvrier, H., Courcoux, P., Della Valle, G. and Lourdin, D. 2008. Structural basis and process requirements for corn-based products crispness. In *Bubbles in food II*, eds. M. Campbell, L. Pyle and M. Scanlon, 369–380. St. Paul: Amer. Assoc. Cereal Chem.

Colonna, P. and Mercier, C. 1983. Macromolecular modifications of manioc starch components by extrusion cooking with and without lipids. *Carb Polym* 3: 87–108.

Colonna, P., Tayeb, J. and Mercier, C. 1989. Extrusion-cooking of starch and starchy products. In *Extrusion cooking*, eds. C. Mercier, P. Linko and J.M. Harper, 247–319. St. Paul: Amer. Assoc. Cereal Chem.

Cuq, B., Abecassis, J., and Guilbert, S. 2003. State diagrams to help describe wheat bread processing. *Int J Food Sci Technol* 38: 759–766.

Davidson, V.J., Paton, D., Diosady, L.L. and Rubin, L.J. 1984. A model for mechanical degradation of wheat starch in a single screw extruder. *J Food Sci* 49: 1155–1158.

de Cindio, B., Gabriele, D. and Pollini, C.M. 2002. Filled snack production by co-extrusion-cooking: 2. Effect of processing on cereal mixtures. *J Food Eng* 54: 63–73.

Della Valle, G., Koslowski, A., Colonna, P. and Tayeb, J. 1989. Starch transformation estimated by the energy balance on a twin screw extruder. *Lebensm. Wiss. u-Technol* 22: 279–286.

Della Valle, G., Barrès, C., Plewa, J., Tayeb, J. and Vergnes, B. 1993. Computer simulation of starchy products transformation by twin screw extrusion. *J Food Eng* 19: 1–31.

Della Valle, G. and Vergnes, B. 1994. Propriétés thermophysiques et rhéologiques des substrats utilisés en cuisson-extrusion. In *La Cuisson-Extrusion*, eds. P. Colonna and G. Della Valle, 439–468. Lavoisier Tec&Doc, Paris.

Della Valle, G., Colonna, P., Patria, A. and Vergnes, B. 1996. Influence of amylose content on the viscous behavior of low hydrated molten starches, *J Rheol* 40: 347–362.

Della Valle, G., Vergnes, B., Colonna, P. and Patria, A. 1997. Relations between rheological properties of molten starches and their expansion by extrusion. *J Food Eng* 31: 277–296.

Della Valle, G., Vergnes, B. and Lourdin, D. 2007. Viscous properties of thermoplastic starches from different botanical origin. *Int Polym Proc* 22: 471–480.

Desrumaux, A., Bouvier, J.M. and Burri, J. 1998. Corn grits particle size and distribution effects on the characteristics of expanded extrudates. *J Food Sci* 63: 857–863.

Donovan, J.W. 1979. Phase transitions of the starch-water systems. *Biopolymers* 18: 263–275.

Eerikaïnen, T. and Linko, P. 1989. Extrusion cooking modelling, control and optimization. In *Extrusion cooking*, eds. C. Mercier, P. Linko and J.M. Harper, 157–204. St. Paul: Amer. Assoc. Cereal Chem.

Fan, J., Mitchell, J.R. and Blanshard, J.M.V. 1994. A computer simulation of the dynamics of bubble growth and shrinkage during extrudate expansion. *J Food Eng* 23: 337–356.

Ficarella, A., Milanese, M. and Laforgia, D. 2006. Numerical study of the extrusion process in cereals production: Part I. Fluid-dynamic analysis of the extrusion system. *J Food Eng* 73: 103–111.

Ganjyal, G. and Hanna, M. 2002. A review on residence time distribution (RTD) in food extruders and study on the potential of neural networks in RTD modeling. *J Food Sci* 67: 1997–2002.

Gaspar-Cunha, A., Covas, J.A. and Vergnes, B. 2005. Defining the configuration of co-rotating twin-screw extruders with multiobjective evolutionary algorithms. *Polym Eng Sci* 45: 1159–1173.

Gibson, L.J. and Ashby, M.F. 1997. Cellular solids, structure and properties. Cambridge Press University, 510 p.

Guessasma, S., Babin, P., Della Valle, G. and Dendievel, R. 2008. Relating cellular structure of open solid food foams to their Young's modulus: finite element calculation. *Int J Solids Struct* 45: 2881–2896.

Guy, R.C.E. and Horne, A.W. 1988. Extrusion and co-extrusion of cereals. In *Food structure – its creation and evaluation*, eds. J.M.V. Blanshard and J.V. Mitchell. Butterworth, London.

Hailemariam, L., Okos, M. and Campanella, O. 2007. A mathematical model for the isothermal growth of bubbles in wheat dough. *J Food Eng* 82: 466–477.

Hayter, A.L., Smith, A.C. and Richmond, P. 1986. The mechanical properties of extruded food foams. *J Mat Sci* 21: 3729–3736.

Launay, B. and Lisch, J.M. 1983. Twin screw extrusion cooking of starches: Flow behavior of starch pastes, expansion and mechanical properties of extrudates. *J Food Eng* 2: 259–280.

Linko, P., Uemura, K., Zhu, Y.H. and Eerikaïnen, T. 1992. Application of neural network models in fuzzy extrusion control. *Trans Ichem E* 70: 131–137.

Madeka, H. and Kokini, J.L. 1992. Effect of addition of zeine and gliadin on the rheological properties of amylopectin starch with low-to-intermediate moisture. *Cereal Chem* 69: 489–494.

Meuser, F., Van Lengerich, B. and Reimers, H. 1984. Extrusion cooking of starches. *Starch* 36: 194–199.

Norton, T. and Sun, D.W. 2006. Computer fluid dynamics (CFD) – an effective and efficient design and analysis tool for the food industry: A review. *Trends Food Sci Technol* 17: 100–620.

Núñez, M., Sandoval, A.J., Müller, A.J., Della Valle, G. and Lourdin, D. 2009. Thermal characterization and phase behaviour of a ready-to-eat breakfast cereal formulation and its starchy components. *Food Biophysics* 4: 291–303.

Pansawat, N., Jangchuda, K., Jangchuda, A., Wuttijumnonga, P., Saaliac, F.K., Eitenmiller, R.R. and Phillips, R.D. 2008. Effects of extrusion conditions on secondary extrusion variables and physical properties of fish, rice-based snacks. *Lebens Wissen Technol* 41: 632–641.

Park, C.B., Behravesh, A.H. and Venter, R.D. 1998. Low density microcellular foam processing in extrusion using CO_2. *Polym Eng Sci* 38: 1812–1823.

Poulesquen, A. and Vergnes, B. 2003. A study of residence time distribution in co-rotating twin-screw extruders. Part I: Theoretical modelling. *Polym Eng Sci* 43: 1841–1848.

Poulesquen, A., Vergnes, B., Cassagnau, P., Michel, A., Carneiro, O.S. and Covas, J.A. 2003. A study of residence time distribution in co-rotating twin-screw extruders. Part II: Experimental validation. *Polym Eng Sci* 43: 1849–1862.

Ralet, M.-C., Della Valle, G. and Thibault, J.-F. 1993. Raw and extruded fibre from pea hulls. Part 1: Composition and physico-chemical properties. *Carb Polym* 20: 17–23.

Redl, A., Guilbert, S. and Vergnes, B. 1999. Rheological properties of gluten plasticized with glycerol: Dependence on temperature, glycerol content and mixing condition. *Rheol Acta* 38: 321–348.

Robin, F., Engmann, J., Pineau, N., Chanvrier, H., Bovet, N. and Della Valle, G. 2010. Extrusion, structure and mechanical properties of complex starchy foams. *J Food Eng* 98: 19–27.

Rosentrater, K.A., Richard, T.L., Bern, C.J. and Flores, R.A. 2003. Economic simulation modelling of reprocessing alternatives for corn masa by-products. *Res Conserv Recycl* 39: 341–367.

Teixeira, C., Covas, J.A., Berzin, F., Vergnes, B. and Gaspar-Cunha, A. 2010. Application of evolutionary algorithms for the definition of optimal twin-screw extruder configuration for starch cationisation. *Polym Eng Sci* DOI: 10.1002/pen.21801.

Valette, R., Bruchon, J., Digonnet, H., Laure, P., Leboeuf, M., Silva, L., Vergnes, B. and Coupez, T. 2007. Méthodes d'interaction fluide-structure pour la simulation multi-échelles des procédés de mélange. *Méca Indus* 8: 251–258.

Valette, R., Coupez, T., David, C. and Vergnes, B. 2009. A direct 3D numerical simulation code for extrusion and mixing processes. *Intern Polym Proc* 24: 141–147.

van Vliet, T. 2008. Strain hardening as an indicator of bread-making performance: A review with discussion. *J Cereal Sci* 48: 1–9.

Vergnes, B. and Villemaire, J.P. 1987. Rheological behaviour of low moisture molten maize starch. *Rheol Acta* 26: 570–576.

Vergnes, B., Villemaire, J.-P., Colonna, P. and Tayeb, J. 1987. Interrelationships between thermomechanical treatment and macromolecular degradation of maize starch in a novel rheometer with preshearing. *J Cereal Sci* 5: 189–202.

Vergnes, B., Della Valle, G. and Tayeb, J. 1993. A specific slit die rheometer for extruded starchy products. Design, validation and application to maize starch. *Rheol Acta* 32: 465–476.

Vergnes, B., Della Valle, G. and Delamare, L. 1998. A global computer software for polymer flows in co-rotating twin screw extruders. *Polym Eng Sci* 38: 1781–1792.

Vergnes, B., Souveton, G., Delacour, M.L. and Ainser, A. 2001. Experimental and theoretical study of polymer melting in a co-rotating twin screw extruder. *Intern Polym Proc* 16: 351–362.

Vergnes, B., Della Valle, G. and Colonna, P. 2003. Rheological properties of biopolymers and application to cereal processing. In *Characterization of cereal and flours: Properties, analysis and applications,* eds. G. Kaletunc and K.J. Breslauer, 209–265. New York: Marcel Dekker.

Wang, L., Smith, S. and Chessari, C. 2008. Continuous-time model predictive control of food extruder. *Control Eng Pract* 16: 1173–1183.

Willems, P., Kuipers, N.J.M. and de Haan, A.B. 2009. A consolidation based extruder model to explore GAME process configurations. *J Food Eng* 90: 238–245.

Willett, J.L., Jasberg, B.K. and Swanson, C.L. 1995. Rheology of thermoplastic starch: Effects of temperature, moisture content and additives on melt viscosity. *Polym Eng Sci* 35: 202–210.

Willett, J.L., Millard, M.M. and Jasberg, B.K. 1997. Extrusion of waxy maize starch: Melt rheology and molecular weight degradation of amylopectin. *Polymer* 38: 5983–5989.

Yeh, A.I. and Jaw, Y.M. 1999. Effects of feed rate and screw speed on operating characteristic and extrudate properties during single screw extrusion cooking of rice flour. *Cereal Chem* 76: 236–242.

Zheng, X. and Wang, S.S. 1994. Shear induced starch conversion during extrusion. *J Food Sci* 59: 1137–1142.

14 Troubleshooting

Galen J. Rokey

CONTENTS

INTRODUCTION

Troubleshooting is the identification or diagnosis of "trouble" in an extrusion system caused by a mechanical failure or misapplication of extrusion processing parameters. Extrusion cooking is a complex process that is often defined as more of an art than a science. There is, however, a systematic approach to understanding and training for extrusion troubleshooting by organizing the process into manageable topical areas. Training shortens the learning curve and reduces time spent in diagnostics and troubleshooting and returning the extrusion system equilibrium.

The subject of troubleshooting and operator training can be divided into the following subtopics:

1. Overview of the extrusion process
2. Process variables

3. Equipment maintenance
4. Actual diagnostics and troubleshooting

OVERVIEW OF THE EXTRUSION PROCESS

The extrusion cooking process can control a wide range of product characteristics such as shape, density, rehydration time, texture, and color. Many process variables impact these final product characteristics and operator training is essential to achieve desired results. When organizing the thought process on troubleshooting, it is necessary to focus on four key areas which include the following:

- Raw material considerations
- Hardware configuration
- Processing conditions or software
- Final product technical qualities

RAW MATERIALS

Ingredient selection has a tremendous impact on final product texture, uniformity, extrudability, nutritional quality, economic viability and ability to accept topicals during a coating or flavoring process. Extrusion converts cereal grain and protein blends into a dough state and, as process temperatures and/or moisture continue to increase, the extrudate transitions into a melt state which allows flow through the die orifice. The common components of a recipe include starch, protein, fat, and fiber. An understanding of each component and how the extrusion process is affected is critical to forming an approach for effective diagnostics and troubleshooting.

Starch

Carbohydrates, in the form of grains, legumes or tubers, are commonly used in the manufacturing of essentially all human foods. Starches, sugars and cellulose are all carbohydrates and are similar in chemical composition, but differ in chemical bonding. These differences in chemical bonding determine the digestibility of these compounds. The use of starches from various plant sources and how they may affect extrusion manufacturing processes is important for organizing troubleshooting efforts.

The common food grains include corn, wheat, rice, oats, barley, and sorghum; while the common root or tuber crops include potatoes, sweet potatoes, yams, and cassava (tapioca). Many legumes are also a major source of starch and are becoming more common ingredients in extruded recipes—often driven by human food marketing initiatives or by food safety concerns. Regardless of the merits, there is increasing interest in alternative carbohydrate sources.

Starches are polysaccharide molecules made up of repeating glucose units linked together by hydrogen bonds to form long, straight chains or long, branched chains. These two types of starches are amylose (long straight amorphous chains composed almost entirely of α-(1-4) linked glucose molecules with 0.2%–0.5% α-(1-6) linkages and typical molecular weight of about 10^6 daltons), and the larger amylopectin (long branched semicrystalline chains with α-(1-4) linked straight chain and 4%–5% α-(1-6) linked branched chain glucose molecules with molecular weights of 10^7 to 5×10^8 daltons) (Hill et al. 2008). The linear amylose molecule is a helical structure and makes up approximately 20%–30% of most native plant starches. Plant starches are essentially all amylose, all amylopectin, or various ratios of these starches. Some starches, waxy corn for example, contain primarily amylopectin.

The interior of the amylose helix structure is hydrophobic, which allows it to form chemical bonds with fatty acids and other lipid-based compounds. Starch-lipid complexes can alter the gelatinization temperature and viscosity and thus impact extrusion processing conditions. Typically, tuber starches will have less lipids and proteins in the starch matrix than will grain starches. Most flavor components are lipid-based, therefore tuber starches have a blander flavor than cereal grain starches and exhibit lower gelatinization temperatures.

Wheat has both large and small starch granules. Granular shape may be disks or spheres, symmetrical or asymmetrical, and have various surface textures. Some starches, such as oats, rice, cassava, sweet potato and sago have compound starch granules with many individual granules bound together in an organized manner. It is somewhat incorrect to refer to this constituent only as starch in the sense that all plant starches also contain variable amounts of amino acids, lipids, fiber components and ash. These nonstarch fractions influence starch reactivity and functionality—especially during the extrusion process.

Starch gelatinization is the disruption of the molecular structure within the granule, including irreversible changes such as granular swelling, melting of the crystalline structure, starch solubilization and loss of birefringence under polarized light. Gelatinization temperature is determined by starch type, concentration, granular shape, uniformity of the granule and the method of determination. Since plant starches typically are composed of a range of granule sizes which do not gelatinize at the same rate, gelatinization temperatures are generally expressed as a range rather than a specific temperature. Both high-amylose and waxy varieties tend to gelatinize at higher temperatures than normal varieties of grains.

Granule size appears to be a contributing factor in how rapidly, and at what temperature range, a starch will gelatinize. Small granules appear to gelatinize at higher temperatures and over a greater temperature range than large granules of the same grain (MacGregor and Fincher 1993). Larger granules have less molecular bonding, tend to swell faster than smaller granules and often gelatinize at lower temperatures. Larger granules may tend to increase viscosity, but this larger physical size also makes them more sensitive to shear (granule breakage) during extrusion.

In general, amylose contributes to gel formation during extrusion, while amylopectin contributes to viscosity. Due to its branched structure, amylopectin is more susceptible to shear during extrusion than the straight chain amylose. Both types of starches are likely to decrease in molecular weight (become shorter chains) under severe extrusion conditions. In extruded products, amylose will provide some crispness (brittleness) in a product, but can also provide much expansion. Some researchers have reported that amylose starches consistently yielded more expansion than amylopectin starches under the same extrusion parameters.

The energy input needed for gelatinization increases as the chain length increases for both amylose and amylopectin. However, the significant presence of other hydrophilic ingredients such as sugars (short-chained saccharides) and alcohols (e.g., propylene glycol and glycerin) will also impact gelatinization by reducing the amount of water available for gelatinization. As the ratio of amylose to amylopectin increases, a higher processing temperature is needed to gelatinize the starch. Linear amylose molecules have more hydrogen bonding, which requires more energy input to break the bonds and gelatinize the starch.

Furthermore, the ratio of amylose to amylopectin determines finished product texture. Amylose, with its greater number of hydrogen bonds, provides the strength, while the amylopectin provides the viscosity (a function of molecular weight). As gelatinization is completed, new hydrogen bonds are formed and the starch recrystallizes or re-associates (a process called retrogradation). In general, amylose will undergo retrogradation much more rapidly than amylopectin. The level of retrogradation is somewhat dependent on amylose to amylopectin ratio, botanical source of the starch, temperature, starch concentration and concentrations of other ingredients such as lipids and salts. The presence of lipids and fiber components within the starch matrix appear to reduce hydrogen bonding in the final product, resulting in extrudates that may be more prone to breakage and increased fines.

In a study conducted by Politz, Timpa, and Wasserman (1994), lower die temperature (160°C vs. 185°C) and moisture (16% vs. 20%) reduced average molecular weights of wheat starch. As a general rule, gelatinization of starches is greater at higher moisture content as long as sufficient energy is present during extrusion to break intermolecular bonding. Shear can be increased by either changing screw configuration or by increasing the screw speed. Figure 14.1 illustrates the impact of screw speed and extrusion moisture on a direct expanded snack made from degerminated corn meal (78% starch).

Starch gelatinization properties will vary as a result of crop growing conditions and can result in variation in the amylose to amylopectin ratio. Although this is not fully understood, this may be part of the reason for drastic changes that are sometimes needed in extruder settings when new crop grains are first used in processing.

Changes in the energy input (revolutions per minute of the extruder shaft), feed rate of material into the extruder, temperature and moisture (added water and steam) will change the rheology of the dough during the melt stage inside the extruder just before expulsion through the die. This dough rheology is somewhat determined by molecular weight degradation, which in turn is determined by gelatinization. Many of these observed changes are related to amylose to amylopectin ratio in the raw grains and starch sources.

Starches play a large and an important role in finished extruded product texture. By altering the type of starch (amylose and amylopectin) and amounts, the amount of added water, temperature and shear conditions, the texture, expansion, crispness and bowl life of the finished product can be altered.

Amylose is usually the primary factor in the strength and crispness of the extruded product. Increasing amylose in a formula makes the dough in the extruder more resistant to shear degradation, and could help improve cutting and shape retention during drying and enrobing.

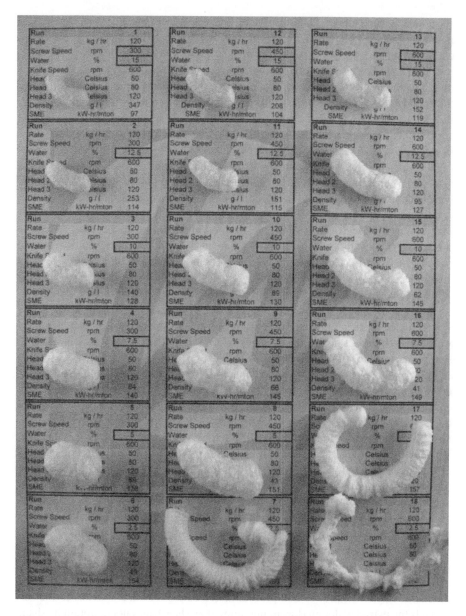

FIGURE 14.1 Impact of screw speed and extrusion moisture on product appearance and density.

Amylopectin is more susceptible to shear degradation, so will break down and gelatinize more easily during typical extrusion conditions. However, increasing amylopectin will increase the stickiness and difficulty in cutting operations as the dough exits the extruder die. Amylose will usually contribute to greater expansion during flash–off of the extruded product. Compared to other common starches, potato starch has a very large granule size and huge swelling capacity. When used as a major component of a recipe, these attributes will likely contribute to this stickiness characteristic.

Under normal extruder conditions (high temperature and high shear) in the production of many food products, native starches in grains are preferred. In general, the following statements can be made regarding extrusion and traditional grains; 1) increasing extruder barrel temperature will usually increase expansion of extruded starches; 2) increasing the percentage of amylose (up to approximately 50%) will usually increase expansion; 3) screw speed alone can influence expansion; 4) changing the screw configuration to increase friction and shear may result in greater expansion; and 5) lower moisture tends to increase friction and shear forces in the extruder barrel and results in increased expansion. Published papers clearly demonstrate that conditions for optimal gelatinization and expansion are highly dependent on processing variables and will change for any given combination of ingredients. Extrusion processing of most foods at less than 20% moisture may result in increased screw and barrel wear as well as increased opportunity for the production of dextrins. Dextrins are produced under high temperature-high shear conditions when there is not enough water to adequately hydrate the starch.

Pregelatinized starches may be used in lower temperature processes, lower added water and lower shear applications, but may form sticky dough that does not convey well nor maintains shape after cutting or forming. Cross-linked starches are also sometimes used for this application.

When extruded products processed under high temperature and high shear conditions do not have the desired crispness or hardness due to poor or weak expansion, consider increasing the amylose content by varying the amount of specific grains or other starch sources. If this same textural problem occurs with lower temperature and shear conditions, consider increasing the amylopectin content in the formula. If the extruded product has poor cutting at the die and/or poor retention of shape, it may be due to low viscosity of the dough during extrusion. In this case, the formulation should be adjusted to increase the amount of amylose subjected to high shear conditions. If poor cutting at the die and/or poor retention of shape occurs during low shear conditions, the amylopectin content in the formula should be increased. Under both high and low shear conditions, water adjustment may also be helpful.

When the extruded product appears sticky after extrusion, it may be helpful to increase the amount of amylose in the formula, adjust the amount of added water, and/or decrease the amount of shear applied during extrusion. Under low shear conditions, it may be helpful to add pregelatinized or cross-linked starches to the formula. If the extruded dough (low temperature, low shear) does not maintain its shape or does not produce relatively uniform thickness of rolled sheets, it could be helpful to adjust the amount of water used, increase the amylopectin content, and/or use a lower water-holding starch.

Increasing the amount of amylose in the formula and/or using cross-linked starches, can remedy a product's tendency to stick to the teeth. Alternatively, the specific mechanical energy inputs during extrusion could be decreased. Some extruded products do not hold together well, or fracture easily and produce excessive amounts of fines. This is likely a result of a recipe that contains one or more of the following: too little starch, too much fat, too many added fines or too much non-functional proteins.

Grains, such as wheat, may have sprouted in the field or during storage. This sprouting may not be readily apparent or visible, and is usually associated with frost damage, wet conditions near the mature stage, or late maturity. However, if this happens, the α-amylase that breaks down starches is already active to provide glucose to the sprouting plant, so the starch molecular weights will be smaller. This breakdown of starch structure changes the physiochemical properties of the grain during extrusion and may actually make these starches more difficult to gelatinize (cook) due to their sticky nature. If this situation is suspected, a falling number test or Rapid Visco Analyzer (RVA) test can provide an indication of starch damage. Relatively large swings in the ampere load on the extruder motor may also be observed. In this situation, reduction in steam may help, along with the addition of small amounts or fat or oil during extrusion to help reduce the shear.

A general understanding of the interactions that occur among ingredients, processing parameters and equipment design and capability is essential in new product development and production of quality extruded foods. In summary, the starchy components gelatinize during extrusion resulting in a substantial uptake of moisture and an increase in dough viscosity. Starch and starch-bearing ingredients contribute to binding and expansion in the extrusion process. Starches also exhibit excellent binding properties and can greatly influence product durability and texture when incorporated at low levels in a recipe. Storage and processing of raw materials prior to extrusion greatly influence their reaction to heat, pressure and shear. For example, cereal chemists recognize "after-ripening" factors which are biochemical changes that occur in grains during storage as a function of time. Grain that has been recently harvested extrudes much differently than grain that has been stored for more than six months. Whole grain is "alive" if it is sound, and therefore, changes with time. The exact mechanics are unknown, but freshly harvested grain is often sticky when heated and hydrated during a particle size reduction or extrusion step. The same grain, when stored for at least six months, may lose much of these sticky characteristics when processed. The extrusion processing conditions may require some adjustment by reducing extrusion moisture and temperature to avoid the stickiness resulting from new crop inputs.

Protein

Protein constituents may impact elasticity and gas-holding properties that are characteristic of hydrated and developed glutinous doughs. Other proteinaceous materials, such as those with low water solubilities, may contribute less to the adhesive and stretchable functional properties. Increasing protein levels in recipes that are to be extruded generally results in less expansion and a more rubbery, or occasionally, a hard texture.

Proteins can be classified as plant and vegetable sources or as animal and marine sources. This method of classification is workable for nutritional and processing purposes. Vegetable or plant proteins are largely water soluble and therefore possess very functional properties during extrusion. The functionality or water-soluble properties of plant proteins can be measured by several laboratory tests. The primary test for potential functionality is the measurement of the protein dispersibility index (PDI).

The PDI is a means of comparing the solubility of a protein in water, and is widely used in the soybean processing industry. A sample of the plant protein to be measured is ground, mixed with a specific quantity of water, and then blended together at specific revolutions per minute for a specific time. The resulting liquid is then measured for protein content, and the PDI is calculated as the percentage of protein in the liquid portion divided by the percentage of protein in the original sample. A PDI of 100 indicates total solubility. It has been shown that the PDI can be affected not only by the source of the protein, but also by the manufacturing processes employed to refine or extract the protein. Heat has been shown to lower PDI and is usually also reflected in color comparisons of the available plant proteins. Lower PDI values correlate with darker colors. During the milling or extraction steps to refine a plant protein for use as an ingredient in extruded products, there are often one or more heating steps which affect the PDI value. These heating or drying operations are usually very mild and do not significantly lower PDI values. A PDI value of greater than 40 will have significant functionality during extrusion, reasonable binding and some expansion potential. Extremely high PDI values (>80) may actually be so functional that, at high levels in a recipe, may contribute to stickiness or tackiness when hydrated that eventually results in unstable extrusion conditions.

Proteins of animal or marine origin may be subjected to much higher temperatures during manufacturing. Higher process temperatures are employed for many reasons—improved extraction and separation from fat and water components, and adequate pasteurization. Where high temperatures have been employed over an extended time period, the resulting protein solubility is quite low and these proteins may be essentially inert during the extrusion process. Inert means that the protein will not contribute to binding or expansion, but may actually reduce expansion. This is in part due to the presence of significant levels of minerals and fat components, but is mainly due to the denatured (nonsoluble) structure of the protein. The high temperature processing of ingredients will be reflected in low PDI values and dark colors.

There are other useful tests to determine the suitability of protein flours or meals for extrusion. One test that has been used in the industry is referred to as the "flow test." A sample of the animal or marine protein is sprinkled from a specified height onto a solid shaft of 5 cm in diameter. The resulting height of the cone-shaped product above the shaft is measured and is an indication of the flowability of that protein ingredient. Heights of more than 5 cm indicate poor flow characteristics. These ingredients will likely have poor transport characteristics and easily bridge in bins and metering devices. If poor flow characteristics are suspected, the flowability can be improved by grinding the material to a smaller particle size, by adding more oil and water during the extrusion step, or by installing a more aggressive extruder configuration.

Animal proteins are supplied to the extrusion system in a fresh (uncooked or lightly cooked) or spray-dried form will have significant solubility and functionality. Protein solubility is an indication of the degree of denaturation of protein ingredients. Denaturation does not necessarily impact protein digestibility. Denaturation does impact extrusion functionality and usually occurs in a temperature range of 55°C–70°C. Further processing by subjecting the protein to excessive temperatures can result in heat-damaged protein which does impact digestibility. Process temperatures in excess of 150°C will begin to affect digestibility and result in the formation of heat-damaged protein. Beyond actual digestibility studies, heat-damaged proteins can be quantified by measuring the amount of nitrogen in the acid detergent fiber fraction. The short retention times commonly experienced during extrusion of most food stuffs will not result in significant loss of available amino acids.

Oil and Other Lipid Components

Almost all ingredients contain some level of oil or other lipid constituents. Oils or derivatives of various fats such as lecithin or mono and diglycerides are often added to recipes to impart specific emulsifying or textural properties. The presence of oil and similar ingredients will act as a lubricant in the extruder screw. Fat addition reduces specific mechanical energy inputs. At lower inclusion rates, lipids can disrupt cell structure and texture by affecting plasticity and viscosity. In most recipes, the addition of lipids will begin to affect expansion and product durability at levels of less than 7% (total crude fat). If an internal fat levels exceed 12% (total crude fat), distinct shapes may not be possible. At moderate inclusion levels, fats will tend to yield large cell sizes and thick cell walls in the extrudate.

Extrusion of starchy foods results in the formation of complexes between starch and lipids and between proteins and lipids. The complexes between starches and lipids are due to the ability of the amylose fraction of starches to bond lipids. Complex formation has been studied by a number of researchers and has been shown to affect water solubility and enzyme susceptibility. Studies have also shown an impact on product bulk density, expansion ratio, and other physicochemical properties of extruded starches (Bhatnagar and Hanna 1994). The starch–lipid complexes formed during extrusion can affect the extraction of lipids from an extruded sample. Crude fat analysis of extruded products will not accurately reflect total crude fat as ether and other solvents fail to extract the lipid portion involved in the starch–lipid and protein–lipid complexes. It is necessary to pretreat the sample with an acid hydrolysis step to free the lipid fractions for subsequent extraction.

Fiber

Fiber is a broad classification of ingredients that includes cellulose, hemicellulose, lignin, and other compounds such as oligosaccharides. Many extruded food products contain fiber as a natural component of various ingredients in the recipe and also as a supplement to promote certain nutritional and health benefits. Fiber sources can be categorized based on their solubility and fermentability. Insoluble fibers have low or only partial fermentation and examples include whole grain bran, vegetables like celery and zucchini, fruit skins, vegetable peelings, and resistant starches. Soluble fibers are readily fermented and include beta-glucan from oats and barley, fruit pectin, psyllium seed, inulin, root vegetables, legumes, and some gums.

At low inclusion levels (less than 5%), fibrous ingredients may not have a noticeable impact on extruded products. Particle size of the fiber is important and if smaller than 400 μm, the fiber may actually increase expansion and reduce bulk density of the extrudate. Large particles of fiber in a recipe usually result in a coarse, fuzzy product surface appearance after extrusion. If the particle size is less than 50 μm, there is less effect on expansion even at higher levels in the recipe. Very fine fiber particles create an extremely small cell structure in the product after extrusion. Insoluble fiber remains nearly inert during extrusion and the individual fiber particulates can serve as nucleating sites during the expansion process at the die. More soluble forms of fiber have less contribution to reduced expansion even at high inclusion levels. Several studies have indicated that extrusion can increase fiber solubility. The extent of this conversion depends on processing conditions.

There are a host of other ingredients—major and minor—that could be considered in a discussion of extrusion troubleshooting. If time and resources allow, the simple research technique of "cause and effect" can be implemented by extruding a product with and without a minor ingredient present in the recipe. The production personnel usually do not have the responsibility to establish recipes and are often not aware of changes or additions to a familiar recipe. Raw materials or recipe components are very rarely selected to be a part of the recipe based on their compatibility with the extrusion process.

Raw materials are selected primarily based on their nutritional and functional contributions. Secondly, economics enters into the selection process. Many recipes are formulated based on least cost formulation software programs. Thirdly, the availability of the raw material becomes a factor.

When purchasing or selecting raw materials, establish a specification range based on desirable characteristics. This range of specifications should include the proximate analysis and other known critical qualities. However, some desirable characteristics are only vaguely recognized and no satisfactory test exists as yet to monitor quality in a reliable manner.

There exist variabilities within a raw material due to influences such as the variety, growing season, and postharvest handling or processing of grains. Different types of grains, legumes, and variations within animal or marine protein sources are reflected in the processability of raw materials.

Many problems can be avoided by developing historical databases that record raw material characteristics that correlate with good processing. Establishing a sample library of acceptable and unacceptable raw materials may be especially useful in maintaining a smooth running extruder and troubleshooting future challenges.

Hardware

An extrusion cooker is essentially a pump and a bioreactor that simultaneously transports, mixes, heats, shears, stretches, shapes and transforms, chemically and physically, the material under elevated pressure and temperature in a short time. The raw material, a fine-ground meal or flour at ambient temperature, can be fed directly into the extruder inlet hopper. Most food extrusion processes today employ a preconditioning step prior to extrusion cooking if extrusion moistures are at 18% or higher. The higher extrusion moistures require more mixing and retention time before extrusion in order to achieve uniform hydration of the recipe components. This can be especially critical where recipe particles exceed 400 μm in size. A preconditioner hydrates, heats and continuously mixes the raw material with steam and water in an atmospheric vessel. The operating parameters exceed the glass transition temperature of the raw material and accomplish as much as 60% gelatinization. The preconditioned material is conveyed into the extruder which further mixes, then softens and gelatinizes and/or melts a plasticized material (dough) which flows downstream in the extruder channel. Shear force within the extruder physically disrupts the starch granules and the starch molecules, and allows much faster transfer of water into the starch molecules resulting in rapid loss of crystalline structure. As this process continues, the material is then shaped by high pressure flow through the die at the end of the extruder. Upon exiting the die, the material can undergo further physical changes as a result of rapid expansion due to moisture flash off. The resulting porosity has a major influence on the product properties. Since extrusion involves simultaneous chemical and physical changes, the process is still not well understood although much effort has been expended to describe or model the process.

An extrusion system, whether a single-screw or corotating twin-screw configuration, must accomplish a number of phenomena in a very short time under controlled, continuous, steady-state conditions. These include tempering, feeding, mixing, cooking, cooling, and shaping. The pressure, temperature, moisture, and resulting viscosity of the extrudate are affected by the recipe, system configuration (hardware), and processing conditions (software).

Selection of the proper system configuration includes making a choice of the following hardware components: a) feed delivery system, b) preconditioning phase, c) extruder barrel components, and d) die and knife configurations (see Figure 14.2). Each component is designed to accomplish a specific function in the process of cooking and forming of food products.

BIN (FEED DELIVERY SYSTEM)

Ingredients, preground and mixed to the appropriate recipe, are delivered to the holding bin above the extruder which is of adequate volume to support the extruder operation for a minimum of 5 min. This minimum material requirement will provide a buffer time for the operator and automatic control network to shut the extruder down in an orderly manner should there be a loss of raw material recipe to the extruder. Typically this holding bin is operated between high and low level limits. In addition, the bin must ensure that the extruder receives a continuous supply of material in a uniform manner and is often equipped with a device to prevent bridging of the dry recipe.

METERING DEVICES (FEED DELIVERY SYSTEM)

Essential to any extrusion operation is the feed delivery system. Hoppers or bins are used to hold the dry ingredients above the feeders. These systems must be able to uniformly meter a blend of ingredients. Generally, when the added fat content of a raw formulation exceeds 12%, the portion of the fat above the 12% level should be introduced into the extrusion system in a separate ingredient stream. The dry recipe portion is delivered to the extrusion system through a specialized metering device

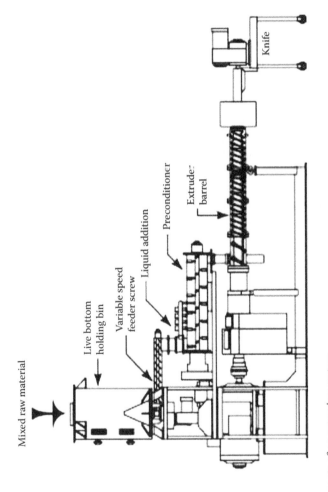

FIGURE 14.2   Components of an extrusion system.

capable of providing uniform flow at the desired extrusion rate. Typically, the rate of extruders is controlled by a screw conveyor or other metering device. Dry ingredients are usually free-flowing and there are a number of capable metering devices which vary in their relative cost and complexity.

Loss-in-weight or gravimetric systems are more complex and expensive, but are preferred as they are not influenced by changes in bulk density of the raw materials nor the height of product in the holding bin. Volumetric systems, on the other hand, are less complex but will deliver feed rates that vary with bin fill due to the material head weight.

The output of the single-screw and twin-screw cooking extruder is independent of extruder screw speed. The metering screw, which delivers dry recipe from the extruder bin, must be equipped with a variable speed drive to allow the screw speed to be set to the rotational speed necessary to deliver dry recipe at the desired extrusion rate. Variable speed augers or screw conveyors can be used to volumetrically meter ingredients. These same devices can be designed and manufactured to act as loss-in-weight (gravimetric) feed systems by mounting the bin and feeder assembly on load cells and adding a microprocessor-based gravimetric controller.

Feed rates of volumetric devices are influenced by degree of fill in the live bin. Dry recipe in live bins that are full exert a "head weight" on the metering screw at the bottom of the live bin. This results in a density increase of the dry recipe which causes the volumetric feeder screw to deliver more recipe with each revolution. The final impact is a net increase in feed rate when the bin is full even though the speed of the metering screw has never changed. The opposite effect occurs when the live bin is nearly empty of dry recipe which reduces "head weight" on the feeder screw resulting in lower feed rates even though the metering screw speed has remained unchanged. This variation in feed rate can alter moisture levels of final extrudate by $\pm1.5\%$ which may impact product characteristics such as bulk density, expansion ratio and texture.

Feed rate controlled by gravimetric feeding devices can maintain accuracies of $\pm0.5\%$ of the desired dry feed rate. Feed rates can be stored in a computer according to product identity as part of a data logging system. The gravimetric metering device continuously records the weight loss of the feed bin as a function of time. This weight loss rate is compared by the computer to the desired feed rate and the speed of the feed screw is then adjusted by the computer to deliver set point feed rate.

PRECONDITIONER

The preconditioning step initiates the heating process by the addition of steam and water into the dry recipe. Uniform and complete moisture penetration of the raw ingredients significantly improves the stability of the extruder and enhances the final product quality especially where recipe particle size is larger than 1 mm or where process moisture exceeds 18%. Preconditioning prior to extrusion may not be beneficial to all extrusion processes. Generally, any extrusion process that would benefit from higher moisture and longer retention time will be enhanced by preconditioning.

The objectives of a preconditioning step are to continuously hydrate, heat, and uniformly mix all of the additive streams together with the dry recipe. Raw material particles are held in a warm, moist, mixing environment for a given time and then are continuously discharged into the extruder. This process results in the raw material particles being hydrated and heated by the steam and water.

Dual shaft preconditioners have improved mixing in comparison to single shaft preconditioners and have a longer average retention time of up to 1.5 min for a similar throughput. As with single shaft preconditioners, they have beaters that are either permanently fixed to the shaft or that can be changed in terms of pitch and direction of conveying. In a double preconditioner, the two shafts usually counter—rotate so that material is continuously interchanged between the two chambers to facilitate mixing.

Of all the preconditioners available today, the differential diameter/differential speed cylinder (DDC) preconditioners are the most sophisticated. The DDC has the best mixing characteristics combined with the longest average retention times of all those available on the market (Figure 14.3). This design offers retention times of up to 2–4 min for given throughputs comparable to the 15 to 45 s possible in double and single preconditioners. As with a double conditioner, the two shafts of a DDC preconditioner usually counter-rotate so that material is continuously interchanged between the two intermeshing chambers.

Unpreconditioned raw materials are generally crystalline or glassy amorphous materials. These materials are very abrasive until they are plasticized by heat and moisture. Preconditioning prior to extrusion plasticizes these materials by the addition of water and steam prior to their entry into the extruder barrel. This reduces their abrasiveness and results in a longer useful life for the extruder barrel and screw components.

Extruder capacity can be limited by energy input capabilities, retention time, and volumetric conveying capacity. While preconditioning cannot overcome the extruder's limitations in volumetric conveying capacity, it can significantly contribute to energy input and retention time. Retention time in the extruder barrel can vary from as little as 5 s to as much as 2 min, depending on the extruder configuration. Average

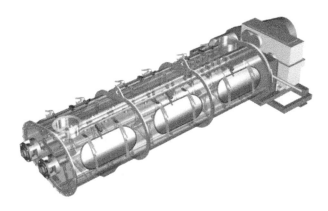

FIGURE 14.3 DDC preconditioner.

retention time in the preconditioner can be as long as 5 min. For some high moisture processes, the energy added by steam in the preconditioner can be as much as 60% of the total energy added to the process.

When hydration, heating, and mixing, the three essential objectives of preconditioning prior to extrusion are adequately satisfied, several results should be expected. First, the raw material particles should be thoroughly hydrated to eliminate the dry core present in the center of raw material particles. This leads to more efficient cooking of starch and protein. Complete hydration of the raw material particles also assists in heat penetration. Second, the raw material particles should be thoroughly heated to eliminate the cool core present in the center of raw material particles prior. This, coupled with complete hydration, results in more complete starch gelatinization and protein denaturation.

EXTRUDER BARREL

The extruder barrel assembly consists of a rotating extruder shaft and elements (segmented screws and shearlocks), a stationary barrel housing (comprised of segmented sections), and a die and knife assembly. Both the length to diameter ratio of the extruder barrel and the actual geometrical design of each individual component can be varied.

Available screw elements vary depending upon the manufacturer and the application. In addition to simply transporting the material from the inlet to the die, screw geometry can influence mixing, kneading, heating and pressure development. An example single-screw configuration is shown in Figure 14.4. Extruder rotating screw and shearlock elements sequentially convey and heat the material via mechanical energy dissipation. Movement and transformation of material within the extruder can be described as three unit operations; feeding, kneading, and final cooking zones (Figure 14.4).

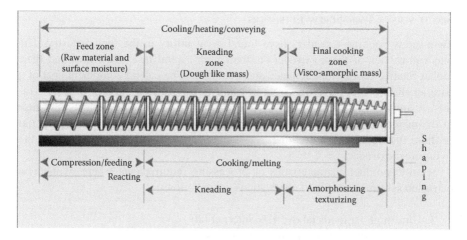

FIGURE 14.4 Various screw element designs and processing zones.

The screw pitch often decreases and the flight angle may also decrease to accomplish more mixing in the cooking/melting zone. Reduced slip at the barrel wall prevents the food material from turning with the screw which is referred to as "drag flow." A continuous screw channel serves as a path for "pressure-induced flow" as the pressure behind the die is usually much higher than that at the inlet. "Leakage flow" also occurs in the clearance between the screw tip and barrel wall. To further increase mixing via leakage flow, the flight of the screw may be interrupted. Energy inputs and moisture levels at this stage are usually present in sufficient amounts so that the material is approaching its melt transition temperature causing the material to exhibit a rubbery texture similar to very warm dough.

Most of the temperature rise in the extruder barrel is from mechanical energy dissipated through the rotating screw. It may be assisted by the direct injection of steam or from external thermal energy sources. The screw profile may be altered by choosing screw elements of different pitch or with interrupted flights, or by adding mixing lobes configured to convey either in a reverse or forward direction. All of these factors affect the conveying of plasticized material down the screw channel as well as the amount of mechanical energy added.

Choosing the proper extruder configuration is critical to a successful operation. The manufacturer of the extrusion equipment should be able to assist in developing configurations tailored for the products to be processed. There are many types of extruders and each has a specific range of applications. An improper extruder selection very rarely results in a smooth running process.

There are two types of extruders most commonly used in the food industry. These are the single-screw and corotating parallel shaft twin-screw. For both single-screw and twin-screw extruders, screw elements of single or multiple-flight geometries may be used. Single flight elements generally yield products of higher bulk densities compared to double-flight screws when operating with the same extrusion parameters because they have less surface area and similar volumetric capacities.

SINGLE VERSUS TWIN-SCREW EXTRUDERS

Twin-screw cooking extruders have found great utility in the food industry. The major drawback of these extruders is their high capital investment cost and their higher relative cost of maintenance and operation. The capital equipment cost of a parallel shaft corotating twin-screw extruder is 1.5–2 times the cost of a state-of-the-art single-screw extruder with comparable hourly production capacity. Because of the increased costs, only those food products with strong value added potential or those that require intermeshing, self-wiping screw flights are processed via the twin-screw extruder.

Specific product characteristics or processing requirements where twin-screw extrusion systems have found applications are as follows:

1. Ultra-high fat foods (above 12% internal fat)
2. Products which have high moisture levels (above 30%)
3. Uniform shape/size product (portioned foods)

4. Ultra-small products (0.6–2.0 mm diameter products)
5. Recipes that become sticky or tacky when moistened and heated (self-wiping screw flighting prevents substantial buildup of recipe on the screw elements)

As fat levels in a formulation are increased above 15%, it becomes increasingly difficult in a single barrel extruder to transmit mechanical energy from the screws into the product. Fat actually provides lubricity and reduces friction within the extruder barrel. However, through more positive transport provided by the two inter-meshing screws, the corotating twin-screw permits internal fat levels approaching 25% while maintaining a cooked product.

Once the proper extruder is selected, it must be assembled correctly and then adequately maintained. Training is essential and the supplier of extrusion equipment must be able and willing to provide this service.

The extrusion chamber is capped with a final die which serves two major functions. It provides restriction to product flow causing the extruder to develop the required pressure and shear. In addition, the final die shapes the extrudate as the product exits the extruder.

The amount of expansion desired in the final product can be controlled by formula manipulation and open area in the die. Unexpanded, but fully-cooked, foods generally require 550–600 mm^2 of open area per metric ton of throughput. Highly expanded foods require 100–250 mm^2 or less of open area per metric ton throughput.

A face cutter is used in conjunction with the die, which involves cutting knives revolving in a plane parallel to the face of the die. The relative speed of the knives and the linear speed of the extrudate results in the desired product length. Blades of the knife run in very close proximity to the die face, and in the case of spring-loaded blades, may actually ride on the surface of the die. Knife blade metallurgy, design, positioning relative to die face, speed, and extrudate abrasiveness determine their life. Many food extrusion applications require changing or resharpening blades every 6–8 h. This is especially critical with intricate shapes. Dull blades distort the product shape and increase the number of "tails" or appendages on the product which later are broken off in drying and handling, resulting in fines.

Knowledge of the reasonable or expected life of components will avoid costly and inconvenient shut-downs. Records are imperative in this endeavor and will greatly reduce the necessity of costly parts inventories.

SOFTWARE

The manipulation, management and monitoring of processing conditions fall under the heading of "software." This critical area reflects the most intimate man/machine interface. Prior to starting an extrusion line, a checklist should be developed for which personnel are responsible and accountable. These prerun tests should include inspection of supporting utilities as consistent and sufficient utility sources are often overlooked.

The selection of a reliable control system is especially important in today's labor environment. Simplicity and repeatability are necessary features of any control system. Modern programmable logic controller (PLC) control systems available today often provide a less than 1.5 year payback. Furthermore, they simplify record keeping. Records are important. There is just too much information to remember with all the variables that are associated with today's sophisticated extrusion system. An expanded discussion of independent and dependent variables associated with software is found later in this chapter.

FINAL PRODUCT SPECIFICATIONS

The final product is the result of the first three areas discussed: raw materials, hardware and software. Final product monitoring usually includes checking moisture content, bulk density, and other visual characteristics. Specific ranges of product bulk density are required for acceptable packaging and storage. There may also be parameters such as color, shape, size, and texture which are necessary to preserve product identity. Labeling requirements must be met and are often dependent on the level of process control. Records of final product characteristics as a function of the extrusion process and recipe are a must. Proper management of raw material formulation, system configuration and processing conditions allows for the profitable manufacture of palatable, functional and tailor-made foods.

On-line devices are available which can continuously measure a number of product technical qualities such as moisture, protein, fat, starch, temperature, and bulk density. These devices are very helpful in troubleshooting the process because they give an almost instantaneous feedback as the product responds to recipe and process changes. The most reliable on-line devices make use of microwave, near-infrared, and load cell technologies. Although most of these devices represent a significant investment, the return in measurement frequency, labor, and process control justifies consideration of their installation.

PROCESSING VARIABLES IN THE PRECONDITIONER

This section focuses on the tools that an operator can use to control and correct the extrusion process. Extrusion processing variables are defined as follows:

1. *Independent variables:* Variables directly controlled by the operator that do not change as a result of changes to other variables.
2. *Dependent variables:* Variables not directly controlled by the operator that change as a result of changes to independent variables.
3. *Responses:* physical or chemical changes in the final product caused by changes of independent or dependent variables.

Examples of independent variables in the preconditioning process prior to extrusion are listed below:

- Dry recipe flow rate (kg/h)
- Water flow rate to preconditioner (kg/h)
- Steam flow rate to preconditioner (kg/h)

- Additive flow rate to preconditioner which can include other liquids or gases (kg/h)
- Preconditioner speed (rpm)
- Preconditioner configuration

When any of the independent variables are adjusted, they will impact one or more of the following dependent variables:

- Moisture content (%)
- Temperature (°C)
- Thermal energy inputs (kJ/kg)
- Average retention time (min)
- Retention time distribution

Changes in the independent and dependent variables will in turn impact certain responses in the final product. Responses commonly monitored are pasteurization, starch gelatinization, bulk density and other product traits such as plasticity.

The extruder operator can adjust any one of the independent variables which, in turn, will change one or more dependent variables. For example, if the operator increases the steam flow rate to the preconditioner (with all other independent variables remaining constant), the temperature and moisture (dependent variables) of the recipe will increase. An increase in moisture and temperature will likely result in an increase in starch gelatinization (product response) as the thermal energy inputs will be higher. Increasing energy inputs to the system (either preconditioner or extruder) will usually result in higher temperatures and higher cook levels and lower bulk densities. Energy inputs are categorized as either thermal (direct or indirect heat inputs such as steam injection) or mechanical (energy contributed by the preconditioner and extruder main drive motors). Preconditioner mechanical energy inputs are usually low enough to be ignored.

Measurement of independent and dependent processing variables in the preconditioning system is essential to understanding, controlling, and troubleshooting the food extrusion process.

1. Dry Recipe Flow Rate (kg/h): The dry recipe is delivered into the preconditioning stage of the extrusion system by a specialized metering device capable of providing uniform flow at any desired extrusion rate. Consistent and uniform metering of the dry recipe is necessary for uniform preconditioning and stable extrusion.
2. Steam, Water, and Other Liquid Flow Rates (kg/h): Flow rates of these inputs to the preconditioner can be controlled manually by adjusting a flow control valve or automatically by a process controller in an automated system. A flow meter is used to monitor these inputs. Pressure of steam supplied to the preconditioner should be 2–2.4 bar (30–35 psi). Higher supply pressures may result in high steam velocities and a reduction of steam absorbed into the dry recipe. This phenomenon is often observed as excess steam blowing out of the preconditioner vent. Recipe particulates can also

be carried out of the vent and result in sanitation and house-keeping problems. Steam flow rates greater than 8% of the dry recipe rate will also often result in excess steam supplied to the preconditioner yielding the same venting problems. Steam must be free of condensate to maintain consistent moisture contributions and must be at the proper pressure in order to accurately measure flow rates. Most flow meters are calibrated for a specific operational pressure. Steam and water flow rates are balanced to achieve a heated, moistened recipe discharging the preconditioner with temperatures up to 95°C and moisture contents of up to 35%. Optimum process water properties for the preconditioner are 40°C–60°C and 3.0 bar (40–45 psi) pressure. Consistent water pressure is critical and variations can be eliminated with proper plumbing. If lipids are added in liquid form as a separate stream into the preconditioner, the point of addition is critical to achieve desired starch gelatinization while maximizing the level of fat. Fat is usually added near the discharge of the preconditioner to allow optimum hydration. Fat tends to coat individual recipe particles hindering moisture absorption and the transfer of thermal energy required for gelatinization. If substantial amounts of fat are to be added (15%–20%) during extrusion, a portion of the total fat may be injected near the discharge end of the extruder barrel. This avoids the deleterious effects of lubricity when the process is dependent on significant inputs of mechanical energy. Typically, extended steam preconditioning and longer extruder barrels to increase product residence time during the cooking phase are employed to overcome the added lubricity of high levels of fat.

3. Preconditioner Speed (rpm): Most preconditioners are constant speed. Occasionally the preconditioner speed can be adjusted to control retention time and mixing intensity. Reduction of preconditioner speed can increase retention time but will also reduce mixing efficiency. The optimum technique for controlling retention time is to vary preconditioner beater configurations as described later.

4. Preconditioner Configuration: Beaters orientation in the preconditioner can be altered to achieve the desired fill. The pitch of the beaters can be set in a forward, reverse, or neutral position. The orientation affects recipe transport and mixing intensity within the preconditioner. If the preconditioner design incorporates threaded beaters, the operator can adjust this pitch by loosening the locknut and rotating the beater to the desired angle.

The dependent variable of recipe average retention time in the preconditioner can be calculated as follows:

Step 1: Operate preconditioner at steady state with all mass flow inputs

Step 2: Simultaneously dead stop all mass flow inputs and preconditioner shaft rotations

Step 3: Remove and weigh all recipe retained within the preconditioner

Step 4: Calculate average retention time: weight of recipe retained in the preconditioner divided by recipe rate

PROCESSING VARIABLES IN THE EXTRUDER

Examples of processing variables in the extruder are listed below:

- Recipe flow rate from the preconditioner (kg/h)
- Water flow rate to the extruder (kg/h)
- Steam flow rate to the extruder (kg/h)
- Additive flow rates to extruder (kg/h)
- Extruder speed (rpm)
- Extruder configuration
- Die configuration
- Heating/cooling of extruder barrel

As in the preconditioning step, changes in these independent processing variables will impact one or more of the following dependent variables:

- Extrudate rate (kg/h)
- Retention time in the extruder (s)
- Extrudate moisture content (%)
- Extrudate temperature (°C)
- Specific thermal energy (kJ/kg)
- Specific mechanical energy (kJ/kg)
- Extruder pressure (bar)

Changes in the independent and dependent variables will in turn impact certain responses in the final product. Product responses commonly monitored include the following:

- Starch gelatinization and/or protein denaturation
- Bulk density (g/l)
- Cell structure or texture
- Shape/size/color

Measurement of processing variables in the extrusion system is essential to achieving the correct product responses.

1. Recipe Flow Rate from the Preconditioner (kg/h): This mass flow rate is a total of all mass flow inputs to the preconditioner. Loss of steam from the recipe during the transition from the preconditioner to the extruder barrel can be ignored unless this is excessive.
2. Process Steam, Water and Other Liquid Flow Rates (kg/h): As with the preconditioner, flow rates of these inputs can be controlled manually or with a process controller. Pressure of steam supplied for direct injection should be 6–8 bar (90–120 psi). Low supply steam pressure may result in plugging of the injection orifices due to product pressure if the barrel is filled in this area. Optimum extruder water and other liquid additive pressure and temperature are the same as for the preconditioner (3 bar

and 60°C). Moisture is a critical catalyst in twin and single-screw cooking extruders. Moisture contributions are made from the direct addition of hot water (60°C) and injection of steam. These additions are made in the preconditioning device and the extruder barrel. If a preconditioner is supplied as part of the food extrusion system, approximately 80% of the steam and water are introduced into the preconditioner and the remainder into the extruder barrel. Extruder steam injection provides additional energy for cooking. This additional energy input can result in capacity increases, more tolerance for high fat levels in the formulations and reduced mechanical energy requirements. Product requirements determine optimum extrusion moisture. Lower moisture levels during processing can contribute to destruction of heat-labile nutrients such as lysine and ascorbic acid. Extruder moisture controls product bulk density. Higher moisture levels are also necessary for the complete hydration and viscoelastic properties of the extrudate. The development of these properties is especially critical in the presence of high levels of fat, which tend to weaken dough strength and hinder final product textural development.

3. Extruder Speed (rpm): The ability to vary the extruder speed gives added flexibility to the extrusion process. Increasing the independent variable of extruder speed also increases the specific mechanical energy input. Producing a given food texture requires a specific input of mechanical energy. As screw speeds increase, the mechanical energy inputs increase unless product throughputs are also increased to maintain desired energy input per unit of throughput. Simply increasing the mechanical energy per unit of throughput will also decrease the bulk density of extruded foods.

4. Extruder Configuration: The extruder configuration consists of an extruder barrel (usually composed of barrel segments which are sometimes referred to as heads) and rotating elements (usually composed of segmented screws and other devices to interrupt flow or to provide restriction). Helical profiles of screw elements are designed to convey the recipe along the extruder barrel and through the restriction of the die assembly covering the discharge. If there were no restrictions to recipe conveyance, limited mixing would be accomplished as the die assembly would provide the only restriction. To improve mixing and the transformation of the recipe from partially–cooked dough to a fully–cooked, visco-plastic extrudate, the conveyance is purposely interrupted by several possible devices.

   a. Interrupted flight screw elements: The helical screw flight is interrupted by small cuts.

   b. Straight-ribbed barrel liners: This liner design interrupts the flow pattern of the recipe through the extruder barrel, resulting in more mixing or shear stress.

   c. Shearlocks and mixing lobes: These devices are collars or rings that have various diameters to restrict product flow. They can also be shaped in a bi-lobal or tri-lobal configuration.

A guideline for determining how changes in extruder configurations can affect this process is as follows: "Any device in an extruder configuration that disrupts conveying will also increase cook."

5. Die configuration: The die assembly, which mounts at the discharge end of the extruder barrel, serves two main functions:
   a. Provide a restriction to product flow
   b. Size and shape the extrudate
6. Heating/Cooling of Extruder Barrel: Extruder barrels are usually segmented into units called head sections. Head sections are often jacketed to circulate heat transfer fluids such as steam, water or oil for heating or cooling. The extruder head jacket temperature profile can be manually controlled or controlled by a computer to assist processing. If the level of cook in a product is very low, heating the jackets via steam circulation, a source of thermal energy input, will increase percent cook and decrease bulk density. In products that are already well-cooked, additional heating may actually increase final product bulk density by decreasing extrudate viscosity and extrusion pressures. Temperature profiles along an extruder barrel are usually established by installing thermocouples that protrude through the head section and mount flush to the inside diameter of the barrel line to measure product temperature. Although these are not a reflection of actual product temperatures, they can be used as set points for process management.

Until approximately 15 years ago, the extrusion process had been almost exclusively manually controlled. Most extrusion operations rely on an individual operator to set the feed rate, steam and water flow rates, and to adjust the product temperature profile. Although the process requires relatively few process variable changes when it is operating at equilibrium, the initial setting of these process variables can make a significant impact on product quality and the economics of the plant. Process control and automation remove the human influence and the inevitable product variability associated with the operation of the extruder.

EQUIPMENT MAINTENANCE

Service and maintenance of equipment is necessary for smooth processing without unplanned shutdowns. The favorable economic impact of a good preventative maintenance program versus emergency repairs has been documented many times. Most of the equipment suppliers offer the following services to assist in this area when required:

- New equipment site supervision
- New equipment startup/commissioning
- 24/7 emergency call-in service
- Emergency onsite service
- System audits
- Refurbish/rebuild assistance
- Training seminars for equipment operators
- Special services (oil analysis and vibration analysis)

Inhouse preventative maintenance programs should include the following:

- Utilization of outside resources for nonroutine circumstance
- Use of top quality replacement parts
- Checklists/records for responsible personnel
- Parts re-order system
- Adherence to safe practices

Safety programs are a prerequisite to the operation and maintenance of any process line. All personnel must be familiar and in compliance with those measures established for their safety. Equipment maintenance checklists should be developed for personnel responsible for maintaining the process line. Assistance is available from the vendors regarding their specific equipment design; however, the following is a generic checklist which can be tailored for each plant.

I. Extruder Live Bin/Metering Device Maintenance
    a. Clean
        – Live bin air-vent sock (monthly)
        – Surge bin air-vent sock (monthly)
        – Feeder/preconditioner transition (weekly)
    b. Check
        – Rotation of equipment (daily)
        – Calibration of metering devices (monthly)
II. Preconditioner Maintenance
    a. Check
        – Vent system (daily)
        – Shaft rotation (daily)
        – Oil level and temperature in drive unit (daily)
        – Beater orientation (annually)
        – Drive shaft couplings (annually)
        – Calibration and flow of all liquid and steam plumbing (bi-annually)
III. Extruder Maintenance
    a. Check
        – Assembly/adjustment of bearings (bi-annually)
        – Shaft alignment and run-out (monthly)
        – Belt tension and alignment (monthly)
        – Sheave installation (annually)
        – Assembly of barrel components (monthly)
        – Wear of barrel components (bi-annually)
        – Die/knife assembly wear and adjustment (daily)
        – Calibration and flow of all liquid and steam plumbing (bi-annually)

Lubrication programs should be developed around the equipment manufacturer's recommendations. General guidelines for lubrication frequency and inspection are as follows:

- Items that rotate at speeds less than 500 rpm should be lubricated monthly.
- Items that rotate at speeds greater than 500 rpm should be lubricated weekly.
- Gearboxes should be serviced (change oil and filters) at least every 10,000 hours.

Extruder barrel and die/knife components are subject to wear. The average life of extruder barrel components and die assemblies is around 3000 h of operation. Knife blades may only have a life of 72 h before they require replacement or resharpening. Many factors influence the rate of wear and the average life of parts in a specific process line is best established by review of historical data which requires accurate record-keeping.

As barrel components wear, they lose their conveying capacity which is initially reflected in higher extruder loads and eventually in unstable processing conditions. Proper measurement techniques for evaluating wear in extruder barrel components are available from extrusion equipment suppliers. Unstable conditions can be expected in the process when the gap between the outside diameter of the screw element and the inside diameter of the head section increases to 2.5 times the distance of new parts. Observing the wear patterns of these components can give an indication of potential process problems.

ACTUAL DIAGNOSTICS AND TROUBLESHOOTING

Troubleshooting the extrusion process can be complex because of the interrelationship of the many independent and dependent process variables. To simplify the approach, the most common challenges facing the extruder operator are listed in outline form with suggested corrective actions and brief explanations. If a process problem persists after implementing corrective actions, the equipment manufacturer's service department should be contacted for additional assistance.

I. Achieving Desired Final Product Bulk Density: The bulk density of the final product is important as it is an indication of proper processing and also determines "bag fill" properties. To reduce final product bulk densities, the following corrective actions should be taken:
   a. Increase thermal energy inputs by implementing one or more of the following:
      - Increase process steam addition to preconditioner or extruder
      - Adjust process water addition to achieve density
      - Increase process water temperature
      - Increase retention time in preconditioner
   b. Increase mechanical energy inputs:
      - Increase extruder speed
      - Reconfigure the extruder to be more aggressive (impart more shear)
      - Decrease dry recipe rate
      - Decrease total open area in the die
      - Increase die restriction by increasing land length

II. Eliminate or Reducing Extruder Instability (Surging): The most common causes of surging are listed in order of frequency of contribution.
   a. Check the wear on extruder barrel components. Excessive wear causes extruder fluctuation and high extruder loads because the screw elements have less positive transport and increased leakage flow.
   b. Nonuniform delivery of dry recipe or any of the other mass flow inputs such as water, steam, or liquid additives.
   c. Extruder barrel temperature is too high requiring the addition of cooling water in the head section jackets.
   d. Open area in the die is too large preventing adequate barrel fill. This phenomenon will also result in long and short product lengths.
   e. Extruder steam injection into the recipe is too high creating restriction to product flow.

III. Eliminate Dark or Light-Colored Stripes or Regions on Product: This color variation is usually the result of nonuniform cook, moisture, or product flow and can be corrected as follows:
   a. Increase steam and/or water flow rates to the preconditioner
   b. Improve mixing in the extruder barrel by using a more aggressive configuration (screw profile that imparts more shear)

IV. Rough Surface on Final Product: Product appearance is important to consumer perception even though quality may not be compromised in any other way. Create smooth product surface by the following:
   a. Reduce extruder barrel temperatures by cooling jackets
   b. Reduce mechanical energy inputs
   c. Increase extrusion moisture
   d. Utilize die materials that reduce the coefficient of friction (Teflon, etc.)
   e. Reduce particle size of raw recipe

V. Improper Cutting of Product at Die: Product appearance and the levels of fines or off-specification product sizes are increased when the extrudate is not cleanly cut at the die face. The major causes of a poor cut are as follows:
   a. Knife blades are dull or damaged
   b. Knife blades are not set close to the die surface
   c. Die opening is worn
   d. Knife blades are not parallel with the die face
   e. Die inserts are not flush with the face of the die

VI. Wedged Product: Product shape can become distorted for a number of reasons and these are listed below:
   a. Extrusion moisture too low (product will fracture easily).
   b. Extrusion moisture too high (product is so soft and tender that it is easily distorted).
   c. Slow knife speed causes wedging as the knife blades are covering the die openings momentarily, resulting in distortion.

REFERENCES

Bhatnagar, S. and Hanna, M. 1994. Amylose-lipid complex formation during single-screw extrusion of various corn starches. *Cereal Chem* 71: 582–587.

Hill, D., Hoke, J., Rokey, G.J. and Grollmes, D. 2008. Carbohydrate sources and their effect on extrusion. Presented at Pet Food Association of Canada Annual meeting. Montreal, Quebec.

MacGregor, A.W. and Fincher, G.B. 1993. Carbohydrates of the barley grain. In *Barley: Chemistry and technology*, eds. A.W. MacGregor and R.S. Bhatty, 73–130. St. Paul, MN: American Association of Cereal Chemists Inc.

Politz, M.L., Timpa, J.D. and Wasserman, B.P. 1994. Quantitative measurement of extrusion-induced starch fragmentation products in maize flour using nonaqueous automated gel-permeation chromatography. *Cereal Chem* 71: 532–536.

Index

For Product Safety Concerns and Information please contact our EU representative GPSR@taylorandfrancis.com Taylor & Francis Verlag GmbH, Kaufingerstraße 24, 80331 München, Germany

Printed and bound by CPI Group (UK) Ltd, Croydon, CR0 4YY

01/05/2025

01858557-0001